Praise for *Hitler's Scientists*

"John Cornwell tears the mask off the myth of the German überscientist. The author does a fine job of focusing on the critical effects that radar, codes, rocket technology and nuclear physics had on both sides of World War II. He also meticulously charts how German research and weapons production became more helter-skelter as the economy buckled, the tide of war turned against the Axis powers, slave labor replaced skilled workmanship and Hitler's orders grew more deranged."
                                                    —*Houston Chronicle*

"The time is right for this book. Over the past thirty years, historians have unearthed new material and provided fresh interpretations on topics such as Fritz Haber—chemical warfare pioneer, Nobel prizewinner and Jew—racial hygiene, medicine, physics, the German atomic bomb, rocket science and the leading German scientific institution, the Kaiser Wilhelm Society. But this is the first book to pull these themes together. Cornwell has written an engaging synthesis of the original research: articulate and intelligent, with an eye for telling detail or anecdote. His lively account is also a damning indictment."   —*Nature*

"The great merit of Cornwell's book is to point out the individual responsibility of scientists for the use that may be made by governments of their work."   —*The Nation*

"[Cornwell's] insights are as unsettling as they are revelatory. Cornwell presents compelling, detailed life stories of several German scientists, juxtaposing their accomplishments against the political and social turmoil of two world wars, anti-Semitism, Adolf Hitler and the Nazis."
                                        —*Milwaukee Journal Sentinel*

"Cornwell's previous book, *Hitler's Pope*, attracted significant controversy. . . . His latest work . . . raises questions about the relationship between scientific progress and warfare that suggest uncomfortable parallels between the past and present. A polemic but a timely one appropriate for audiences beyond war and science buffs."   —*Booklist*

"The book's broadness dilutes the indictment against any single scientist [but] the broad indictment is no less damning as a result. A noble effort to seek the truth."   —*San Jose Mercury News*

PENGUIN BOOKS

HITLER'S SCIENTISTS

John Cornwell, the prizewinning author and journalist, is in the department of history and philosophy of science at Cambridge University. He is a regular feature writer on science at the *Sunday Times Magazine* (London) and *New Scientist* and author and editor of four books on science, in addition to the *New York Times* bestseller *Hitler's Pope*. He lives in England.

Hitler's Scientists

Science, War and the Devil's Pact

JOHN CORNWELL

PENGUIN BOOKS

PENGUIN BOOKS

Published by the Penguin Group

Penguin Group (USA) Inc., 375 Hudson Street,
New York, New York 10014, U.S.A.
Penguin Group (Canada), 10 Alcorn Avenue, Toronto,
Ontario, Canada M4V 3B2 (a division of Pearson Penguin Canada Inc.)
Penguin Books Ltd, 80 Strand, London WC2R 0RL, England
Penguin Ireland, 25 St Stephen's Green,
Dublin 2, Ireland (a division of Penguin Books Ltd)
Penguin Group (Australia), 250 Camberwell Road, Camberwell,
Victoria 3124, Australia (a division of Pearson Australia Group Pty Ltd)
Penguin Books India Pvt Ltd, 11 Community Centre,
Panchsheel Park, New Delhi – 110 017, India
Penguin Group (NZ), cnr Airborne and Rosedale Roads,
Albany, Auckland, New Zealand (a division of Pearson New Zealand Ltd)
Penguin Books (South Africa) (Pty) Ltd, 24 Sturdee Avenue,
Rosebank, Johannesburg 2196, South Africa

Penguin Books Ltd, Registered Offices:
80 Strand, London WC2R 0RL, England

First published in the United States of America by Viking Penguin,
a member of Penguin Group (USA) Inc. 2003
Published in Penguin Books 2004

1 3 5 7 9 10 8 6 4 2

THE LIBRARY OF CONGRESS HAS CATALOGED THE HARDCOVER EDITION AS FOLLOWS:
Cornwell, John 1940–
Hitler's scientists : science, war, and the devil's pact / John Cornwell.
p. cm.
Includes bibliographical references and index.
ISBN 0-670-03075-9 (hc.)
ISBN 0 14 20.0480 4 (pbk.)
1. Science and state—Germany—History—20th century.
2. World War, 1939–1945—Science—Germany. I. Title.
Q127.G3C67 2003   2003057650

Printed in the United States of America
Set in Bembo

In memory of Max Perutz

Contents

List of Illustrations

29. Messerschmitt Me-262.
30. Experiments on a prisoner at Dachau.
31. Primo Levi.
32. Heisenberg's reactor at Haigerloch.
33. Farm Hall.
34. Max von Laue.
35. Carl Friedrich von Weizsäcker.
36. Paul Harteck.
37. 'Fat Man' nuclear weapon.
38. General Groves and J. Robert Oppenheimer.
39. Wernher von Braun surrenders to US counterintelligence personnel.
40. President John F. Kennedy and Wernher von Braun.

Photographic acknowledgements

1: Hulton Archive/Getty Images; 2, 3, 4, 5, 6, 8: Max Planck Archive, Berlin-Dahlem; 9: photo by A. B. Lagrelius & Westphal, courtesy AIP Émilio Segrè Visual Archives, Weber Collection; 10: photo by F. D. Rasetti, courtesy AIP Emilio Segrè Visual Archives; 11: Niels Bohr Archive, Copenhagen; 12, 28: Smithsonian Institution; 13, 14, 16: Christopher Hurndall; 15, 37: Bettman/CORBIS; 17: Time Pix/Getty Images; 18: Magnum; 19: UK Atomic Energy Authority; 20: Bundesarchiv Koblenz; 21: Hulton Archive/Getty Images; 22: Hulton-Deutsch Collection/CORBIS; 23, 24, 26, 27, 29, 38: Imperial War Museum; 25: CORBIS; 30: Dachau Museum; 31: Gente; 32: AIP Emilio Segrè Visual Archives, Goudsmit Collection; 39, 40: NASA.

Every effort has been made to contact all copyright holders. The publishers will be glad to make good in all future editions any errors or omissions brought to their attention.

Acknowledgements

Drawing together a history of science in Germany in the first half of the twentieth century with special reference to the period of the Third Reich involves acquaintance with a wide circuit of scholarship over the past three decades. I have profited from the work of many historians, but especially from the published work of Jeremy Bernstein, Alan Beyerchen, Michael Burleigh, David Cassidy, Ute Deichmann, Helge Kragh, Christie Macrakis, Benno Müller-Hill, Michael C. Neufeld, Thomas Powers, Robert Proctor, Monika Renneberg, Richard Rhodes, Paul Lawrence Rose, Ruth Sime, Mark Walker and Paul Weindling. I also owe a special debt of gratitude to those who attended the *Copenhagen* symposium at Jesus College, Cambridge in November 2002, and especially to Michael Frayn.

Others I wish to thank for information and insights in various ways are Paul Alexander, John Barrow, Antony Beevor, Herman Bondi, Tom Bower, Andrew Brown, Soraya de Chadarevian, Robin Donkin, Marcial and Marie-Louise Echenique, Peter Glazebrook, Walter Gratzer, Henning Grunwald, David Hanke, Stephen Heath, Jana Howlett, Tim Jenkins, Gerry Kearns, Jonathan Keeling, Peter Lipton, Austyn Mair, Marta Mazzoro, Michael Minden, Juliet Mitchell, Veronique Motier, John Naughton, the late Max Perutz, Nicholas Ray, David Reynolds, Simon Schafer, Kirstin Shepherd-Barr, Brendan Simms, Derek Taunt, Adam Tooze and Lewis Wolpert.

I owe a special debt of gratitude to Jeremy Bernstein, Ray Freeman and Walter Gratzer for reading through the manuscript, and for making many useful suggestions; any errors that persist are entirely my responsibility. It also gives me much pleasure to record my gratitude to William Saslaw for guiding me through the world of physics, and to Peter James Bowman, who tirelessly handled my

German sources and assisted me in a wide range of bibliographical and research tasks.

I wish to thank the librarians of the Imperial War Museum, the British Library, Churchill College Library, the London Library, the Whipple Library in the Cambridge University Department of History and Philosophy of Science, Cambridge University Library, the Niels Bohr Archive in Copenhagen and the Max Planck Institute Archive of the History of Science in Berlin. I am indebted to Kate Barker of Viking-Penguin in London, for her help with pictures, and to David Watson for the copy editing. I am grateful to my agents Bob Lescher in New York, and Clare Alexander in London, and, as ever, to my editors Wendy Wolf in New York and Juliet Annan in London. I thank the Master and Fellows of Jesus College, Cambridge and the Department of History and Philosophy of Science in the University of Cambridge, for providing ideal auspices and stimulus for research and writing.

To Crispin Rope I owe the original inspiration that led to this project, as well as the constant encouragement and enthusiasm that kept me going.

'Science without conscience is the ruin of the soul.'

(Rabelais)

Hitler's Scientists

Introduction: Understanding the Germans

I have an early impression of my father holding me up to glimpse a black growling angel trailing fire in the moonlit sky across London: it was a V1 – a long-range pilotless 'pulse-jet' flying bomb, what today we would call a cruise missile. It was early summer of 1944 when the V1 rockets, known to Londoners as doodlebugs and flying bombs, first dived on the capital, ripping through buildings and shattering windows for hundreds of yards around. They made a hideous sound not unlike a powerful, faltering motor bike with a broken exhaust; when the engine cut out, people braced themselves for the explosion that would follow within about twelve seconds. Later came the V2s, ballistic missiles, more menacing because they hurtled through the stratosphere silently, carrying a larger warhead. It was possible to hear the whoosh and thunder of their supersonic atmospheric entry after the explosion on the ground. On the way back from primary school on the bus with my mother, I saw the devastation caused by a V2 rocket that landed on the borders of Wanstead and Woodford in the eastern suburbs of London. The missile had laid waste several acres of woodland and made an immense crater. A woman had been walking her baby, and the pram still hung high from the branch of a tree. Mother and child, and several other hapless pedestrians, had been killed outright.

For a boy born in London in May of 1940, conceived in the month when Britain declared war against Germany, it was as if the war had always been and always would be. War for a child was a permanent, puzzling crisis, as well as an occasional adventure: the race to the corrugated iron shelter when the sirens started up their melancholy chorus; finding silvery shards of shrapnel in the streets the morning after an air raid; gazing up at glittering barrage balloons which blotted out the sun like sailing whales. War for our parents

and elder siblings was the devastation of the Blitz (which left 43,000
British civilians dead); the news of military casualties on land, sea
and in the air; nights spent underground. We grew up associating
these terrors and miseries with Germany and the German people;
and, understandably, we saw the war with Hitler as *our* war.

In the summer of 1945 my mother took me to see German
prisoners of war interned behind barbed wire at a transit camp on
the Wanstead Flats, an expanse of parkland close to where we lived.
They were lounging around, sun-tanned and relaxed; some wore
gay kerchiefs around their necks, dark glasses and field grey forage
caps at jaunty angles. One of them winked at me and pulled a funny
face. It was hard to demonize Hitler's fighting men in quite the
same way after seeing them in the flesh. Yet, even after the war had.
ended, and the Victory party bunting had been taken down, the
impression that there was something congenitally malevolent about
German people persisted, exacerbated and deepened by a growing
conversational acquaintance with an older generation that had
fought in the Great War. My grandfather kept used brass shells from
that conflict over the fire place and would take them down and
polish them every day of his life. He talked of the days before
the Great War when Germany's navy was building battleships to
threaten the Royal Navy. He spoke with awe of the vast stacks of
German goods 'dumped' on the docksides of London, flooding
England with cheap manufactured consumer products. There was
a connection, he told me, between those battleships, and the stacks
of cheap German goods – toys, tools, pens, kitchen utensils, lamps,
scissors, sewing machines, typewriters – and the stacks of bodies
heaped up behind the trenches. 'The Germans,' he used to say,
'were too clever and too wicked by half.' The remark anticipated
another aphorism that echoed through my boyhood and youth:
'The only good German is a dead German.'

German Science

Can we by studying the history of science in Germany in the first half of the last century draw significant conclusions about the relationship between science and the good society? Does doing science make human beings more rational, sceptical, internationalist, objective? And do we expect science to flourish better, and the discoveries of scientists to be used more responsibly and ethically, under democracies than under dictatorships?

Exploring the story of the German science communities during a period that spans two world wars is an essential part of attempting to understand the nature of science and the conduct of scientists in the twentieth and the new century. Recounting the drama of ideas and discoveries, probing the behaviour of Germany's researchers – towards their disciplines, towards their governments and regimes, towards their fellow human beings – cannot be separated from the task of understanding science and technology and its impact on all of us since World War II.

This chronicle of German 'science', a word I frequently use in this narrative to include the natural sciences, medicine and technology, where the range of categories is implied, is written against the background of the current war on terror, the enlargement of the European Union and the ever-growing importance of Germany. Today, Germany, split for almost half a century, is assuming an increasingly powerful role at the centre of a Europe that stretches from the Atlantic to the borders of Russia and from the North Sea to the Aegean.

Understanding German science, however, is fraught with pitfalls, since the victors of the last war have tended to see themselves, understandably, in morally superior contrast to things German.

The indebtedness of Europe and North America to German influence – before the impact of Hitler, Goebbels, Himmler – has been powerful, complex and often ambivalent. The German term for science, *Wissenschaft*, incorporated traditionally a huge circuit of intellectual disciplines in which the German-speaking peoples, never forgetting Austria, excelled, and often took the lead,

including history, literary and scriptural criticism, philosophy, the-
ology and psychology, with far-reaching influences on Western
thought. The roll-call of sublime genius – Bach, Goethe, Beethoven, Kant – was succeeded, beyond Hegel, Fichte, Schelling,
Frege, by the protean influences of Marx and Nietszche, thinkers
fatally appropriated by the founders of Soviet communism and
fascism. The early part of the twentieth century saw the dynamic
influence of Max Weber in the social sciences, Ludwig Wittgenstein
in philosophy, the unsettling impact of Freud, Adler and Jung, and
the seminal if baleful musings of Martin Heidegger, who anticipated
existentialism and the deconstructions that drove much of French
post-war critical thought – Sartre, Ricoeur, Derrida.

Meanwhile, existing for the most part in philosophical and political limbo, while giving impetus to new technologies with the
power to transform the world, the success of the natural and medical
sciences in Germany was, before 1933, prodigious: Wilhelm
Konrad Röntgen, discoverer of X-rays; Fritz Haber, who fixed
nitrogen from the air; David Hilbert, who set for the world's
mathematicians sufficient tasks to keep them busy for a century;
Max Planck, a founding father of quantum theory; Albert Einstein
and his epoch-making theories of relativity; Werner Heisenberg
and quantum mechanics. After more than half a century in which
Germany built an unrivalled power-house of chemistry, organic,
inorganic and industrial, German nationals, in the first two decades
of the twentieth century, walked away with more than half the
Nobel awards in every discipline of the natural sciences and
medicine. In the meantime, a constituency of biologists and anthropologists had claimed that populations, society and history could
be explained by a historico-biological theory of race, as spurious as
it was menacing. Thus Darwin's and Mendel's explicators were
appropriated by Ernst Haeckel and H. F. K. Günther, who in turn
spawned ideas that were used to justify Hitler's anti-Semitism and
so-called racial hygiene.

Rediscovering what we nevertheless owe, and therefore share,
with Germany is an essential part of understanding some of the

leading ideas we have employed to understand ourselves and the nature of the world through the twentieth century. Germany's decisive contributions to the natural sciences have been a crucial part of that process.

My generation, however, growing up in Britain in the immediate post-war era routinely characterized German science as Nazi science: at best, brutally efficient and militarized – the *Blitzkrieg* combination of Stuka dive-bombers, Panzer tanks and motorized infantry; the V1 and the V2 rockets, and U-boats – at worst, monstrously sadistic – experimentation on humans in the death camps. As late as the early 1960s the German scientist was still depicted in popular film culture as Peter Sellers's Doctor Strangelove.

In contrast, Britain's scientists, who led the scientific war against Hitler until America joined the conflict in 1942, were characterized in a gamut of movies, documentaries and popular war histories as ingenious boffins, effortlessly superior, and modest with it – eccentric civilians in threadbare tweed jackets creating wizardry like radar out of sheer native brilliance in the garden sheds of England. We were at our most offensive when Barnes Wallis invented his bouncing bombs to breach Germany's hydro-electric dams on the Ruhr. 'Bomber' Harris's 'area bombing' was portrayed as a strategic necessity. The story of British participation in the creation of the atom bomb, which Germany had failed to achieve, was not a celebrated chronicle, except to hammer home the point of failed Nazi science. Naturally, we assumed, the Nazis would have dropped an atom bomb on us if they had managed to create one.

Through the 1950s questions were increasingly raised about the political and moral dimensions of the science of the atomic bombs dropped on Japan. The debate continues to this day.

In time my generation, who were students in the late 1950s and early 1960s, came to ponder, and debate, the details of the British night raids on the civilian populations of Dresden, Hamburg, Lübeck and other German towns and cities. There was the exploitation of technology in a carefully judged combination of pressure and phosphorus bombs to create firestorms that incinerated hundreds of thousands of innocent men, women and children in order to break German morale as much as to hit industrial targets. There were the careful inquiries by RAF bombing experts of specialists like the architectural historian Nikolaus Pevsner about ancient buildings: not to preserve such structures, but to discover their combustible properties. The RAF raids on Hamburg which began on the night of 24 July 1943, known as Operation Gomorrah, culminated in an inferno four nights later – a fireball that surged two kilometres into the night sky, imploding oxygen and raising furious wind storms strong enough to uproot trees. Sugar boiled in cellars, glass melted and people were sucked down into the asphalt on the streets. On that one night an estimated total of 45,000 Germans died compared with a similar number killed in Britain during the course of the entire war. German estimates of civilians killed by Allied bombing of cities and towns in the Reich are 450,000 killed and 600,000 injured. It took another generation of Germans to speak loudly and clearly about the horrors that had been visited on their population.[1]

By 1962 my generation had shared the eyeball-to-eyeball terror of the Cuban missile crisis and taken views on the nuclear arms race, the doctrine of tactical nuclear weapons as a response to a conventional Soviet invasion of West Europe. We had learned of the global effects of radioactive contamination from nuclear weapons testing, both East and West. By the 1970s we were learning about the indiscriminate and long-term effects of anti-plant agents and napalm in Vietnam. It is estimated that the United States was responsible for the destruction of some 30 per cent per year of food production in Vietnam in the course of a decade.[2]

Meanwhile, historians had long since taken a more detached and considered view of the imperial rivalries that prompted World War I, and the effect of the Versailles Treaty on the upheavals in Germany of the Weimar period that contributed to the growth of National Socialism. By the 1980s, as cruise missiles were shipped to West Europe and Ronald Reagan planned his Star Wars initiative, it was possible to disengage from the stark contrast between German or Soviet military science and military science under the auspices of the democracies. While bearing in mind the unique deeds of Nazism and its historical contexts – how can one forget them? – it was possible to rethink the political history of scientists under Hitler not just as Nazi or German scientists, but as scientists.

Science under Hitler

No account of scientists under the Nazis can be understood in the absence of the narrative that forms the first part of this book: the growth and success of the natural sciences, medicine, mathematics and technology in Germany in the second half of the nineteenth century and the early decades of the twentieth. By the end of the first decade of the twentieth century, Germany had become the international Mecca of science. Researchers, basic and applied, flocked to German universities from all over the world; learned German to read the leading science journals and to participate in conferences and seminars. Germany was well placed to take a leading role in the development of a new physics that would transform the technology of the century, involving from the outset Max Planck, Albert Einstein, Max Born, Werner Heisenberg and Erwin Schrödinger, German-speakers all, alongside scientists from Denmark, the Netherlands, France and Britain. In turn, the new physics led to quantum mechanics and, ultimately, to nuclear physics, the science of the atom and the hydrogen bomb.

Yet by the beginning of World War I the reputation of Germany's scientists had been tarnished in the eyes of its Western enemies. Germany was the first to torpedo a civilian ocean liner (even the great German novelist Thomas Mann rejoiced in the wondrous display of technology employed against the *Lusitania*). A year into the war, a large constituency of German scientists along with academics and intellectuals declared their subservience to the state and to the military. In April 1915, moreover, one of Germany's most distinguished scientists, Fritz Haber, encouraged the army to use poison gas against the Allies, and provided the means to make it a reality. Germany's scientists readily turned over their civilian

...ence institutions, the Kaiser Wilhelm Institutes, to poison gas research. Were these cases of Germans behaving according to type as Germans? Or scientists in Germany behaving according to type as scientists?

In the post-war era, despite lack of funding, boycotts on the part of the victorious nations and political and economic upheavals, science continued to prosper in Germany. The Weimar period saw remarkable developments in the new physics and collaboration with scientists from many nations. There were other, darker developments in the scientific milieu: an increase in antagonism towards the work of Jewish scientists and the growth of 'racial hygiene' encouraged by distorted versions of medicine, anthropology and evolutionary theory. But victimization of Jews within scientific communities occurred also in France, Britain and the United States.

When Hitler came to power in 1933, science, medicine and technology were pressed into the service of the new regime. The Third Reich called for a spirit of *Gleichschaltung* (marching in step): all the resources of the new regime were to work in coordination. Education, the media, psychology and communications were adapted to serve Nazi ideology in order to channel and shape public opinion in the National Socialist 'revolution'. Scientists, with remarkably few exceptions, speedily acquiesced under those pressures. As the historian Joseph Haberer puts it, the scientific leadership engaged in 'expediency and compliance' and colluded with 'victimization of members of the community'.[3] Yet some groups – notably doctors and anthropologists – not only acquiesced but took a lead in promoting racist policies, and, in some cases, one segment of the scientific community oppressed and coerced another: experimental physics, in the view of some influential Nazi scientists, was more authentic than theoretical physics, which was deemed to be 'Jewish'.

Hannah Arendt expounded famously in *Eichmann in Jerusalem* (1961) her thesis of 'the banality of evil': the proposition that evil in the Third Reich arose out of the class of banal, amoral bureaucrats. Arendt's perspective has been deepened, and at the same time challenged, by more recent insights and interpretations, shedding light on the behaviour of scientists, medical doctors and engineers in the Third Reich. One perspective argues that despite the apparent rhetoric of conformity in the *Gleichschaltung*, the inchoate nature of Nazi programmes encouraged forms of 'auto-coordination'. The circumstance was augmented by the fear that hovered continually on the edge of each individual's consciousness 'by encouraging doubt over not belonging to a compelling movement of fellow citizens'.[4] At the same time, and by no means contradictory to such a thesis, Michael Wildt's study of what he terms the 'generation of the unbound' points to the careers of those who suddenly prospered after Hitler's rise to power, and who, with the outbreak of war, saw no limit to their scope of action.[5] Such a diagnosis may better explain a Wernher von Braun, and the ease with which he exploited slave labour, than Arendt's analysis.

Nazi propaganda gave an impression of technocratic modernity matched with a restored and bracing naturalism – exercise, healthy living and outdoor pursuits. The making of the Volkswagen, the people's car, and the great autobahns (known as 'Adolf Hitler's roads') were symbols of a modernizing nation state, signalling, in the propaganda of the time, a marriage between motorized transport for all and the beauty and freedom of the landscapes of the Fatherland. Fritz Todt, who was in charge of the project, accepted the call for landscaped highways as crucial to the new *Deutsche Technik*, which transcended selfish, consumerist, profit-orientated capitalist motives. The pretensions to modernity were superficial; Nazi technocracy was seldom unmixed with mawkishly bucolic dimensions.

The rallies, using searchlight sky patterns; the clean lines of the new Berlin stadium for the 1936 Olympics; the neo-classical Nazi pavilion (designed by Albert Speer) out-frowning the Soviet pavilion at the 1937 International Exhibition in Paris; the emphasis on callisthenics and health programmes – like the Nazi anti-smoking, anti-drinking campaigns – all reinforced the bracing, clear-eyed image of *Deutsche Technik* to the outside world. Communications and new media, including the first use of magnetic tape (to record

Hitler's voice for posterity) and television (to broadcast the Berlin Olympics), flourished.

The veneer of modernization, however, did not indicate the flowering of a technocratic state, nor an ideal world for any form of intellectual endeavour, as the dismissals of Jewish researchers and the bonfires of books on 12 May, 1934 amply demonstrated. Basic science, deprived of many leading researchers, and oppressed from within and without by ideological pressures, did not entirely collapse, but neither did it thrive.[6] As for applied science, technology and medicine, attempts on the part of certain groups to politicize science, creating such movements as German physics, a German biology and even a German mathematics, marked an era of division and decline, while the competition between overlapping power centres of industry, the SS, the armed forces, the civil service would result in prodigious waste and ineptitude. The Third Reich saw the misuse and misapplication of innovation, loss of freedom and diversity, neglect and decline in some branches of the natural sciences, physics in particular, and marking of time in others.

The strongest science-based imagery within the Nazi *Weltanschauung* was a bogus borrowing from anatomy, with an ominous continuity between the symbolic and the real – the body politic expelling the unwanted pathogen. The pseudo-scientific vision of National Socialist Germany was of a racially hygienic 'body', invoking the researchers and data of an expanding community of doctors, anthropologists and eugenicists. In the first wave of anti-Semitic measures hundreds of scientists were dismissed – expelled as if they were germs rejected by the body's immune system. A generation of researchers faced the choice of condoning racist legislation or leaving Germany: they acquiesced. It would not be long before the ideologues of racial hygiene promoted sterilization and 'euthanasia' – the elimination of 'lives not worth living'.

As the provisions of the Versailles Treaty were broken one by one, and Germany rearmed, powerful traditions of engineering and inventiveness gave the new Reich many apparent advantages in military applications, raising the familiar wartime spectres of science

and technology as Pandora's Box and Faustian bargain. In the 1930s
and through the war German scientists and engineers developed an
impressive list of innovations: proximity fuses, infra-red night sights,
jet engines, radar, mechanical encryption, synthetic fuels, synthetic
rubber, ballistic missiles, the snorkel and hydrogen-peroxide assisted
submarines. From 1939 to the war's end, scientists working under
military control began research on nuclear chain reaction with the
prospect of arming Hitler with an atomic bomb. In most of these
activities, and especially missile research, development and production, exorbitantly resourced high-tech rational methods were
employed for irrational goals. By 1943, however, and after years
of selective suspension of the rule of law, few areas of science,
technology, medicine and industry had not been tainted by brutality, slave labour, torture, human experiment without consent and
casual murder.

Was the science of the Third Reich, oppressed by totalitarianism,
secrecy, wasteful overlap and racist exclusions, therefore doomed
to failure? Could this poisoned tree not bear the occasional good
fruit? In recent years, historians have claimed, without attenuating
their indictments of atrocious Nazi experimental research, that
some areas of science flourished under National Socialism – for
example, the regime's war against cancer which saw pioneering
epidemiological research of the highest quality.

But is there a clear demarcation – a discontinuity – where Nazi
science starts and ends? Historians are revising their views. There
was an assumption at the end of World War II that the dark shadow
cast over Germany's science during the period of the Third Reich
was no more; that following the punishment of guilty individuals
at the Nuremberg Trials, and the restoration of democracy, West
Germany's scientists made a fresh start. Historians of science, however, have brought to our attention more recently the issue,
formerly taboo in German historical circles, of 'fellow travelling',
arguing, even, that fellow travellers did more damage than card-
carrying Nazi scientists, since they failed to challenge the consciences of the uncertain and the fence-sitters. According to this
argument, the majority of Germany's scientists carried a burden of

guilt beyond the war's end. Some critiques take the argument further, claiming that the taint was inherited in the West, as in the Soviet Union, through the scientific plunder carried out by the Allies after the war, in particular the evacuation of the Peenemünde team of rocketeers under Wernher von Braun, which had exploited slave labour. By the same token, the history of West Germany's economic miracle, the zero hour growth of its industries under democracy, has been revised, especially in relation to companies like Mercedes Benz which switched speedily from war production to domestic products. In the words of Chris Keller in Arthur Miller's *All My Sons*, 'what you have is really loot, and there's blood on it'.

Science and Individual Conscience

Beyond all the collective political and moral issues, there are instructive matters of individual conscience and behaviour. A case in point is the head of the German atomic research project, Werner Heisenberg. Did he deliberately impede progress on a German atom bomb? Or did he simply fail through inadequate physics, then claim a retrospective moral superiority after the war for not succeeding? Whatever the case of his intentions, heroic or hypocritical, it is clear that the failure of the German atomic bomb was due at least in part to a reluctance on Heisenberg's part to take responsibility for pressing for the priority of a highly speculative venture that would put an unbearable strain on Germany's resources. That, as some historians have noted, was a pragmatic rather than a moral decision.

The significance of Heisenberg's story, however, is its relevance to the question: how *should* any scientist behave when drawn into involvement with weapons of mass destruction? The issue was highlighted by the best-selling book by Robert Jungk, *Brighter than a Thousand Suns*, first published in English in 1957, which suggested that Heisenberg and his colleagues may have behaved more morally than those scientists, many of them Jewish émigrés, who had worked on the Allied bombs that were dropped on Hiroshima and

Nagasaki. 'It seems paradoxical that the German nuclear physicists, living under a sabre-rattling dictatorship,' wrote Jungk, 'obeyed the voice of conscience and attempted to prevent the construction of atom bombs, while their professional colleagues in the democracies, who had no coercion to fear, with very few exceptions, concentrated their whole energies on production of the new weapon!'[7] The view, which caused consternation and indignation for another forty years (although Jungk eventually recanted his verdict), was given a further boost in 1993 by Thomas Powers's *Heisenberg's War*. Powers's conclusion at the end of a 500-page book was that:

from the first weeks of the war until the last, Heisenberg was at the very heart of the German bomb programme; no physicist enjoyed greater respect from his colleagues or greater trust by the authorities. These considerations, taken as whole, lead me to conclude that Heisenberg consciously did what he could to ensure that there would be no ambitious German effort to develop atomic bombs.[8]

It was the apparently imponderable nature of Heisenberg's true intentions that prompted Michael Frayn, in his play *Copenhagen*, to draw a parallel between Heisenberg's uncertainty principle in quantum physics and the uncertainties of history and biography. For the historian, however, Heisenberg's story demonstrates the importance of evidence in understanding the behaviour of scientists faced with extraordinary dilemmas and choices in the midst of war. In recent years the publication of the Farm Hall tapes – records of the conversations between detained German physicists, including Heisenberg, at the end of the war – have done much to dispel the 'uncertainty' that once shrouded the truth about Heisenberg's state of mind and the German atomic bomb programme. The status of that evidence, along with recently released letters of the Danish physicist Niels Bohr, with whom Heisenberg had a wartime meeting in Copenhagen, forms a concluding chapter of this narrative.

Ethics and Science

One of the toughest criticisms Michael Frayn encountered in the aftermath of his play was the charge that it exhibits a fashionable moral relativism. Some critics complained that by giving the audience a variety of versions of Heisenberg's thoughts and motivations, Frayn appeared to be saying that it is impossible to arrive at a final judgment on Heisenberg's intentions and behaviour. Frayn denies the charge.

On the face of it, scientists are no different from other human beings caught in complex moral dilemmas. But there is a problem. Scientists themselves frequently claim that basic science is morally and culturally neutral. At the level of molecules and particles, they argue, there are no ethics, no politics, no culture: water boils at the same temperature in Peking as it does in Berlin. Real scientists, basic scientists, they argue, generate knowledge; technologists, industry, governments apply it. Basic research under Hitler, according to this view, is the same as research under any other government or regime.

In Britain, for example, the one-time chairman of the Committee for Public Understanding of Science, Professor Lewis Wolpert, writes: 'It is not for scientists to take moral or ethical decisions on their own: they have neither the right nor any special skills in this area. There is, in fact, a great danger in asking scientists to be more socially responsible.'[9]

How does this view operate applied, for example, to those German academic researchers during the war who were involved in exploiting data provided by human experiments in the death camps? As Joseph Rotblat puts it: 'A scientist is a human being first, and a scientist second.' A clear example of the depraved values of Wernher von Braun's version of value-free science was a remark he made after the war: that he did not care whether he worked for Uncle Joe or Uncle Sam: 'all I really wanted was an uncle who was rich'.[10]

In a history of the behaviour of scientists in wartime, moral judgments are inescapable. Overall I have been inclined in this narrative to adopt a utilitarian approach, judging the actions of scientists by the known, or at least probable, consequences of their discoveries and published knowledge. Focusing exclusively on consequences, however, may nevertheless fail to resolve the complexity of choices that scientists routinely face. Integrity for a scientist is not confined to choices made in isolation, socially and in time, but involves choices that affect the conduct of a scientist's entire life project. A moral life involves a committed pattern of behaviour, beliefs and principles, leading to feelings and convictions of self-respect. In other words, how does a scientist look himself, or herself, in the mirror every morning?

How, moreover, do scientists behave, as members of a community of specialists? In the course of this narrative we constantly encounter the pressures and demands that are special to scientific disciplines, forming the context in which scientists struggle to attain integrity. Being a scientist involves an unusual degree of *competition*: for funding, academic and professional recognition, employment, promotion, publication. Competition to be *first* with a discovery, or publication, is uppermost in a scientist's mind.

The *loyalties* of scientists are multi-layered and frequently in conflict – between family, institution, discipline, nation. Scientists are unusually *dependent*, compared with artists, writers or composers, say. They are dependent on superiors, patrons, fund-holders, paymasters of every kind.

Scientists, moreover, are constrained by high standards of intellectual integrity, strict codes of procedure in constructing experiments, devising research programmes, collecting data, reporting results. They inhabit a social framework which obliges them to cite, acknowledge, assess and endorse the work of colleagues.

Under the Third Reich, the pressures of competition, dependence and maintenance of standards were increased and exacerbated often to an unbearable degree by a regime determined to exploit every aspect of science and education to its own ends. At the same time, academic and professional integrity were distorted and corrupted by the need to survive, by the desire, in some cases, to thrive, under a depraved regime in which many norms of law – the treatment of 'lives not worth living', for example – had been suspended. At every stage of this narrative we shall see the presence of pressures, temptations and compromises, involving spouses, families, fear of loss of opportunity to do science; desire to receive credit, recognition, resources; conflicting loyalties, towards academic discipline, colleagues, institutions, student bodies and Fatherland. The choices made by scientists, moreover, had consequences not only for the circumstances of the moment but also for their 'projects' as integrated human beings and the morale of their milieu. These are the moral dilemmas 'internal' to their practice. But what of the extrinsic or external moral dilemmas?

Science and Democracy

There is a widespread view, first promoted by the Western Allies in the early stages of World War II, that science reaffirms the values of Western liberal democracy. At first sight the notion of a symbiosis between science and democracy presents a powerful alternative view to the 'irresponsible purity' endorsed by many scientists under the Third Reich. The danger is, however, that scientists working within more or less democratic societies are inclined to abdicate responsibility by assuming that their democratically elected leaders know what is best.

To assume that more or less democratic governments have tended to behave with integrity in the funding, administration and application of science and technology is naive, as the final chapters of this book will seek to demonstrate. Free and well-ordered science, albeit under the auspices of democracy, hardly made a return to the ideals of freedom and diversity in the post-war era, when East and West opposed each other in a climate of secrecy, suspicion, a nuclear missile arms race and the wholesale militarization of science, aided and abetted by industry and the universities. Nor have we seen since 1989 the emergence of a golden age for science as a post-Cold War peace dividend. Today, fourteen years on from the fall of the Berlin Wall, we face alarming trends in the new era of biotechnology and genetics, the continuing pollution of the planet, new generations of nuclear weapons for pre-emptive use rather than deterrence, a multi-trillion-dollar strategic missile 'shield' and increasing incursions on civil liberties in the global war against terrorism.

Challenges to well-ordered science, responsive and responsible to society in an environment of democracy, include the collapse of ideals of free access to information, the spread of intellectual property ownership in the interests of profit rather than freedom of knowledge, the aggressive patenting of nature, attempts to break morally accepted guidelines on the exploitation of human embryos and cloning, continued abuse of the global environment and ecology, the further impoverishment of the developing world, continued research into biochemical weapons.

Scientists cannot ignore the auspices under which they work and receive funding, nor relax their moral and political vigilance, even if they believe that a democracy is the best of all possible worlds for the conduct of science. Pondering, as this book does in its frankly historicist conclusion, the moral and political predicament of scientists today, in the light of Germany's scientific prostitutions in the first half of the twentieth century, prompts not only contrasts and parallels with the past but consciousness of present and future danger.

PART ONE

Hitler's Scientific Inheritance

1. Hitler the Scientist

On his twenty-seventh birthday, 23 March 1939, Wernher von Braun, Germany's brilliant young rocket engineer, met Adolf Hitler for the first time. The Führer had agreed to be briefed on the progress of the army's advanced ballistic missile programme at Kummersdorf West, a research facility south of Berlin.

Walter Dornberger, von Braun's superior, has left an eye-witness impression of Hitler's encounter with one of the most significant high-tech inventions of the century.[1] It was, he reported, 'a cold, wet day, with an overcast sky and water still dripping from the rain-drenched pines'. Hitler's thoughts seemed elsewhere. 'His remarkably tanned features, the unsightly snub nose, little black moustache and extremely thin lips showed no sort of interest in what we were to show him.' Dornberger put on a series of demonstrations of roaring rockets and guidance systems to impress his Führer: he demonstrated the power of a 650-pound thrust rocket motor, then showed off one with a 2,200-pound thrust for comparison. But Hitler 'kept his eyes steadfastly fixed on me', wrote Dornberger. 'I still don't know whether he understood what I was talking about.'

Next the young von Braun, a fleshy-looking young man of Junker stock, gave a presentation of the internal workings of an A3 rocket using a cutaway model; Hitler apparently listened, closely at first, but then stalked off shaking his head as if uncomprehending. Another static demonstration took place, this time with an A5, which was to precede in development a much larger and more sophisticated missile – the A4, the army's missile of choice as a long-range weapon.

At lunch Dornberger sat diagonally opposite Hitler. 'As he ate his mixed vegetables and drank his habitual glass of Fachingen mineral water . . . [Hitler] chatted with Becker about what they had seen,' wrote Dornberger. 'I couldn't tell much from what was said, but he seemed a little more interested than during the demonstration or immediately after.' Later Hitler made the laconic remark, 'Es war doch gewaltig!' (That was tremendous). Dornberger remained puzzled. The visit had seemed 'strange' to him, 'if not downright unbelievable'. Dornberger had been used to visitors being 'enraptured, thrilled, and carried away by the spectacle', like Luftwaffe chief Hermann Goering, who, on being shown the rocket hardware, leaped about laughing and slapping his thighs with unrestrained glee.[2]

Reflecting on the episode after the war, Dornberger wrote that Hitler did not grasp the significance of missile technology for the future. 'He could not fit the rocket into his plans, and what was worse for us at that time, did not believe the time was ripe for it. He certainly had no feeling for technological progress, upon which the basic conditions for our work depended.'

The episode encapsulates Hitler's approach to new technology: his tendency to make decisions in isolation, depending on the certitude of his personal intuition and inspiration, rather than on the basis of careful inquiry and the conclusions of committees. As it happened, Hitler was right to be suspicious of the imminent effectiveness of ballistic missiles in 1939; nor did his apparent lukewarm reaction indicate an unwillingness, as Dornberger infers, to fund further research, at first on a medium level of priority. In time, however, the story of the Führer's decisions and ambitions for the Nazi rocket programme – a technology in which Germany was a generation ahead of the rest of the world – would reveal profound flaws in his capacities as leader of one of the most advanced scientific nations. Hitler became seriously interested in rockets only at a point when defeat seemed inevitable: the deployment of the V2 was to be no more an act of ritualistic vengeance, a gesture of what the novelist Thomas Mann described as 'technological romanticism', than a rational strategy that could help win the war.

Hitler's Bio-political Rhetoric

Hitler's knowledge and appreciation of science and technology were warped, degenerate and profoundly racist. At the Nuremberg trials of the Nazi leadership, Albert Speer, Hitler's architect and his Armaments Minister from February 1942, proclaimed that he, Speer, was 'the most important representative of a technocracy which had showed no compunction in applying all its know-how against humanity'. In a statement to the judge, Speer commented that in a mechanized age dictatorships required, and had produced, a type of individual who took orders uncritically. 'The nightmare of many people that some day nations will be dominated by technology,' he declared, 'almost came true in Hitler's authoritarian system. Every state in the world is now in danger of being terrorized by technology. But this seems inevitable in a modern dictatorship. Hence: the more demanding individual freedom and the self-awareness of the individual.'[3] The former Nazi minister had revealed no such refined ratiocinations while serving the Third Reich, yet faced with the hangman's noose he admitted the insidious exploitation of science and technology in Hitler's totalitarian state, while intimating future dangers for the victors of World War II.

What was absent, however, from his 'confession', which alludes principally to weapons technology, communications and the media, was an acknowledgement that Adolf Hitler's view of science, at its most influential at the outset of the regime, featured crude borrowings from the ambit of pathology and racist 'genetics', to articulate his notion of the German nation state and its destiny. Hitler's favoured rhetorical metaphors, as he rose to power, have been described as 'bio-political'. Hitler subscribed to the idea of the German nation state as a type of anatomy, subject to circumstances of health and disease like the human body.

Hitler betrayed a profound ignorance of Mendelism and particulate inheritance. His 'biological' notions of race evidently found their origins in Joseph Arthur de Gobineau, the French nineteenth-century man of letters and early exponent of racial theory, and a tradition of latter-day racist 'philosophers': Houston Stewart Chamberlain, Erwin Baur, Eugen Fischer and Fritz Lenz. Hitler believed that the purity of the Germanic-Aryan race had been compromised through a 'blending process'.[4] The task ahead was to encourage and preserve uncontaminated stocks of Aryan blood.

By 1925, as Hitler completed his political testament *Mein Kampf*, the racist epithets of Teutonic supremacy, culled from the pamphlets of his lean days in Vienna, were giving way to a vulgarized version of geopolitics, *Lebensraum* – the quest for living space, allied to pseudo-scientific quasi-medical imagery. He harped on the introduction of undesirable hereditary strains into the healthy Nordic body, the *Volkskörper*, and extraneous factors operating like pathogens. Jews were invaders, undermining the integrity of the German organism – bacilli, cancers, gangrene, tumours, abscesses. His political programme was seen in terms of cures, surgery, purgings and antidotes. He lamented in 1925 that the state did not have the means to 'master the disease' which was penetrating the 'bloodstream of our people unhindered'.[5] Such ideas, bogus as they were pernicious, culled from the so-called discipline of racial hygiene, contained inevitable propensities towards solutions which saw the German *Volk* as a patient, the Jew as a sickness and Hitler as the beneficent physician.

The images of Jews as a disease were all too familiar by the mid-1930s as the ideological bio-political content merged with Nazi medical science. The cofounder of the Nazi Physicians League, Kurt Klare, talked of the 'decomposing influence of Jewry'.[6] The *völkisch* body was in need of 'cleansing', according to the physician and Nazi plenipotentiary Dr Gerhard Wagner. Hence the race laws of 1935 were underpinned by images of immunity and calls for radical therapy, the 'cauterizing of the tumour'. By 1940, Hitler was seen as the great 'healer'. In a basic text, explaining the necessity of the invasion of Poland, the Nazi publicist Ernst Hiemer declared that from Poland 'these Jewish bacilli crossed over to us, bringing the Jewish sickness to our land. Our people almost died from this sickness, had Adolf Hitler not delivered us in the nick of time.'[7] As the war progressed, the bio-rhetoric saw the convergence of images that argued a continuity between medical metaphor and prophylactic realism, hastening to an inevitable conclusion. Jews were not only a parasitical invasion of the host body of Germanhood, they were responsible, it was claimed, for actual current epidemics in the East requiring immediate isolation and quarantine – degenerate euphemisms for the ghettos and the camps. In the pathological paradox that frequently attends science as salvation, the purveyors of death thus become those who respect and preserve human life. Just as a physician acts to cut away an infected appendix from a patient, the 'Jew', as declared by Auschwitz physician Fritz Klein, 'is the gangrenous appendix in the body of mankind'.

Hitler and the Bomb

As Hitler's thoughts turned to the conquest of Europe, however, his need to understand the power and scope of applied science and technology for war-making assumed a practical urgency. He was keenly interested in weapons and quick to grasp how a piece of equipment worked. It was often remarked that he could rephrase a long-winded technical account with a terse, highly accurate summary. Speer wrote that Hitler 'was antimodern in decisions on armaments'. Hitler opposed the machine gun because, according to Speer, 'it made soldiers cowardly and made close combat impossible'.[8] He was against jet propulsion, because he thought its extreme speed was an obstacle to aerial combat, and distrusted German attempts to develop an atomic bomb, calling such efforts, according to Speer, 'a spawn of Jewish pseudo-science'.

On 23 June 1942, Albert Speer discussed the atomic bomb with Hitler. Speer wrote in his memoirs that the Führer's intellectual capacity was quite obviously strained by the idea, and that 'he was unable to grasp the revolutionary nature of nuclear physics'. Speer noted that out of 2,200 points raised in his conferences with Hitler, nuclear fission was raised only once and then only briefly. Hitler, it seemed, had acquired a garbled version of atomic science from his photographer, Heinrich Hoffmann, who in turn had picked it up from a minister who was sponsoring an atomic research project for the Post Office. Speer, meanwhile, reported that the head of the official nuclear research programme, Werner Heisenberg, had been unable to confirm that a chain reaction could be controlled 'with absolute certainty'. There had been suspicions among the scientists that a chain reaction, a release of massive energy in fissile material by the instantaneous splitting of its atomic structure, once started, would continue on through the material of the entire planet. Speer wrote that in consequence Hitler was 'plainly not delighted with the possibility that the earth under his rule might be transformed into a glowing star'. Hitler, Speer went on, liked to joke that the scientists 'in their unworldly urge to lay bare all the secrets under heaven might some day set the globe on fire'.

Yet when Hitler invaded Poland in September 1939 there were physicists in Germany who knew at least as much, if not more, than the Anglo-Americans, and who were organizing research programmes for harnessing the power of the atom as a weapon. In fact, it had been a German, Otto Hahn in Berlin, assisted by Fritz Strassmann, with crucial input from Lise Meitner and her nephew Otto Frisch, who first discovered nuclear fission, or the splitting of the atom, in December of the previous year, even though it had probably been first achieved, unwittingly, by Enrico Fermi in Italy.

At the same time, at Peenemünde on the Baltic coast some 200 miles north of Berlin, the German army had by 1939 gathered hundreds of scientists and engineers with unprecedented research and development facilities to create and mass-produce supersonic rockets to enable Hitler to strike at his enemies hundreds of miles distant. In the last year of the war the rocket scientists were drawing up plans for booster rockets that would carry payloads as far as 400 miles and even beyond. Had the Third Reich been first to construct an explosive nuclear device, or even a 'dirty bomb' composed of conventional explosive and radioactive materials,[9] it is likely that its first employment against an enemy would have involved delivery by long-range guided missile, and history would have been very different. There can be little doubt that Hitler would have used an atom bomb had he possessed one. Albert Speer remembers Hitler's reaction to the final scene of a newsreel in the autumn of 1939. In montage a plane dives towards the British Isles: 'A flash followed, and the island blew up in smithereens.' Speer wrote that Hitler's enthusiasm was unbounded. Similarly, when Walter Dornberger, head of the German rocket development project, spoke with Hitler about the potential of ballistic missiles in the summer of 1943, a 'strange, fanatical light' came into the Führer's eyes. Hitler declared: 'What I want is annihilation – annihilating effect.'

Historians of science have argued to this day about the feasibility of a Nazi atomic bomb. It is clear that Hitler's scientists had not overcome the main technological problems by the end of the war; it is also apparent that Germany lacked the matériel, the manpower and economic resources necessary to develop such a weapon during the war. Hitler's racist policies, moreover, had resulted in the dismissal of hundreds of key Jewish physicists, skilled in theoretical and nuclear physics. Hitler's ignorance of science and technology, scientists and engineers, as well as the grotesquely inefficient and corrupt 'polycratic' nature of the power structures of the Third Reich, undermined Germany's ability to win a long-term war based on sophisticated science and technology calling for massive resources. The Manhattan Project, the American atom bomb programme, involved two separate paths – a uranium bomb and a plutonium bomb – while the research and development involved some 150,000 personnel and an expenditure of $2 billion at the time. America could call on these vast resources without strain. With Germany, lacking capacity in every area of weapons production, the case would have been different.

But Germany's failures in science and technology were systemic and wide-ranging. When Hitler went to war in 1939, Germany's education system, once the envy of the world, was in chaos, along with the country's national policies for the fostering and exploitation of science and technology. Some members of the Reich leadership even spoke of closing down the universities until the war was over. Thousands of highly educated technicians, engineers and science students were drafted into the armed forces irrespective of the stage of their education or their usefulness to the war effort; it took three years for the regime to reverse the process. Links between academia and the military, and hence the bias of research programmes, were pursued and fostered within academic institutions out of self-interest rather than rationality, while warring elements of the military vied with each other to appropriate academic personnel; laboratories and institutions as a matter of prestige and inter-service rivalry. In the absence of a rationalized, centralized executive, science and technology in the Third Reich were at the whim of competing warlords and commercial and bureaucratic fiefdoms. At the same time, Hitler deliberately generated rivalries between the main armed services and the SS, while failing to establish a policy for priorities in the complex mobilization and rearmament of a technologically advanced nation state.

Hitler's Scientific Education

Hitler grew to maturity against the background of remarkable advances in science and technology at the turn of the last century in Germany and Austria. Germany's scientific research centres, industry, university science departments, its *Technische Hochschulen* (tertiary-level technical colleges of high academic standing with excellent research capabilities) and later its Kaiser Wilhelm Institutes, were regarded as among the best in the world.

Hitler was, in the words of the historian Ian Kershaw, 'not unintelligent', and he had a formidably retentive memory. He did not attend school beyond the elementary level, and he was by nature indecisive and lethargic. By the time he came to power he had evidently acquired only a remote and patchy knowledge of the natural sciences, engineering and mathematics. Not inclined to rue his lack of secondary and tertiary education, Hitler told the readers of *Mein Kampf* that at school he 'sabotaged' those aspects of the curriculum he thought unimportant and unattractive. Given, nevertheless, to an extraordinary degree of self-assurance, he claimed that at his *Realschule* in Linz, in Austria, his best subjects were geography and history. But he lied. These were among his worst subjects; although his marks in mathematics were even lower.[10] At school, he claimed, he provoked his teachers by exposing the contradiction between religion and science in the curriculum: 'At 10 a.m. the pupils attend a lesson in the catechism, at which the creation of the world is presented to them in accordance with the teaching of the Bible; and at 11 a.m. they attend a lesson in natural science, at which they are taught the theory of evolution.' He said that this contradiction forced him to 'run his head against the wall'. Often, he wrote, he complained on this score to his teachers, 'and I remember that I drove them to despair'. By a curious quirk of fate, Hitler overlapped with Ludwig Wittgenstein, who would become an eminent philosopher; but whereas Wittgenstein was a year ahead of his age, Hitler was a year behind, despite their being a few days apart in age.[11]

According to Speer, who was well educated, a mathematician manqué, an architect and, as Hugh Haffner once commented, a 'pure technician', Hitler did not choose direct routes to obtaining information on science and technology from responsible people, 'but depended instead on unreliable, incompetent informants to give him a Sunday-supplement account'.[12] It was proof, Speer added, 'of his love for amateurishness and his lack of understanding of fundamental scientific research'. Hitler nevertheless liked to pontificate on science, technology and medicine, as if he had got their measure by a process of intuitive genius. He had placed science within his own grand vision of things and would lecture visitors on its history, the physical laws of nature and their applications.

In 1942 we hear him haranguing one of his admirals on the relationship between the shape of fish and the design of ships. The tone of disquisition and its dogmatic appeal to first principles give an impression of Hitler's invincible egotism. 'The current design of ships certainly does not conform to the laws of nature,' he told the admiral.

If it did, then we should find fish furnished with some sort of propulsive element at the rear, instead of the lateral fins with which they are endowed. Nature would also have given the fish a stream-lined head, instead of that shape which corresponds more or less exactly to a globule of water . . . by thickening the prow you reduce by so much the pressure produced from in front on a pointed bow. It is only quite recently, too, that it has been realised that a pointed spade is not the best spade.[13]

In mid-1944, when Luftwaffe senior officers were urging the use of the Me-262 jet aircraft as a fighter, Hitler, insisting that jets should only be used as bombers, lectured them on the physiology of high-speed flight:

Fighter jets, fighter jets – you've got them spinning your heads around – I don't want to hear any more about it! That's not a fighter, and you won't be able to fight with it. My doctor told me that parts of your brain simply shut down in the dog fights which you have to do when in combat.[14]

While projecting an image of healthy living and energetic leadership, Hitler was given to periods of indolence, spending half the night watching banal movies and rising at lunchtime. He espoused an eclectic medley of medical fads, centred, as in his political and administrative life, in a constellation of orbiting and competing influences, from which he chose as if at random so as preclude the monopoly of a particular brand of science or medicine.[15] He subscribed to strict dietary fads (which he occasionally broke) – uncooked fruits, grains and vegetables – based on anecdotal 'science'. At the same time, he imbibed bromides and a highly toxic gun-cleaning fluid, a 'medicine' he believed did his stomach good in the trenches of World War I. Never would he allow an examination of his naked body, as if unwilling to surrender his authority to professional scrutiny. Shortly after becoming Chancellor he went to see an old woman to consult her on vegetarian diet. When his Gestapo chief attempted to broach an administrative matter with him, Hitler snapped that there were 'far more important things than politics – reforming the human lifestyle, for example. What this old woman told me this morning is far more important than anything I can do in my life.'[16] Hitler's vegetarianism, which dates from 1931, was said to be due to the influence of the composer Richard Wagner, who had claimed that human beings had been contaminated by racial mixing and the consumption of meat. Hitler believed that the human lifespan had shortened as a result of sterilization of food through cooking and consequent *Kulturkrankheiten* (diseases of civilization), including cancer. He was convinced that humans started to eat meat when forced to do so by the Ice Age. The promotion of raw fruits and vegetables, in his view, would reverse this process. Anxious about contracting cancer, Hitler was convinced that there might be some danger from the influence of 'earth rays', *Erdstrahlen*, in the chancellery, and had a physician, Gustav Freiherr von Pohl, test for 'emanations' with a dowsing rod.

Amongst the luggage of superstition and half-baked knowledge he carried through life was a trust in astrology. Even in the final days of the Reich, early in April of 1945, he was inclined to place his confidence in a file of astrological nonsense, portending favourable circumstances for his personal horoscope, placed in his hands by his Propaganda Minister, Goebbels. On learning of the death of Roosevelt on 12 April, Hitler was beside himself with glee. 'Here we have the great miracle that I always foretold,' he told his cronies. 'Who's right now? The war is not lost. Read it! Roosevelt is dead.'[17]

In his vision of the thousand-year Reich he placed will over education as the driving force of the national state: science in schools, he declared, was to blame for the fragmentation and chaos of the Weimar Republic. Writing on education policy in *Mein Kampf*,[18] he cited scientific and technical education as a reason for 'the plague of our present-day cowardly lack of will'. But his grasp of what science was and what scientists do was narrow and ill-informed. He vaguely understood that scientific propositions, unlike those of metaphysics, were provisional, that scientific research should be allowed freedom unhampered by the need to teach; but he had a poor understanding of the nature of experimental and empirical method, and tended to identify scientific training with mere accumulations of facts, with scant appreciation of how those facts were obtained.

Once in power, he wrote, he would change the curriculum: 'The scientific school training which today is really the beginning and end of all state educational work can with only slight changes be taken over by the *völkish* state.'[19] The *völkish* state, he emphasized, 'will have to put general scientific instruction into an abbreviated form, embracing the essentials'. He was convinced that Germany was passing through a 'materialized epoch', meaning that 'our scientific education is turning increasingly towards practical subjects – in other words, mathematics, physics, chemistry etc.' These preoccupations, he warned, were 'dangerous' since they implied a neglect of an 'ideal' education and the renunciation of 'forces which are still more important for the preservation of the nation than all technical or other ability'.[20]

All the same, although he would change his views about technical innovation during the war, he believed through the 1920s that the Fatherland's best defence 'will lie not in its weapons, but in its citizens . . . a living wall of men and women filled with supreme love . . . and fanatical national enthusiasm'. There was a proper role for 'scientific education':

Science . . . must be regarded by the *völkish* state as an instrument for the advancement of national pride. Not only world history but all cultural history must be taught from this standpoint. An inventor must not only seem great as an inventor, but must seem even greater as a national comrade.[21]

The reflection was no mere spasm of reactionary waffle, for it echoed precisely the Fulda manifesto signed by ninety-three German intellectuals and scientists, who at the outset of the First War had combined to declare that science and knowledge should be entirely at the service of the nation state in arms. The signatories of the Fulda document had expressed this sentiment at a time when physicists in Britain, France, Germany and Denmark had also been reconsidering many of their discipline's cherished structures and preconceptions, questioning whether indeed science could be collaborative and internationalist during wartime. Hitler, who claimed that he had carried Schopenhauer in his greatcoat pocket through the First War, only to reject him later for the 'genius of Nietzsche',[22] had formed a highly nationalistic overview of science; but he was not alone in his failure to perceive the dependence of new physics on pluralist, internationalist collaboration. But Hitler, it seemed, had not progressed much in his thinking beyond potted versions of seventeenth-century astronomy and distorted farragoes of Darwinism.

Sitting over cakes and tea on long evenings in his headquarters during the war years, Hitler often turned to the remarkable transformation effected by the inventions of the microscope and the telescope. His gift to the nation, he liked to say, would be the construction of observatories to be placed like municipal libraries in the towns and cities of Greater Germany. In this way, he claimed, he would overcome 'a world of superstitions'.[23] Among his 'astronomical' interests was an inclination to credit a bizarre theosophical theory known as 'glacial cosmogony', Hanns Hörbiger's mix of myth, imagination and pseudo-science. The theory, enthusiastically espoused by Heinrich Himmler, the SS leader, claimed that the universe was formed out of ice interacting with an original sun that was believed to have been 30 million times bigger than our contemporaneous sun. Hitler commissioned the architect Professor H. Gieseler to design a planetarium and observatory in Linz, the city of his childhood, with a gallery in which Hörbiger would take his place alongside Copernicus, Galileo, Newton and Kepler.[24] Ironically, the famous 'Einstein Tower', the spectacular futuristic observatory built by the astronomer Irwin Finlay Freundlich and the architect Erich Mendelsohn at Potsdam to explore the implications of Einstein's theory of relativity, ceased to do serious basic astrophysics under Hitler. Freundlich was removed as director for questioning the order to give the Nazi salute within the building. The new director, the Nazi Hans Ludendorff, banished the name of Einstein and called it the Institute for Solar Physics. Research based on Einstein's relativity theory was dropped, and the observatory became a centre during the war years for research with military relevance – the link between turbulence in the solar atmosphere and radio interference.[25]

An early decision taken shortly after he came to power was the dismissal of Jews from the civil service. Since most scientists and academics were civil servants in Germany, this meant the loss to the country of hundreds of distinguished physicists, chemists, biologists and mathematicians. The eminent physicist Max Planck went to Hitler and argued the case for retaining Germany's top Jewish scientists despite their non-Aryan predicament. In particular, Planck had in mind Fritz Haber, Jew, patriot, dean of the German science community and director of the world-famous Kaiser Wilhelm Institute for Physical Chemistry in Berlin. Haber had discovered the chemical process for the fixation of nitrogen from the air which provided a blockaded Germany in the First War with potentially unlimited and cheap artificial fertilizers and explosives. Had it not been for Haber, who in deference to prevailing prejudices had converted to Christianity in his early twenties, Germany could not have prosecuted the war for more than a year. Hitler was well aware of the epoch-making difference Haber had made to Germany's industrial and military fortunes, but he did not allow this knowledge to affect his wider prejudice. His notorious response to Max Planck's pleading was this: 'If science cannot do without Jews, then we will have to do without science for a few years.' Through the mid-1930s some of the finest scientists in the world consequently made their way from Germany and Austria to Britain and the United States, with inevitable consequences for both sides in the coming world conflict. But the racist notions underpinning the act had other implications for the outcome of the war against Russia. Blinded by racial prejudice, Hitler was convinced that the Russians were incapable of conducting a technological war of the twentieth century.

Hitler believed that certain races, like the Russians, were incapable of making things: 'The Russians never invent anything,' he liked to say. 'Give them the most highly perfected bombing-sights. They're capable of copying them, but not of inventing them.'[26] Nor were they capable, in his view, of systematic thought, work and organization. Shortly after the invasion of the Soviet Union, in the summer of 1941, he opined over supper: 'The Russian will never make up his mind to work except under compulsion from outside, for he is incapable of organizing himself.' He added: 'And if, despite everything, he is apt to have organisation thrust upon him, that is thanks to the drop of Aryan blood in his veins.'[27] Destroy their factories, he declared, and 'the Russians cannot rebuild them and set them working again'. The Japanese, on the other hand, were capable of 'improving something that exists already, by borrowing from left and right whatever makes it go better'. Lumping together the Russians, the Chinese and the Japanese, he pronounced: 'These people are inferior to us . . . In the sphere of chemistry, for example, it's been proved that everything comes to them from us.'

By the time he had reached his prime, Hitler had constructed a confident, self-taught 'philosophy' in which he had relegated science, the laws of the universe, race and religion to their proper spheres and scales of significance. During those evening monologues at his headquarters he was given to expatiating about science and its relationship to politics and religion. He believed that 'progress in science' had led 'liberalism astray into proclaiming man's mastery of nature'. He advocated a holistic, intuitive acquaintance with 'the laws of nature', but science, he counselled, cannot deal in ultimate questions – 'as for the *why* of these laws, we shall never know anything about it'. And yet, the ultimate questions, he declared, completing the illogical coda, cannot be answered by religion.

Religion, science's enemy, is doomed by science, he insisted. 'The best thing is to let Christianity die a natural death,' he told Himmler early in the war.

The dogma of Christianity gets worn away before the advances of science. Religion will have to make more and more concessions. Gradually the myths crumble. All that's left is to prove that in nature there is no frontier between the organic and the inorganic. When understanding of the universe has become widespread, when the majority of men know that the stars are not sources of light but worlds, perhaps inhabited worlds like ours, then the Christian doctrine will be convicted of absurdity.[28]

Amidst such rants he was given to propounding mind-numbing nostrums to his captive audiences. 'Let us compare science to a ladder,' he liked to say. 'On every rung, one beholds a wider landscape. But science does not claim to know the essence of things.'[29] He had grasped the inability of science to answer meta-physical questions, but science was providing the world with some-thing more valuable: the ability to gaze upon the very small and the very large. 'Science has taught us,' he declared, 'that we are hemmed in not only by the infinitely great, but also by the infinitely small – the macrocosm and microcosm.'[30]

In contrast, although Roosevelt, Churchill and, later, Truman were not educated in science, the Allied leadership maintained an astute respect for the experts. They listened, they took advice, they learned, sometimes painfully, and they consulted and formed expert committees. Stalin, prejudiced in favour of a Lamarckian view of inherited characteristics, supported the bogus plant science of Trofim Denisovich Lysenko, but he evidently showed presence of mind in opting for the robust engineering projects that provided the Red Army with the war-winning T4 tank; after the war his scientists kept pace with America in pursuit of the hydrogen bomb.

Roosevelt, despite an extraordinary aberration early in the Pacific war, involving his plan to drop thousands of bats over Tokyo in the belief that they would terrify the Japanese (the bats died in transit), was open to extensive advice from committees dealing with scientific matters as well as being accessible to relatively recent migrant scientists like Albert Einstein and Leo Szilard, the Hungarian-born physicist. Churchill, who as First Sea Lord expended vast sums in the early months of World War II on useless mechanical trench-diggers to cross what he envisaged would be a World War I no man's land, was nevertheless fascinated by science and technology and asked searching questions of the experts. One of his closest friends during the 1930s and during the war was the physicist Professor Frederick Lindemann.

Hitler, however, portrayed himself as a war leader who had grasped all there was to know about the importance of inventive genius and technology. He found it difficult to bridge the gap between his acts of will, his fiats and the technical feasibility of his decisions. 'What's important,' he told a group of dinner-table cronies early in 1942, 'is to have the technical superiority in every case at a decisive point. I know that; I'm mad on technology. We must meet the enemy with novelties that take him by surprise, so as continually to keep the initiative.'[31] His war against the Allies, he declared, was 'technological'. He spoke as if technical weapons wizardry could be conjured out of thin air; that he could exploit wonder inventions by the application of unassailable logic. 'Ten thousand bombs dropped at random on a city are not as effective as a single bomb aimed with certainty at a powerhouse or a waterworks on which the water supply depends.' At the same time, he told Dornberger, the leader of the rocket plant at Peenemünde, that Germany must build tens of thousands of rockets, at once, and this at a time when the scientists were finding it difficult to make one successful prototype.

From the day Hitler assumed the leadership of Germany in January 1933 nothing he said, or did, indicated that he understood the prodigious scientific and technological legacy of Germany, how it was ordered and organized, how it could be harnessed, or how its ongoing programmes could be ordered according to priority. There is no evidence that he appreciated or understood anything of the nature of its origins and the history of its development.

2. Germany the Science Mecca

In the first three decades of the twentieth century Germany stood pre-eminent in many areas of the natural sciences, mathematics and technology, despite an intervening ruinous war. In 1900 Berlin had failed in its bid to host the great International Exhibition – the honour went to Paris – but this did not prevent the German exhibits, which stretched over fifteen locations from the Tuileries to the Eiffel Tower, outstripping those of other nations in wonder and inventiveness.[1] Millions of Germans made their way to Paris to glow in pride before the scientific and technological achievements of the Fatherland. On the Quai des Nations the Germans had erected a pavilion in the form of a sixteenth-century *Rathaus*, with a steeple that towered above its rivals. Apart from the national pavilions, the participating countries had been encouraged to exhibit side by side in thematic clusters of technology – the emphasis being on the usefulness of products to all nations. The Germans, however, had decided against showing their wares according to company origins, hence emphasizing the national origins of their advanced technology rather than mere trade names.

The Germans had brought together such advances as liquefaction of gases, production of electrical power and electrochemistry.[2] One exhibit revealed the rare earths essential for the 'illumination industry', a feature of intense interest in the palace of electricity, which displayed the brilliance of Walther Nernst's patent lamps. The purest thorium, crucial to the burgeoning electrical lighting industry, was produced by Germany, according to the accompanying promotion; and the German price had been reduced from 2,000 marks per kilo to 50 marks per kilo in the space of six years. In the science and technology exhibition situated on the Champs de Mars visitors gazed at the largest and most powerful dynamo on earth, constructed by Germany's Helios Company. As for the instrument section, Britain's science journal *Nature* reflected that the prodigious array of German precision instruments was not 'one which brings great pleasure to an Englishman, and if he moves on to examine the English exhibit his thoughts cannot fail to be very grave'.[3] The German exhibition catalogue was a major publishing event in itself, a 250-page volume incorporating the soaring, unstoppable hubris of German science and technology. Published in three different language editions, German, English and French, and lavishly illustrated with colour illustrations and Art Nouveau design, it was bound in woven beige muslin. The publication covered the history, development and applications of the German exhibits. The introduction contained a survey of Germany's population expansion, with facts and figures on female employment, divorce, racial and religious groups and emigration. Between 1895 and 1899, the population had risen from 52 million to 55 million, having doubled between 1816 and 1900. Germany was a young country, 61 per cent of the population aged under thirty. There were 4.7 children in every family, and only one in eighty marriages ended in divorce.[4]

And as if the tangible results of Germany's pre-eminence in so many fields was not overwhelmingly evident at the exhibition, there was a striking reminder in Paris that summer of German leadership in the realm of mathematics. The organizers of the second International Congress of Mathematicians had invited the German mathematician David Hilbert to deliver the keynote address in a lecture hall of the Sorbonne University. Hilbert, described that day as 'wiry, quick, with a noticeably high forehead, bald except for wisps of still reddish hair', had about him 'a striking quality of intensity and intelligence'.[5] He chose to use the centenary celebration as an opportunity to set the mathematicians of the world a series of unsolved problems; these were not just mind-teasers, but the profound unfinished business of mathematics. Hilbert told his audience that these problems

show how rich, how manifold and how extensive mathematical science is today . . . the organic unity of mathematics is inherent in the nature of this science, for mathematics is the foundation of all exact knowledge of natural phenomena. That it may completely fulfil this high destiny, may the new century bring it gifted prophets and many zealous and enthusiastic disciples![6]

It was a prophetic statement, and it seemed only natural that a leading German exponent of the discipline should make it.

Twenty years on, despite the loss of a terrible war which devastated the nation's wealth and morale, the promise shown at the great Paris exhibition – that the twentieth century would be *Germany's* century – still seemed possible. By 1921, twenty years into the institution of the Nobel awards, Germans, or at least German-speakers, had won half of all the prizes awarded for the natural sciences and medicine. One of the earliest awards in the post-war era went to Albert Einstein for the photoelectric effect: Einstein, born in Germany, had renounced his German citizenship to become stateless, before acquiring Swiss citizenship; but there were still many Germans, and non-Germans, who regarded him as one of them.

Despite Germany's reputation as a pariah nation in the aftermath of a devastating war, despite boycotts against its participation in conferences, and cancellations of periodical subscriptions, Germany was still viewed as the Mecca of science and technology, its language a crucial condition of scientific education and advancement. The roll-call of Nobel greats, between 1901 and 1921, gives an impression of the range of German genius in the first two decades of the century. Wilhelm Konrad Röntgen for the discovery of X-rays; Emil Adolf von Behring for serum therapy; Adolf von Baeyer for work on organic dyes; Wilhelm Ostwald for chemical equilibria and rates of reaction; Philipp Lenard for cathode rays; Max von Laue for diffraction of X-rays by crystals; Max Planck for discovery of energy quanta; Fritz Haber for the synthesis of ammonia from its elements; Walther Nernst for his work on thermodynamics. And then there was Einstein.

What was it about Germany, its education system, its history, its national characteristics even, that made it so pre-eminent and wide-ranging in original research from the late nineteenth to the early decades of the twentieth century?

Germany's Dyestuffs Challenge

The daunting enterprise of German science into the second decade of the twentieth century had been founded on the discipline of chemistry; in particular, the talent of the Germans for exploiting the potential of new chemical processes in the manufacture of dyestuffs. These chemical technologies freed Germany from dependence on the British Empire for many raw materials and products. That new freedom, in turn, combined with expanding iron, steel and coal industries, had enabled Germany to rise from the status of an agricultural economy to challenge France, Britain and the United States. The consequent competition between Germany, as a burgeoning European great power by the first decade of the twentieth century, and Britain would be a contributing factor in the tensions that led to World War I.

The foundations of Germany's expertise for exploiting chemical processes had been laid early in the nineteenth century and were synonymous with the name of Justus von Liebig, who in 1824 founded his great laboratory at Giessen, where he trained large numbers of chemists. To this day he is associated with nitrogen cycles, and the law named after him, which states that plant growth is limited by the element present in soil in the least adequate quantity. The science of nitrogen cycles would grow to supreme importance as Germany's young population multiplied. Liebig's genius anticipated Fritz Haber's great discovery, made in 1908, that would free Germany from dependence on nitrogen imports.

In 1828 Friedrich Woehler succeeded in converting ammonium cyanate into urea, the product of animal kidneys. Before this it was believed that molecules of living organisms could only be synthesized by living organisms. Chemists now realized that they were on the brink of making a host of new substances that do not occur in nature. Then in 1841 a student at Justus von Liebig's laboratory, August Wilhelm von Hofmann, received his doctorate for a thesis on the derivatives of coal tar (a waste product of coal burning), a substance that would prove crucial in the new world of chemical transformations. Hofmann found that coal tar could yield aniline, which had in the past been found in the distillation of indigo. Could artificial dyes, he wondered, be obtained from coal tar?

Four years later Queen Victoria and Prince Albert, visiting Bonn to celebrate the seventy-fifth anniversary of Beethoven's birth, met Hofmann in his laboratory, a set of rooms where Prince Albert had once lodged as a student. The royal couple were impressed; in consequence Hofmann was invited to head the Prince's new Royal College for Chemistry in London's Oxford Street. In London Hofmann brought together a group of chemists, both German and English, selecting and directing their research projects. An entire new industry was born in Hofmann's laboratory, when in 1856 a seventeen-year-old student chemist called William Perkin, while seeking a means of making quinine from coal tar, stumbled on a process for making a rich artificial dye – the colour widely referred to as 'mauve'. Seeing the commercial possibilities of such an artificial dye, Perkin's father persuaded his son to leave the laboratory, against Hofmann's advice, and to go into business with him. Perkin and son achieved rapid but modest success, considering the potential they had unleashed. Mauve became the rage in the streets and salons of Europe's capitals. The French Empress thought mauve matched the colour of her eyes, and Queen Victoria wore a mauve dress to a wedding.[7]

Meanwhile Hofmann developed other good artificial dyes, especially a range of violets. There followed a rainbow of brilliant colours – reds, blues, greens – yet the industrial and trading advantage was to be reaped not in Great Britain but in Germany, which invested extensively in new technology aimed at exploiting the techniques, while pursuing canny patenting policies and marketing strategies to ensure German monopoly. Many chemistry graduates migrated to England to gain work experience in industry; but most of them returned to Germany to take part in the Fatherland's expanding industries.[8]

Perkin himself, who grieved over the loss to Britain of the rewards of his invention, complained for the rest of his life about the British tendency to seek only short-term gains from investment. Equally decisive was the early symbiosis between science education and industry in Germany. It had been noted as early as 1853, by a fact-finding mission despatched to Germany by Prince Albert, that Britain, by comparison, displayed 'an overweening respect for practice and contempt for science'.

Germany could back the new developments with academic research. In the latter part of the nineteenth century some twelve *Technische Hochschulen* linked with technical secondary schools known as *Realschulen*. Unlike Britain, where education expansion lagged, the German school system developed ahead of industrialization. Meanwhile chemistry departments in German universities were already flourishing in Giessen, Göttingen, Heidelberg and Bonn. At Bonn, for example, the Bohemian-born chemist August Kekulé had discovered the benzene ring, a revelation that had come to him in a dream. He saw a vision of chemically linked carbon atoms, like a serpent, which eventually grabbed its own tail in its mouth. The insight was essential to the understanding of chemical formulae of the known carbon compounds.

A host of discoveries, techniques and processes flowed from Germany's laboratories, the greatest of which was Bayer's organic chemistry laboratory in Munich. Some of the developments came directly from dye technologies. F. W. Beneke succeeded in staining plant and animal cells with aniline dyes in the 1860s. In the following decade Paul Ehrlich in Frankfurt realized that coal-tar staining could highlight parts of the cell. Methyl green, for example, allowed the cell nucleus to stand out in green while the cytoplasm stained red. The staining processes revolutionized biology and heralded the science of biochemistry. Ehrlich's cousin, Carl Weigert, showed that methyl violet highlighted certain bacteria, which assisted Robert Koch's research in Berlin on the bovine anthrax bacillus and tuberculosis. Meanwhile Carl Zeiss had set up his optical instrument workshop in Jena, giving impetus to cell biology, anatomy, histology, embryology and pathology by means of improved microscopy. Typical of the marriage of optical instruments and the birth of biochemistry in the second half of the nineteenth century was the work of Hermann von Helmholtz, physiologist and, as he was known, 'the Reich Chancellor of Physics', the first to measure the speed of nerve impulses in frogs (20 metres per second) and detect with his new ophthalmoscope a live human optic nerve. By the end of the nineteenth century the German optical instrument industry, led by the Zeiss and Leitz companies, was supreme in the world. German students of science took it for granted that they should possess a microscope; this increased production and made Germany's instruments irresistibly cheaper than the products of her trading rivals.

The German university system and the *Technische Hochschulen*, unembarrassed by entrepreneurial associations, forged ever closer ties with industry, and industry with the universities.[9] In the course of the nineteenth century three kinds of relationship reinforced those connections: personal ties between industrial chemists and their contacts in the universities, formalized by chemistry research organizations; the exchange of research based on personal contacts; and the flow of trained chemists from academe into industry. By the first decade of the century ad hoc arrangements between individual researchers and individual companies had developed into collective relationships between laboratories, and within the profession as a whole. Well-funded research centres flourished, and graduate students and post-docs were encouraged to sample, like wandering scholars, courses and research programmes in a variety of universities and industrial laboratories. University experience was intense, with hard work, frequent bouts of heavy drinking, duelling and rigid codes of behaviour towards women. Vacations were spent climbing mountains in Germany, Austria and Switzerland.

Mechanistic, mathematical and empirical, German research appeared to have abandoned any lingering association between the natural sciences and philosophy by narrowing the scope of scientific inquiry. Justus von Liebig is credited with laying the foundations of a reductionist approach to scientific research in biology – understanding the mechanistic whole by scrutiny of the smallest parts. His approach to studying living organisms gave full encouragement to scientific materialism, thus repudiating the metaphysical vagaries of *Naturphilosophie*.[10] The brigades of students trained by von Liebig entered the teaching and research communities, influencing future generations of German students in their rejection of vitalism and speculations about the meaning and purpose of living things, so as to concentrate on the strictly physico-chemical components of biological processes. The spirit of the reduction of vital to inanimate is captured by Thomas Mann in his *Doctor Faustus*, when the narrator, Zeitblom, meditates on the researches of Adrian Leverkühn's father, in which the margins of the organic and the inorganic appear interchangeable:

If I understood my host aright, then what occupied him was the essential unity of animate and so-called inanimate nature, it was the thought that we sin against the latter when we draw too hard and fast a line between the two fields, since in reality it is pervious and there is no elementary capacity which is reserved entirely to the living creature and which the biologist could not also study on an inanimate subject.[11]

There seemed no end to the possibilities as Germany led the way in laboratory biomedical science and pharmacology to form the basis of a global pharmaceutical industry: in the 1890s the invention of aspirin at the huge Bayer plant transformed everyday treatment for pain and fever, while at Hoechst am Main the invention of Novocaine revolutionized local anaesthetic therapy, banishing the more grievous tortures of the dentist's chair. Ehrlich won international fame, and eventually his Nobel prize, for Salvarsan, a synthetic pharmaceutical product for the treatment of syphilis. Between 1870 and 1900 the cost of synthetic dye produced in Germany dropped from 60 marks per kilo to 1 mark per kilo. In the years running up to World War I German products would flood the world's markets: soaps, detergents, paints, printing inks, glazes, laboratory staining dyes, pharmaceuticals, chemical processes for iron and steel production, photographic materials, explosives and above all fertilizers. During the same period Germany's output of iron and steel outstripped even Great Britain's and was second only to that of the United States.

3. Fritz Haber

Fritz Haber, destined to become one of the greatest chemists of the early twentieth century, was a product of the German system, typical of the kind of scientist who stood to make the twentieth century 'Germany's Century'.[1] Fritz Haber's story embodies in striking fashion science's Janus-faced power for good and evil, its potential for peaceful beneficence as well as for violence and atrocity, within a single formula or set of equations.[2] The discovery that bears his name, and the name of Carl Bosch, was a process that culminated in plentiful supplies of artificial nitrogen fertilizer at low cost and thus allowed the world's population in 2000 to reach about 6 billion rather than the 3.6 billion which would have been the estimated maximum without such a discovery. Yet, by the same token, and the same formula, he enabled a blockaded Germany to produce high explosives so as to prolong the hideous trench warfare of 1914–18. Haber's discovery was a powerful exemplar for the notion that science was value-free, neutral, apolitical: the scientist discovered the laws of nature and invented applications; the good and evil perpetrated by those applications was on the consciences of others. And yet, Fritz Haber himself would one day demonstrate the aptitude of scientists for taking a lead in warfare, of blurring the distinctions between basic and applied science. He not only initiated the technological means to make gas warfare in the First War which, by some estimates, killed 1.3 million troops, but enthusiastically led the first attacks from the front.

Fritz Haber was born into a family that had prospered on the success of chemical dyestuffs; his father, Siegfried Haber, was a wealthy Jewish businessman of Breslau, who traded in the chemical processes and products that were making Germany the pre-eminent nation state in Europe. Fritz came up through one of Breslau's top gymnasia, and had benefited from an all-round education typical

of the era, including literary and classical studies. While Germany's
secondary education in mathematics and the natural sciences was
second to none in the world, the educationalists of the new
Germany insisted on the need for a balanced curriculum. At school,
Fritz developed a love for poetry.

His father had set his heart on Fritz entering the dyestuffs business,
but he encouraged him to lay the foundations of an academic
career in the natural sciences. Haber studied chemistry at the University of Berlin under the same August Wilhelm von Hofmann
who had supervised William Perkin in the chemistry laboratories
in Oxford Street in London and who had returned to Germany in
1865. Haber then went to Heidelberg and worked under Robert
Wilhelm Bunsen, one of the great pioneers of spectroscopy, before
returning to Berlin to register at the *Technische Hochschule* in
Charlottenburg.

After a year's military service he spent a brief period in his father's
business, where he managed to make some significant losses as a
result of impetuous transactions. After two more university moves,
to Zurich and to Jena, Haber finally settled at the *Technische Hochschule* in Karlsruhe, where he would stay for seventeen years, publishing some fifty papers on a wide range of topics including organic
and inorganic chemistry, combustion chemistry, electrochemistry
and engineering.

In 1892, aged twenty-four, he converted from Judaism to
Lutheranism, and dropped the middle name, Jacob, to evade associations that could blight the prospects of a scientist's, or any high
official's, career in Germany at that time. Photographs of him reveal
a caricature, posturing Prussian – he sits sideways on to the camera
in military jacket, square, shaven head, a deep scar from lip to jaw
(obtained, it seems, in a duel),[3] severe pince-nez, thrusting chin –
übermenschlich: his admirers would refer to him as a 'superman'. Yet
his darkly intelligent eyes and the wide and full sensual mouth
reveal a passionate, dreamy dimension. He appears an embodiment
of German efficiency, obstinate energy and science; but he had a
capacity for friendship, an accident-prone weakness for women (his

first wife killed herself, his second divorced him) and an aptitude for poetic sentiment.

In 1901 he had married Clara Immerwahr, a girl from his home-town of Breslau. She was a chemist and the first woman Ph.D. at Breslau University. She too had converted from Judaism to Protestantism to further her scientific career. She had nourished hopes of sharing a full intellectual life with her brilliant husband; but she soon discovered that his workaholic ways left little time for conversation and collaboration. In 1902, one year after their wedding, and shortly after the birth of their son Hermann, Haber departed for a four-month tour of the United States to study the country's scientific and engineering education systems. When he returned, he toured Germany, lecturing on what he had seen. In 1909 Clara wrote to her friend Richard Abegg, bemoaning Haber's behaviour:

It has always been my attitude that a life has only been worth living if one has made full use of all one's abilities and tried to live out every kind of experience human life has to offer. It was under that impulse, among other things, that I decided to get married at that time . . . The lift I got from it was very brief . . . and the main reasons for that was Fritz's oppressive way of putting himself first in our home and marriage, so that a less ruthlessly self-assertive personality was simply destroyed.[4]

According to colleagues Clara Haber became a 'Hausmütterchen' (little domestic matron), never out of her apron. As Clara became diminished, Haber flourished.

By the first decade of the century, the widespread enthusiasm for science and technology in Germany had translated into a key Prussian decision to encourage research projects unhindered by industrial aims or teaching duties. Germany had prospered by applying science and technology directly to industry: it had not lost sight of the need for pure scientific research to maintain its pre-eminence, yet it had no independent research institutes to rival the Pasteur Institute in Paris, or the American foundations – the

Rockefeller Medical Institute, the Carnegie and Cold Spring Harbor. In the final years of the nineteenth century Friedrich Althoff, the Prussian minister responsible for university affairs, had pondered a response to American competition, convinced that Germany should take advantage of the fact that the obligation to teach could hamper research, and that in any case not all researchers were good at teaching.

In consequence, by 1910 the German Emperor had assumed patronage of a research society and associated specialist centres named after him – the Kaiser Wilhelm Society and its institutes (after World War II, renamed Max Planck). The institutes, incorporating offices, laboratories and living quarters, were mainly situated in a leafy suburb of south-west Berlin with the hope that the district might one day become a 'German Oxford'. In 1908 some fifty hectares of land had been purchased for the erection of six institute buildings in Dahlem, sometimes known as Berlin's Hampstead. Close to the lakes and forests of Wannsee and Potsdam, the district would one day boast impressive villas by noted German architects: Peter Behrens, Hermann Muthesius and Walter Gropius. In 1895 there were only 153 people living in Dahlem; by 1914 there were 5,500. The first group of institutes was dedicated to chemistry, physical chemistry and biology.[5]

The spirit of the Kaiser Wilhelm Society and its institutes was admirably expressed in an essay penned in 1909 by Adolf von Harnack on the state of German science. Despite the widely held respect for German research, von Harnack suggested that there were many reasons to fear for its future.[6] He appealed to the ideal, widely ascribed to Wilhelm von Humboldt, for the unity of research and teaching, as well as independent institutes funded not solely by the state but by private philanthropy. At the same time he emphasized that Germany's greatness was based on the pillars of two key German drives: *Wehrkraft und Wissenschaft* – military might and science in the widest sense of knowledge.

The shaping of the individual institutes, as it turned out, involved eminent scientists and the views of industrialists. One of the Society's most generous supporters was Leopold Koppel, a Jewish banker, who funded (to the tune of 700,000 marks) the Kaiser Wilhelm Institute for Physical Chemistry in Berlin, on condition that Fritz Haber should run it. Haber accepted and was soon joined by Max Planck, Walther Nernst and Albert Einstein in neighbouring institutes (most of the KWI buildings still survive in Dahlem incorporated into the Free University of Berlin; the society headquarters was in Berlin Palace, which was razed to the ground by the Soviet occupiers in 1945). Preparing in his colleague Richard Willstätter's home for a possible audience with the Kaiser on his appointment, Haber practised walking backwards while bowing, and managed to break one of his host's vases as he did so.[7]

Haber now applied his more characteristically assertive tendencies to reach a solution to a long-standing problem in chemistry. In the late eighteenth century it had been discovered that ammonia, essential for fertilizer and explosives, was composed of one atom of nitrogen and three of hydrogen. From that point onwards chemists had been attempting to synthesize ammonia from these two plentiful gases in nature, without success. The problems were technically formidable as the synthesis involved the application of pressures 200 times the atmosphere at sea level and temperatures of 200°C (390°F). When this was first achieved in difficult laboratory conditions by Haber and his English assistant Robert Le Rossignol, the process yielded only a trickle of ammonia over long periods of time, precluding any hope of industrial production. Haber realized that a catalyst would have to be found to speed up the reaction. After a trial of countless metals, Haber discovered that the powder of the rare metal osmium (only 220 pounds of which existed in the world) produced the desired result. On one of chemistry's great red letter days, 2 July 1909, Haber made a demonstration of his ammonia production at the rate of seventy drops a minute in the presence of the two technical heads of the great German chemical concern Badische Anilin- und Soda-Fabrik (BASF), Alywin Mittasch and Carl Bosch. During the demonstration one of the bolts of the pressure apparatus burst, delaying matters for several hours. Twenty years later one of Haber's students remembered the stress that attended the event: *Tücke des Objeckts*, he called it – 'the inherent malevolence of inanimate objects'.[8] BASF now pressed into service Carl Bosch and Alywin Mittasch to improve and make an engineering reality of Haber's process. Although the company had bought up the entire available stock of osmium in the world, the team was determined to find a catalyst that was both more rapid and more plentiful. After 4,000 trials they came up with an ideal catalyst composed of iron, oxides of aluminium, calcium and potassium. The recipe is virtually unchanged to this day.

The Birth of the Giant IG Farben

The bid to be first to produce and exploit artificial ammonia occurred against the background of a major transformation in Germany's chemical industries. By the first decade of the century the success of Germany's major dyestuffs businesses was a brake on profits in foreign and domestic markets. The industry was fraught with price wars, patent wrangles, bribes to major clients and industrial espionage. The warring companies were locked in hostile competition and a trust or cartel arrangement appeared the only solution. The impetus came from Carl Duisberg, the chief executive of Bayer, a keen pan-German, an inexorable Prussian and an advocate of the leadership principle (*Führerprinzip*) long before Hitler endowed the term with its violent, dictatorial implications. Duisberg was to survive and thrive through the reign of the Kaiser, the era of the Weimar Republic, and into the regime of Adolf Hitler, leading and shaping the increasingly dark and culpable fortunes of IG Farben. IG Farben would emerge in the period of the Third Reich as one of the most formidable and easily the most corrupt multinational in the world.

Duisberg's inspiration to draw together the leading chemical companies into a cartel had been prompted by a trip he made to the United States in 1903. His aim was to establish a Bayer-owned plant at Rensselaer, New York, in order to overcome American protective tariff laws. While in America he studied the trust movement – cooperation to overcome competition – which was flourishing despite the 1890 Sherman Antitrust Act. Duisberg looked closely at the organization and advantages of the Standard Oil trust, which had succeeded in the United States in neutralizing many of the problems encountered in Germany by the chemicals industry. In consequence Duisberg negotiated a formula which brought together the German chemical 'Big Six' – Bayer, BASF, Agfa, Hoechst, Cassella and Kalle – into an *Interessengemeinschaft* (community of interests), hence 'IG' Farben. The aim was to reduce competition between the members and to establish a means of profit sharing. Each company nevertheless maintained its independent identity, each was free to research and develop other products outside the core activity of dyestuffs. Agfa, for example, specialized in photographic materials, Bayer and Hoechst concentrated on pharmaceuticals.

By the autumn of 1913, BASF, which had invested heavily in the nitrogen-fixing formula, was producing up to 5 tons of synthetic ammonia a day. By July of the following year, on the brink of war, the plant was capable of producing 40 tons a day, mostly for fertilizer. The ammonia production plant at Oppau on the Rhine with its rows of tall chimney stacks would soon become one of the industrial wonders of the world and production would expand to include six ammonia-producing factories in Germany.

Although on the declaration of war the British navy was blockading the import of Chilean saltpetre stocks to Germany, it took the War Ministry in Berlin some months to understand the importance of the ammonia process for the German war effort in the production of explosives. In the second week of September 1914 the Wehrmacht's juggernaut towards Paris was stopped in its tracks by a French counter-attack which exposed the German lack of gunpowder. Bosch was ordered to Berlin and a deal was struck, involving a substantial subsidy, to exploit existing ammonia plants for the conversion of ammonia, via nitric acid, to explosives. In May of 1915 Bosch announced that Oppau was set up to mass-produce nitric acid. Germany was now free of those Chilean imports and was set to make the Royal Naval blockade impotent. Bosch soon persuaded the government to subsidize a massive expansion in the

BASF's nitrate production, which led to a new high-pressure plant at Leuna. Meanwhile, under Haber's direction, the Kaiser Wilhelm Institute for Physical Chemistry began to collaborate, with industry, the military and the government, anticipating both the phenomenon of the industrial-military complex and 'Big Science' – as Fritz Stern has commented, 'a kind of Manhattan Project before its time'.[9] The phenomenon of 'Big Science', denoting huge funding and investment, from industry, the state and the military, large complex plant and machinery and large teams of scientists and engineers, was to become a key feature of science later in the century. Researchers within the ambit of Big Science would be diminished, with a consequent reduction of personal responsibility for its products.

Haber reported to the Imperial Civil Cabinet in 1916 that the production of chemicals suitable for fertilizers and explosives had reached 2,400 tons a month, making it possible for continued prosecution of the war.

Britain's Dreadnought Project

Was Germany, then, the instigator of the industrial-military complex and the Big Science that would dominate Western science and technology in World War II and the Cold War? Germany had not been alone in showing such propensities. As the Fatherland expanded to the status of an industrial giant and a great European state in the last decade of the nineteenth century, bursting the framework of the continental balance of power, tensions had arisen with Great Britain, which was traditionally reluctant to permit Europe to be dominated by a single nation state. Germany's power beyond Europe was insignificant, whereas Britain's empire accounted for a third of the populations of the world and a vast trading enterprise secured by the all-powerful Royal Navy. The development in the late nineteenth century of steam-driven, heavily armoured warships, dependent on new technologies, was set to exacerbate Anglo-German rivalries as Germany aspired to become a superpower. The threat became palpable in 1898 when Germany passed its first Navy Law challenging Britain's naval primacy in its own home waters by planning the construction of a fleet of battleships to be stationed in the North Sea.[10] Britain responded. The first Royal Naval Dreadnought keel, laid on 2 October 1905 at Portsmouth Dockyard, heralded the most demanding programme of ship construction ever conducted by any nation in history. Admiral Sir John Fisher, Britain's First Sea Lord during this period, was thus responsible for prompting an arms race that would stretch Britain and the Empire's resources to their limits. Since the Dreadnoughts were superior to all earlier warships, including all the ships of the Royal Navy, and since what now counted was not numbers of ships but the numbers of ships of similar design to the Dreadnought, Britain's rivals could now compete only by emulating the Dreadnought standards. Admiral of the Fleet Sir Frederick Richards conceded in Parliament: 'the whole British fleet was scrapped and labelled obsolete at the moment when it was at the zenith of its efficiency and equal not to two but practically all the navies of the world combined'.[11] Over a period of fifteen years Britain's shipyards were to build thirty-five modern battleships and thirteen cruisers.[12] By the beginning of World War I a Royal Navy battleship was arguably the most complex object on earth, comparable to the launch of a major manned space satellite today, and, at £7 million per ship in actual costs at the time, easily the most expensive. To achieve its targets the Admiralty commissioned vessels from the Royal Dockyards as well as inviting competitive tenders from private shipyards.

The expense of building a fleet that outstripped in numbers and firepower the two next largest navies in the world (the 'two power standard' as it was known) drew exorbitantly on the nation's wealth, diverting financial resources from other needs – in particular the provision of free education. The popular 'naval vote' always won the day over every other sacrifice; and yet, compared with Germany, there was an inherent weakness in Britain's ability to exploit its scientific and technological expertise even in such an important national aim.[13]

Arthur Pollen, a British engineer of genius, had worked on the problem of how to set adjustments to the range and deflection on gunsights relative to the movement of a target. At sea, of course, missiles were affected by wind, and rapidly changing distances between aiming and target vessels while both pitched and rolled. Pollen employed synchronous, gyro-stabilized range and bearing observations in conformity with the firing ship's course and speed. A 'clock' computer – an instrument that worked as an 'analogue' rather than a digital computer – integrated the rapidly altering data and transmitted the required adjustments to the guns from a central point on the ship. A massive cabinet the size of a cathedral altar, operated by up to eight men, Pollen's prototype gunfire-control range finder, built in 1909, was an instrument of prodigious complexity and sophistication for its time. But distrust of complex technology and the preference for in-service personnel, with gunnery experience, led to the rejection of Pollen's system in favour of a less sophisticated compromise with far-reaching consequences for the sea battles of the First War.

The German military harboured no such distrust of civilian scientists.

War, Science and German Chauvinism

At the outbreak of World War I Fritz Haber was forty-six years of age, too old for active service and ineligible for a reserve commission because he was Jewish. But he eagerly put himself and his institute at the service of the military. In the first weeks of the war he responded to a military request to develop an antifreeze substance that would not require scarce toluene, essential for TNT. He speedily found alternatives in xylene and other petroleum derivatives. In the meantime the German military had been seeking alternative means of achieving a breakthrough on the Western Front to conventional firepower, due to shortage of gunpowder. Major Max Bauer, who was the Supreme Command's liaison with industry, was intrigued by the potential of poisonous gas as a weapon. After an abortive attempt to use a Bayer-produced bromine on the Russians, Fritz Haber, who also ran a bureau in the

War Raw Materials Office in Berlin, promoted the advantages of chlorine, a substance that was plentiful in the dyestuffs plants. Chlorine was already stored at BASF in metal cylinders, suitable for the battlefield, as opposed to the more traditional glass containers. Haber set up a collaboration between his Kaiser Wilhelm Institute and the IG companies to speed up the production of chlorine poison in secrecy. The development to weapons grade was not without its dangers. One of his young chemists, Otto Sackur, a friend of Haber's wife Clara from Breslau, was killed in a laboratory explosion at the institute. Clara heard the explosion from their house, which was in the grounds of the Kaiser Wilhelm Institute. Another scientist, Gerhard Just, lost his right hand. Haber and Willstätter wept bitterly at Sackur's funeral. Clara, utterly opposed to the exploitation of science for weapons of war, was distraught.

For Haber, however, German scientists were as much involved in the service of the Fatherland as any frontline soldier. His enthusiastic chauvinism was certainly shared by an impressive constituency of his colleagues and led to the tawdry ritual that would earn Germany, for all its pre-eminence in science, a reputation as the pariah among civilized nations early in the Great War. Haber joined with ninety-two of his academic and science colleagues in signing the Fulda manifesto (so called as it was drafted by Ludwig Fulda, a popular writer and a Jew), titled 'The Appeal of the Ninety-three Intellectuals' (otherwise, in German, known as *Die Kulturwelt! Ein Aufruf* – The World of Culture! A Wake-up Call!). The document repudiated out of hand German responsibility for the war, defended the invasion of Belgium, denied alleged atrocities, and insisted that German culture and militarism were one. 'Were it not for German militarism,' continued the document, 'German civilisation would long ago have been destroyed . . . The German army and the German people are one.'[14] The manifesto was published in a crop of German newspapers on 4 October 1914 and reported around the world: its signatories included many of the top names in German science – Max Planck, Emil Fischer, Paul Ehrlich, Richard Willstätter, Wilhelm Ostwald, Walther Nernst, all of them, in time,

Nobel prize-winners. Albert Einstein, Haber's longstanding friend, notably refused to sign the manifesto. Instead Einstein became signatory to a counter-manifesto launched in the same month by G. F. Nicolai, a Berlin doctor, seeking the influence of intellectuals to promote internationalism and peace in the form of 'an organic unity out of Europe'.

Yet Germany had not been alone, nor first, in promoting the idea of science and chauvinism. The charge that there had been a golden age of universalism in science until it was distorted and corrupted by Germany in the first and second decades of the twentieth century is hardly borne out by the facts. Chauvinism had been rife during the nineteenth century, especially between France and Germany.[15] A French minister of education declared in 1852:

does not our tongue appear especially suited to the culture of the sciences? Its clarity, its sincerity, its lively and at the same time logical turn, which shifts ever so rapidly between the realm of thought and feeling – is it not destined to be not merely [scientists'] most natural instrument but also their most valuable guide?[16]

After the defeat of France in the Franco-Prussian War of 1870, Franco-German rivalry became increasingly incendiary. Louis Pasteur renounced in a letter his honorary degree from the University of Bonn, whose dean of the Medical Faculty replied that, in the light of the 'insult which you have dared to offer the German nation in the sacred person of our august emperor', the letter, with its 'expression of total contempt', was being returned so as 'to protect its archives from pollution'.[17]

German pre-eminence in the chemical industries, according to the French, was simply a token of the Germanic tendency to militarize its organizations and to plunder the ideas of others. In response, Hermann Kolbe, a noted and particularly intemperate German chemist, gave vent to public explosions of Francophobe bile, such as: 'I have a deep hatred and contempt for the French, but I have never considered them so uncivilized, vile and beastly, as we now recognize them to be.'[18]

Amidst the anti-German mud-slinging at the outset of the First War were sentiments destined to become stock anti-Teutonic prejudices through the rest of the century. The French physicist Pierre Duhem, borrowing mainly from Pascal, set the tone, caricaturing German science as typical of the unthinking regimented *Volk*. He liked to contrast the French and the German mind, by contrasting a French *esprit de finesse* with a German *esprit géométrique*. German learning is authoritarian; its laboratories are like factories; dedicated to geometrical thinking, their scientists lack common sense. The 'mathematical mind of the Germans, so fitted to deduce all the consequences from a given principle, is marvellously adapted to extract an industry of extraordinary power from our machines, our physics and our chemistry, the moment they reached a deductive and mathematical stage'. Denying German science the least benefit of the doubt, Duhem also denounced Einstein in terms that would become familiar within Germany itself, when Aryan physicists would condemn theoretical physics, or 'Jewish physics' as they termed it, as bogus. Duhem wrote: 'For a supporter of the principle of relativity to speak of a velocity greater than that of light is to pronounce words bereft of sense.'[19]

As the First War continued, the denunciation of Germany and all things German gathered momentum in Britain and France. But the great and the good of German academia responded by issuing further manifestos – one signed by 3,000 German academics, the other by the rectors of twenty-two German universities. The American physicist Michael Pupin wrote to a colleague, the astronomer George Hale: 'Science is the highest expression of a civilization. Allied science is therefore entirely different from Teutonic science.' In a series of articles in *Le Figaro* German science was characterized as being subservient to the German state, 'reduced to profit, usefulness to Germany, to its supremacy, to its domination'. The American journal *Science* had stated on 1 August 1914, 'German is, without doubt, a barbarous language only just emerging from the stage of the primitive Gothic character, and . . . it should be to the advantage of science to treat it as such.'[20]

In Britain the chemist Sir William Ramsay of University College,

London, a Nobel laureate, excoriated German science in the journal
Nature, with an interesting aside about German Jewish scientists:
'Much of [German scientific] reputation has been due to Hebrews
resident among them; and we may safely trust that race to persist in
vitality and intellectual activity.'[21] Jewish scientists were nevertheless serving the Fatherland enthusiastically during the First War.

Meanwhile Haber's chemical weapons research programme culminated in the first deployment of a weapon of mass destruction,
inflicted on French Algerians on 22 April 1915. The taint on the
reputation of the German army and German science would last
through much of the twentieth century.

4. The Poison Gas Scientists

Known to his scientific colleagues as 'the *Geheimrat*' – privy coun-
cillor – Haber was a distinctive figure in the German trenches as he
went about poisoning the Allied enemy in 1915. A uniformed
academic, pince-nez perched on his nose, pockets bulging with
papers, a cigar stuck permanently between his teeth, he was
accompanied by a team of young, eccentrically attired researchers,
among them the bluff Rhinelander Otto Hahn (one day to
co-discover fission, for which he would win the Nobel prize), and
future Nobel laureates James Franck and Gustav Hertz, as well as
Erwin Madelung and Wilhelm Westphal. These were the new
German warriors: scientists who calculated injury and kill rates with
graphs and equations, and employed toxic gases produced by their
chemical formulae as weapons. As the German-Jewish author
Alfred Döblin would write, by the end of the conflict: 'The decisive
assaults upon mankind now proceed from the drawing boards and
the laboratory.'[1] Gas warfare had brought together civilian scientists
and the military in a new partnership. It was seen as a wonder
weapon.[2]

At 5 p.m. on the afternoon of 22 April 1915, Haber and his gas
troop known as Pionierkommando 36 were dug in along a four-
mile front on the German lines at Ypres, facing the French terri-
torials and an Algerian division of the French Army. His scientists
were in charge of some 5,730 cylinders, each weighing 200 lbs, of
chlorine gas in liquid form. When the order came to attack, the
operators – wearing protective masks – opened the valves and
released the entire contents within ten minutes. A spectacular
blanket about five feet high of thick green-yellow gas swept on a
westward breeze across no man's land to the Allied trenches. Those
troops that were not suffocated broke from the trenches and fled,

abandoning fifty guns. The German infantry then crossed no man's land and took the Allied trenches.

Fritz Haber, now a professor at Berlin University and director of the Kaiser Wilhelm Institute for Physical Chemistry, had broken atrociously the acceptable norms of warfare in the early twentieth century. He claimed, as would other gas poisoners, like the British physicist J. B. S. Haldane, who swiftly emulated Haber's example, that new technology in weapons had the power to save lives since it could achieve swift victory. Haber believed, or at least said so, as did Haldane, that gas warfare was 'a higher form of killing' – that to be injured by gas was better than being blown up by a conventional shell.

Chlorine attacks the mucous membranes of the nose, mouth and throat, causing asphyxiation and blindness. In his diary, Captain J. W. Barnett of the Indian Medical Service recorded: 'Saw gassed men – blue faces choking and gasping.' He added: 'How we hate the Germans.'³ A soldier serving with a Canadian division in a neighbouring section of the front gave his impressions from a less exposed area.

We did not get the full impact of the gas, but what we got was enough for me, it makes your eyes smart and run, I became violently sick, but it passed off fairly soon. By this time the din was something awful where we were, we were under the crossfire of rifles and shells, we had to lie flat in the trenches. The next thing I noticed was a horde of Turcos (the British nickname for Algerians) making for our trenches, some were armed, some were unarmed. The poor devils were absolutely paralysed with fear.⁴

The British Commander-in-Chief, Sir John French, cabled London that day, reporting that hundreds of men had been thrown into a comatose or dying condition, and that within an hour the whole position had been abandoned with some fifty guns. By nightfall some 5,000 troops, it was claimed by the Allies, had lost their lives, and 10,000 were injured (disputes about the casualty numbers continued down the years, until, according to German sources, the figures settled at around a third of the original claims).[5]

The action broke the Hague Conventions of 1899 and 1907 (and it was to be linked, as a major act of inhumanity, with the sinking of the civilian passenger ship *Lusitania* two weeks later by a German U-boat). Despite the kill and injury rate, however, Haber was that evening mightily downcast. According to his colleagues, the *Geheimrat* had hoped to win the war 'scientifically'. He had argued for an attack with a far larger volume of gas, to produce a knock-out blow rather than a mere 'experiment'. He later commented that if the military had followed his advice and made a large-scale attack instead of the experiment at Ypres, the Germans would have won the war.

On 12 June the Germans used gas against the Russians in Galicia, using a lethal mixture of phosgene and chlorine. Phosgene is highly toxic, more so than prussic acid (hydrocyanic acid): its inventor died after one sniff. Soldiers remarked that as phosgene was released the birds dropped from the trees and even rats stopped dead in their tracks. Otto Hahn, who personally led the German infantry across no man's land, described in later years how he felt 'profoundly ashamed and perturbed' on seeing the consequences of the attack, and that he attempted to revive Russian soldiers with his scientific officers' masks – in vain.[6]

Predictably the first use of gas as a weapon only served to deepen the wrath and violence of Germany's enemies, prompting retaliation in kind and the violation of more humane conventions of war. According to Britain's Imperial War Museum historian,

In the days and weeks following the introduction of gas – a new and frightening advance in the culture of war – it is hardly surprising that the precepts of the Hague Convention on the treatment of enemy prisoners carried little or no clout and were completely ignored . . . As well as rage during the battle, the intensive high-octane commitment of a trench raid – particularly if the deaths of comrades during it produced, as it could so easily do, the urge to extract instant equivalent revenge – could also mean the thumbs down for prisoners; they were sitting targets, obvious tokens of reprisal.[7]

On 23 April Sir John French insisted that 'immediate steps be taken to supply similar means of most effective kind for use by our troops'. He added: 'It is also essential that our troops should be immediately provided with means of countering enemy gasses.'[8] By the end of the war some twenty-two different agents had been used on both sides, ranging from lung and skin irritants to blood-poisoning chemicals, delivered by shells, mortars, grenades and aerial bombs. Erich Maria Remarque's *All Quiet on the Western Front* sketches a harrowing description of German casualties stricken by the very weapon Haber had initiated:

We found one dug-out full of them, with blue heads and black lips. Some of them in a shell hole took off their gas masks too soon; they did not know that the gas lies longest in the hollows; when they saw others on top without masks they pulled theirs off too and swallowed enough to scorch their lungs. Their condition is hopeless, they choke to death with haemorrhages and suffocation.

There seems to have been no civilian outcry in Germany against their army's first use of gas; but German troops deplored its use, since the prevailing winds in Flanders came from the west.[9] We find, moreover, private detestation even at the level of the generals. The German General Carl von Einem, supreme commander of the 3rd Army, writing to his wife in 1917, told her: 'I am furious about the gas and its deployment, which has disgusted me from the outset.' He went on to describe gas as 'very unchivalrous too, otherwise used only by blackguards and criminals'.[10] The commander of the 6th Army, Crown Prince Rupprecht of Bavaria, wrote in his diary for 1915:

I made no secret of the fact that the new gas weapon seemed not only disagreeable, but also a mistake, for one could assume with certainty that, if it proved effective, the enemy would have recourse to the same means, and with the prevailing winds he would be able to release gas against us ten times more often than we against him.[11]

. A few days after Ypres, Haber returned to Berlin to visit his family. On 1 May he threw a celebration for friends in the accommodation close to the Institute to celebrate his promotion to captain, an unprecedented honour for a scientist. That same night his wife, herself a scientist of distinction, one of the few women in Germany with a Ph.D., took his service revolver out into the garden, fired one shot in the air and put the next bullet into her breast. Two hours later she died in her son's arms (he too, in time, was to commit suicide). Speculation has persisted down the years that she committed suicide to protest against Haber's role in the initiation of gas warfare: it seems certain that she had argued vehemently against it in previous months as he laid plans for his chlorine tactics. The very morning of her death Haber left the house and made the journey to the eastern front to initiate a gas attack against the Russians.[12]

As the war progressed, Haber led a series of research and development projects for further exploitation of gas as a weapon. One type, eventually favoured over others, was known as *Buntkreuz*, which involved a mixture of a phosgene-type gas and an irritant that could penetrate gas-masks. Otto Hahn, Haber's younger colleague, who experimented on these substances in Berlin in 1916, reported that 'those attacked were forced to tear off their gas-masks, leaving themselves exposed to the poison gas'.[13]

Hahn continued to ponder his participation in gas warfare long after the war was over. Half a century later he commented:

As a result of continuous work with these highly toxic substances, our minds were so numbed that we no longer had any scruples about the whole thing. Anyway, our enemies had by now adopted our methods, and as they became increasingly successful in this mode of warfare we were no longer exclusively the aggressors, but found ourselves more and more at the receiving end. Another factor was that we front-line observers rarely saw the direct effects of our weapon.[14]

Interviewed in the 1960s, Hahn said, 'Haber told me the French already had rifle bullets filled with gas.'[15] He added:

You might say that Haber put my mind at rest. I was still against the use of poison gas, but after Geheimrat Haber had put his case to me and explained what was at stake, I let myself be converted and I then threw myself into the work wholeheartedly. As you know, many other renowned scientists also put themselves at his disposal . . .[16]

As for the ethics of war, Hahn commented: 'At first the English were very surprised by our disregarding the Hague Convention, but from 1916 onward they used at least as much poison as we did.'[17]

These were unprecedented moral conundrums for scientists, and it is arguably unfairly anachronistic to reproach Hahn for his latterday reflections; but on every count his exculpations rehearse the arguments of treaty breakers and perpetrators of atrocities down the ages. Even the physicist Lise Meitner, who would live to regret the fact that she remained working with Hahn until 1938 in Berlin, sympathized with Hahn's rationalizations about poison gas work during World War I. '[I] can well understand your misgivings,' she wrote to Hahn, 'yet you are certainly justified in being an "opportunist". First, you were not asked [but ordered] and second, if you do not do it, someone else will. Above all, any means which might help shorten this horrible war are justified.'[18]

After Germany lost World War I, the Allies sought to bring Haber to trial as a war criminal. For a time, according to Hahn, he went around with a beard 'so that he would not be immediately recognized'. He disappeared to Switzerland; but he was eventually allowed to return to Germany to participate in the reconstruction of his defeated country. To the disgust of the Western scientific community, Haber won a Nobel award in 1918 for his ammonia discovery and continued to prosper as a prominent and honoured leader of German science for more than a decade. Otto Hahn would win the award in 1944, and James Franck, who assisted Hahn in experiments on phosgene, won his in 1925 for demonstrating that atoms in collision gain or lose energy in quantum steps.

Haber continued to be absorbed by his work on chemical weapons. Even after the war he tried to persuade Koppel, the Institute's benefactor, to fund a laboratory for weapons technology as well as a Kaiser Wilhelm Institute for Chemical Warfare to be directed by himself. Six million marks were voted to the project and Fritz Haber, Walther Nernst (chemist and old rival to Haber, as well as an enthusiastic poison-gasser) and Emil Fischer were made board members in prospect.

The Institute of Chemical Warfare was never realized, but after returning from Switzerland Haber continued work on poison gas despite the restrictions of the Versailles Treaty. Through a go-between, Dr Hugo Stoltzenberg, Haber was involved in several significant chemical warfare operations: the use by the Spanish army of poison gas to put down Abd el Krim's revolt in Morocco; a secret deal with the Soviets to produce poison gas; the establishment of a poison gas factory near Madrid; and the construction of a chemical warfare plant near Wittenberg. Haber also encouraged the development of hydrocyanic acid: it had a dual use – as a pesticide, and as a lethal gas against humans in enclosed spaces. It would be known as Zyklon B, one day to be used as the principal means of killing Jews in the death camps.

Even after Haber ceased to be a leading figure in gas warfare research, the Kaiser, living in exile in Holland, continued to consult him in anticipation of a future war of revenge against the Allies. In June 1927 the Kaiser wrote that he was especially interested in the scope for 'total gassing of large cities'.[19]

A correspondence between Haber and the pacifist chemist Hermann Staudinger (a generation younger than Haber) offers insight into Haber's ethical views on gas warfare. Staudinger had published an article in French at the time of the Versailles Treaty negotiations which he sent to Haber in order to initiate a dialogue on the subject. Haber wrote back accusing the young chemist of 'smiting Germany in the back at its hour of greatest need and helplessness', and providing propaganda that could only prompt more punitive measures. A patriot, Haber pointed out, should concentrate on combating foreign slander, not exacerbating it.[20]

Staudinger countered that Haber had not understood his drift. His main concern, he told the senior scientist, was not to apportion blame but to warn that an advanced science in the form of chemistry had exerted an evil effect on warfare. Haber replied that these ethical considerations were a matter for political and military authorities rather than scientists, but in any case he was utterly against international treaties banning the use of poison gas: morality was for individuals. This prompted Staudinger to write again, pointing out that the warfare of the future, if things went on like this, would end in total destruction.[21]

A token of Haber's influence was his invitation to give a paper to officers of the Reichswehr Ministry, published on 11 November 1920, the second anniversary of the armistice. It was a response to the resolution passed by the League of Nations on gas warfare on 28 October, reaffirming that gas warfare broke the Hague Convention and was hence in contravention of international law. Haber declared that poison gas weapons were no more cruel than conventional weapons, and that international condemnation of gas warfare was illogical. He stressed, moreover, the continuing development of gas weapons in other countries after the end of the war and encouraged German officers to learn the technical aspects of gas warfare and its potential.[22] Haber urged elsewhere in his correspondence to Staudinger the homely reflection that 'permanent peace cannot be assured in terms of the technical side of things. The amicability of a husband and wife in a marriage must issue from their basic convictions and self-discipline, not from the denial of access to sticks and pokers to brawl with.'[23]

But if Haber and his influential voice had remained intransigent on the merits of gas warfare, as well as fervently patriotic towards the Fatherland, a voice of internationalism, peace and anti-militarism had persisted in Germany in the person Fritz Stern has described as Haber's 'fraternal opposite'.[24] Not long after the end of the First War, Albert Einstein paid a visit to the battlefields of France, accompanied by Maurice Solovine, a Romanian Jewish physicist of philosophical bent, and Paul Langevin, the eminent French physicist. They wandered through blasted woods and cratered fields, and stood at the graves of French and German soldiers lying side by side. 'We ought to bring all the students of Germany to this place,' Einstein remarked, 'all the students of the world so that they can see how ugly war is.'[25] There was some doubt in the minds of Einstein's companions as to how the local French would take the presence of a 'German' scientist. At lunch, a party of French officers and a woman recognized Einstein. When Einstein's group rose from their table, the French party also rose without saying a word, all moving together, and bowed low and respectfully to Einstein. It is entirely likely that these citizens of France had remembered Einstein to be a leading figure among the handful of academics who had refused to sign the Fulda manifesto supporting the view that 'German' science should be at the unqualified service of the Fatherland and the military.

Haber after the War

Haber had met his second wife, Charlotte Nathan, in 1914 at the Deutsche Gesellschaft, a society for the great and the good of Germany. She was the secretary of the organization, a Jewess and twenty years his junior. She appears to have taken a liking to his poetry, which expressed a youthful infatuation with her. They married in the autumn of 1917 in church, she converting to Christianity in order to do so.

Marriage and two children born to him by 1920, however, could not shift the deep depression Haber felt after the war. Even the Nobel award failed to lift his spirits, but his acceptance speech was memorable. He said, among other things: 'It may be that this solution is not the final one. Nitrogen bacteria teach us that Nature, with her sophisticated forms of chemistry of living matter, still understands and utilizes methods which we do not as yet know how to imitate.'[26]

One of his friends noted that he seemed '75 per cent dead'. In this state of melancholy he now embarked on a strange quest to help his Fatherland in its period of post-war peril. The reparations payments exacted by the Versailles Treaty sought 20 billion gold marks from Germany by May 1921, and some 132 billion in subsequent payments. This gargantuan ransom was equal to two-thirds of the gold reserves of the entire world. To make matters worse, the Allies had declared Germany's carefully protected patents null and void, including the famous Haber-Bosch process, depriving the country of the means of earning revenues to pay those reparations. Could science rescue the situation and save the nation from impending economic and political chaos?

Working on the calculation that a tonne of seawater contains several milligrams of gold, Haber calculated that the oceans of the world could yield countless millions of tonnes of gold. With a fourteen-strong team of researchers Haber set sail for New York from Hamburg in July 1923 on the ocean liner *Hansa*, in which he had installed a laboratory to test the waters for gold traces. In October he set sail for Argentina in order to test the seas of the Caribbean. The following year he sailed from San Francisco to Honolulu, Yokohama, the China Seas, the Indian Ocean, and through the Suez Canal to the Mediterranean. He also prevailed on friends to send him appropriate specimens of sea water from different quarters of the globe. Meanwhile his research team tested some 5,000 samples of sea water in great secrecy in Berlin, and finally reached the conclusion that the original estimate of several milligrams of gold per tonne of seawater was very wide of the mark. The actual concentration of gold per tonne was 0.008 milligrams, a thousandth of the original estimates, indicating that hopes for commercial extraction were in vain.

Through the 1920s Haber's depression deepened; but he never ceased to promote science and to travel far and wide. By 1927 his second marriage broke down. Later he wrote to Willstätter: 'I am fighting with declining strength against my four enemies: sleeplessness, the financial demands of my ex-wife, my unease about the future, and the feeling of having made grave mistakes in my life.'[27] The greatest blow in his life, however, awaited in 1933, when Hitler came to power.

5. The 'Science' of Racial Hygiene

Throughout his life Haber had busied himself in the wide-ranging ambit of chemistry. He had seen himself as a science patriot for his Fatherland. He had, as we have seen, adopted Christianity and abandoned his Jewish birthright. He was born into an educated generation in Germany that saw Jewishness principally as a religious or cultural affiliation that could be dropped or repudiated as easily as submitting to Christian baptism. But there had been movements within science, or claims for science, throughout the course of his life that were to make his Jewishness, and the Jewishness of many of his colleagues, a deep and dangerous matter.

While Germany emerged as the chemistry power-house of the world in the second half of the nineteenth century, there had been important developments in biology, anthropology and the study of race, with far-reaching consequences for Jews in Germany. In the year 1900 the new century was celebrated in Germany with the announcement of a controversial essay prize to be awarded in the field of biology. The organizers sought to exploit theories of evolution in order to undermine revolutionary socialism and encourage nationalist, conservative views in the public domain. The point was to encourage and reward biologists who could promote an equivalence between social and political history and the influence of biological inheritance. The initiative was master-minded in secrecy by a major figure in German industry, the steel and armaments tycoon Friedrich 'Fritz' Krupp, the grandson and sole heir of the mighty firm's founder, who was forty-six years of age in 1900. The circumstances of the award reveal the scope for manipulation of theories of evolution to draw political and social conclusions. The part played by Krupp reveals the vulnerability of Darwinism to pseudo-scientific conclusions. Krupp had already decided how the winning essay should be cast before it had been written; his insistence that distinguished academics should 'front' the publicity and adjudication of the prize reveals the tainted nature of the initiative from the very outset.

Fritz Krupp was close to Wilhelm II, and by 1900 was profiting prodigiously from the Kaiser's passionate determination to build a navy. Krupp had purchased the Germania shipyards at Kiel in 1896 in order to create the specialized capacity necessary to build a fleet of battle cruisers for Germany; but his foundries at Essen, with their jealously patented state-of-the-art armour-plating production, were also providing guns, shells and armour for countries all over the world, including Britain, and even China and Japan, who were at war with each other.

In common with other industrial firms, the Zeiss lens company and Siemens engineering, Fritz Krupp had sought to develop partnerships between industry and science. In 1898 he funded a chemical and physical institute for experimental research with obvious advantages for steel and weapons. His passion for biology, however, was more a token of the influence of his wife, who had put in his way a book by the German Darwinian biologist Ernst Haeckel, *Natural History of Creation*, expounding hierarchical evolution in human populations. Krupp developed an interest in welfare programmes under the strong direction of company management, and advocated the virtues of gradual 'progress' as opposed to the cataclysms of revolution: a socio-political direction, of course, that favoured his goals as an industrialist. These interests now merged at points with his amateur fascination for evolutionary biology.

He became a serious student of marine biology and equipped two of his private yachts for research at sea. From his villa on the island of Capri he made friends with Anton Dohrn, director of the Naples Zoological Station, a renowned marine laboratory to this day. Krupp donated 100,000 marks to Dohrn's institute and gave it the run of one of his research yachts. He also funded a fresh-water laboratory at Plon, where he spent much time gazing at the patterns made by plankton.

There were sixty-six entries for the Krupp biology and society prize (valued in total at 30,000 marks): forty-four from Germany, the rest from Austria, Switzerland, Russia and the United States. Krupp had anonymously written the original manifesto, pointing out that he was looking for a formula to revitalize conservative policies in the government of the state.[1] The academic members of the prize commission and judges' panel attempted to restrain Krupp's input, advising him that the science should be kept separate from history. In the event they failed, but the announcement of the prize at least succeeded in recommending a separation of 'natural inheritance' from 'inheritance of tradition'. Krupp chose the figures responsible for organizing and judging the prize: conservatives all, including the Darwinist Haeckel.

Most of the prize money was split four ways, with several lesser prizes. The noted Aryan racist among the entries, Ludwig Woltmann, did not win first prize (the judges criticized his entry for manifest errors in biology); he came fourth, and as a result withdrew his essay and refused to accept the money. The first prize, in the event, went to Friedrich Wilhelm Schallmayer, whose history indicated the eugenic direction of medicine at the turn of the century.

Schallmayer, who was a psychiatrist, had been influenced by the French psychologist Théodule Ribot, who was convinced of the hereditary nature of both mental disease and personality disorders. As historian of science Paul Weindling comments, Ribot's notions 'marked the transition from liberalism to professional elitism, once the medical profession claimed the power to decide who were the degenerates'.[2] Schallmayer, as he expounded it in his prize essay and elsewhere, recommended the abandonment of the deregulatory liberalism which had dominated psychiatry from the mid-nineteenth century in Germany (in short, letting inmates out of their asylums), and the promotion of proactive policies designed to combat not only mental illness but social deviancy. He called for mental health assessment panels, composed of doctors who, as officials of the state, would sit in judgment on the status of patients and deviants of various kinds. People would be universally issued with health passports to identify at all times the sound from the unsound. He believed that the mentally sick should not be released back into society, where they could produce offspring. Here were

powerful indications of eugenic policies of future years. His prize
essay, moreover, argued for a strong nation state, based on hereditary biological principles and ubiquitous state intervention. In this
way Germany would enjoy an advantage in the competitive struggle
for survival with other nation states.

Fritz Krupp did not live to see how his initiative had served to
fuel the influences that culminated in National Socialism. His end
was imminent and tragic. In 1902 newspapers in Germany and Italy
accused him of homosexual and paedophile crimes committed
while he stayed on the island of Capri. His wife asked the advice
of the Kaiser, who recommended, with consummate irony, that he
should be placed in an asylum. Krupp's friend Admiral Friedrich
von Hollmann, however, countered this scheme by raising doubts
about Frau Krupp's state of mind. With the assistance of four
eminent psychiatrists, Krupp had his wife locked up in an asylum
in Jena. The bizarre circumstances resolved themselves when Fritz
Krupp died suddenly in November 1902. The death certificate
averred stroke; but suicide was more generally suspected.

Fritz Krupp would be remembered for the welfare schemes that
made his business, as Jérôme Bonaparte commented, 'a state within
a state', the homosexual scandals, and the legacy of a hugely profitable company which had employed 21,000 workers when he started
and 43,000 when he died. Less well known, as he had kept his
name a secret in the enterprise, was his seminal influence on the
growth of racial hygiene and eugenics in Germany, a tendency that
had been gathering strength in Europe and North America during
the second half of the nineteenth century.

Charles Darwin's Idea

In the early sixteenth century, Spanish philosophers and theologians
came together in *juntas* to debate the natural status of Amerindians
against the background of the Spanish occupation of the New
World and quarrels about the legitimacy of taking indigenous
people into slavery. Were these creatures human, or were they not?
On one side of the argument there were those who invoked the

idea, based on Aristotle, of the existence of 'natural men' – to explain the existence of savages who were slaves by nature, inclined to cannibalism and lacking full souls. Opposed to this view were theologians who insisted on the Judaeo-Christian belief in the unity of the human race as descended from Adam and Eve. Human beings, according to this doctrine, possessed without exception immortal souls and were capable of evangelization and salvation. Francisco de Vitoria, for example, in his *De Indis* (1539), invoked alternative texts in Aristotle to argue that American Indians were 'true men', whose rational faculties were potential rather than actual.[3] The Spanish disputes, like most arguments about race in Europe and North America, continued until the nineteenth century, based on philosophical and theological rather than biological considerations.

The earliest attempt to characterize race as the principal dynamic of human affairs was made by the French man of letters Arthur Comte de Gobineau, whose *Essay on the Inequality of the Human Races* (1855) argued that 'racial vitality' impelled the great shifts in human history, including the progress from the Stone Age to the Bronze Age, and the Fall of Rome. Gobineau maintained that races are innately unequal, and that the white races, especially the Aryan ones, are at the pinnacle of a hierarchical racial pyramid. 'Contamination,' he believed, resulted from inter-breeding between the white and other races and even between white races that are themselves unequal. Gobineau's thesis was scorned by de Tocqueville, who commented that such ideas would find a suitable reception only among the plantation owners of America's South. But the principal interest of Gobineau's essay is its emphasis on science as a means of distinguishing between human populations.

Science was by then the emerging universal explanation as well as offering increasing means of social control. Typical of such arguments were the theories of the Italian Cesare Lombroso, who thought it possible to identify criminals by the shapes of their heads. Then there was the American George Morton, who thought that cranial evidence proved the low intelligence of Indians, Blacks and women. 'Scientific racism,' claims historian of science Robert

Proctor, 'was an explanatory program, but it was also a political program, designed to reinforce certain power relations as natural and inevitable.'[4]

Meanwhile Charles Darwin's *Origin of Species* (1859) profoundly undermined the biblical basis of understanding human groups by declaring that human beings had evolved not since Adam, a mere four millennia past, but over hundreds of millennia, and by suggesting that races had evolved by a process of adaptation to local habitats. Darwin offered the prospect of understanding the human race biologically, and it was a short step for certain of his followers to invoke natural selection and survival of the fittest as the basis of human behaviour and racial characteristics. In the United States there were early Darwinists who appealed to the theory in support of white racial superiority in the form of a competitive capitalist spirit. In Germany, however, Darwinism took a rather different direction: calls for social intervention that would control selection in order to avoid the degeneration of human groups.

In October of 1866 Charles Darwin greeted at Down House, his home in Kent, the man who had become his most enthusiastic German supporter, the zoologist Ernst Haeckel. The encounter had its difficulties, since Haeckel was so overcome with exuberance that Darwin could scarcely comprehend him. Soon afterwards Haeckel's *Generelle Morphologie* (in two 500-page volumes) was delivered to Darwin's door. In a painful process of word-by-word translation aided by dictionaries, Darwin began to get the gist of the German professor's ambitions for the scope of evolutionary theory. The two volumes, it turned out, were a mere fragment of a theory of everything, a grandiose proposal for the universal relevance of Darwinism ranging from the finer details of embryology to excursions into the formation of the liberal nation state. Beyond this lay the promotion of the superiority of the Germanic peoples – all the more superior for becoming united within one nation state – and the need to combat Christianity, the priesthood and its 'gaseous' God. Haeckel called his evolutionary philosophy monism since, in his view, it was the only viable explanatory principle in science. The nation state, according to Haeckel, was comparable to an evolutionary biological organism, struggling towards progress and constrained by natural laws. Only the fittest racial types would survive and prevail; and the fittest race must be on its guard against disease and deterioration.

Haeckel, a devotee of Goethe, combined a curious blend of evolution and German 'nature philosophy', which lent a powerful notion of teleology, or sense of inherent goals and purpose, to his interpretation of Darwin's theory. Evolution, according to this view, was driven relentlessly onwards and upwards by an inexorable propensity towards the complexity and genius of human beings as the pinnacle of nature. Haeckel's blend of nature philosophy and evolution survives to this day among certain professional biologists.

German academic opinion on Darwinism was by no means united even among its enthusiasts. Haeckel studied medicine at Würzburg University under Rudolf Virchow, who had exploited staining techniques, using dyestuffs technology, and founded modern pathology, laying down the principle that cells derive only from cells ('omnis cellula e cellula'). Haeckel, while still a student, was already under the spell of evolutionary theory, having read The Voyage of the Beagle. Virchow disagreed with Darwin on some basic principles of the theory. Mutation of individuals which gave rise to the evolutionary process, he argued, was not the result of random agents of change, but cellular alterations that were precursors of disease. By the time that Haeckel had been appointed to a chair at the University of Jena, where he lectured to large numbers of enthusiastic students, Virchow and he were clashing publicly over the political implications of social Darwinism. Virchow maintained that leftists influenced by social Darwinism would plunge society into violence and bloodshed similar to that of the French Revolution. The dispute popularized the issues, and Haeckel, in the eyes of many Germans, appeared to be getting the best of the argument. At the same time, Haeckel's views began to merge with ideas of Gobineau, whose essay had found widespread resonance in

Germany, where the theme of Teutonic superiority was in time developed by the Englishman Houston Stewart Chamberlain. Chamberlain had become a friend of Wagner and an ardent devotee of mystical pan-Germanism, and was to exert a powerful influence over Adolf Hitler. Chamberlain, born in 1855, was the son of a British admiral and a German mother. In his book *Foundations of the 19th Century* he opined that Germans were superior to other races and should be considered the very foundation of advanced societies, since they were responsible for industrialization. Focusing on Darwin's suggestion that the evolution of homo sapiens was a result of enlargement of the human brain, Chamberlain became preoccupied with brain size within the human species as a key to race and evolutionary progress. He incorporated phrenology into his theory, the idea that various bumps on skull and cranial con-figurations denoted certain moral and intellectual characteristics in individuals.

Here were the origins of the notion of *Rassenhygiene* (race hygiene), a term coined by Alfred Ploetz, a medical practitioner, who, in 1936, aged seventy-six, would be appointed by Hitler to a professorship at Munich University. Among Ploetz's leading ideas, based on a mish-mash of evolutionary theory, was the notion of counter-selection leading to degeneration. He maintained, for example, that advances in medicine would only encourage the survival of degenerate human stock by sustaining them artificially. He advocated sending only inferior specimens of the race to the front in war as cannon fodder, sparing the best specimens. The driving force of Ploetz's racial hygiene was the fear that inadequate people were going to multiply faster than the fit and the gifted. He also urged the enhancement of human germ plasm (thought to be the biological 'substance' of heredity) and its protection from damaging agents such as alcohol and venereal disease. In this way, he believed, the process of natural selection of the fittest could be attained by manipulation of human breeding rather than by war and struggle.

His notion of racial hygiene focused on the good of the race rather than the individual. In the furtherance of racial hygiene he recommended a three-pronged thrust of anthropology, socio-politics and medicine. In 1895 he published his magnum opus, *The Vigour of our Race and the Protection of the Weak*,[5] followed in 1904 by his notorious journal for *völkish* enthusiasts, *Archives of Racial and Social Biology* (*Archiv für Rassen- und Gesellschaftsbiologie*). In 1905 he founded the Society for Racial Hygiene (Gesellschaft für Rassenhygiene) with local branches in Berlin, Munich and Freiburg. His stated aim was to investigate 'the principles of the optimal conditions for the maintenance and development of the race'.[6] In 1909 the society agreed that membership should be restricted to members of the Nordic 'race'. He was also a member of a secret club that promoted the supremacy of the Nordic type, and he helped found a club called the Nordic Ring, which promoted Germanic racial traits through sports and self-discipline.

Ploetz was not at this early stage a radical anti-Semite; in fact, he put Jews a little lower than the Aryan and cited Jesus, Spinoza and Marx as evidence of Jewish talent. He was convinced, however, that the aptitudes of Jews arose from Aryan racial mixing. He was convinced that the Caucasian was better adapted to variations of climate and terrain than the Negro and claimed that it was a fact 'known to every American school child' that Negroes learn more slowly than Whites.[7] The intelligence of a White compared with that of a Negro, he declared, offered the same contrast as the intelligence of a Negro compared with that of a gorilla.

Racial Hygiene after the First War

Up till the mid-1920s racialist and eugenic tendencies were to be found on both the political left and the right. There were socialists who saw eugenics as a means of rationalizing the means of production and favoured state-organized eugenic programmes. There were socialists who claimed that the heroes of their movement were mostly of Nordic stock. Karl Valentin Müller, a member of the German Social Democratic Party (SPD), argued that 'the goals of the workers' movement extend only to *white* workers: socialism simply cannot mean the same thing for a Chinese and an Indian, for an "Alpine" Frenchman or a Nordic Swede'.[8]

Through the 1920s the link between Nordic supremacy, right-wing politics and racial hygiene, however, grew stronger, particularly under the influence of the publisher Julius Friedrich Lehmann. Lehmann, who was enthusiastic for the ideas of Ploetz, was one of the most successful publishers of medical text-books and journals in Germany, and exerted widespread influence over the medical profession.

Another influential racist, under the spell of Ploetz, was the self-taught anthropologist H. F. K. Günther, known as 'Rassen-Günther' (Race Günther), who published in the early 1920s his *Rassenkunde des deutschen Volkes* (*Racial Hygiene of the German People*), a book which Hitler plundered. With Nazi backing Ploetz and Lehmann managed to secure Günther a chair in anthropology at Jena. He was finally appointed in 1932 despite vehement objections on the part of the majority of the members of the university senate. Hitler came to hear Günther's inaugural lecture.

Ploetz himself was nominated, unsuccessfully, for a Nobel peace prize for his work in racial hygiene. His anti-war rhetoric, however, was fundamentally racist, arguing that people of Nordic stock were particularly affected by war, since they were strong and large and therefore put at the front of battle; also, Nordic people, he argued, were 'more willing to fight for their ideas',[9] whereas Jews 'tend to suffer least from war, partly due to their smaller physique and weaker constitutions, and partly due to their lack of solidarity with fellow citizens and the state'. Hitler, under the influence of such ideas, proclaimed before the German Parliament in May of 1935: 'Every war goes against the selection of the fittest,' and hence National Socialism was 'profoundly and philosophically committed to peace'.[10]

Two other figures influential in the early formation of National Socialism as a 'political expression of biological knowledge' were Eugen Fischer and Otmar von Verschuer. Fischer, born in 1874, came from a prosperous family and originally studied medicine, his father having refused to allow him to become a zoologist. But as part of his medical degree he wrote a thesis on a famous orang-utan which had died in Stuttgart Zoo, concentrating on its urinary and reproductive organs. As he worked his way up through the academic ladder on the borders of zoology and medicine, he began to make a name for himself developing new approaches to anthropology by working on soft organs and tissues in different human races. For example, he 'demonstrated' that the facial muscles of Europeans were much finer than those of Papuans, whereas the Papuans showed far greater similarities to the apes. He did similar work on nose shape, eye pigment and tongue muscles.[11] Fischer later studied the consequences of inter-breeding in Germany's colonies, and of bastard children in the Rhineland fathered by French African troops of occupation after the First War, concluding that 'inferior' breeds weaken the stock.

The foremost prophet to emerge from this school of thought was Erwin Liek, a medical doctor of Danzig, who argued that pain was a benign secretion of a disorder and an essential part of the healing mechanism: to suppress it impeded recovery. Endurance of pain was, moreover, a prime virtue and one with which the superior races were better provided. Liek believed that illness was due to lack of moral fibre, a conviction that in time would add impetus to the influences within professional medicine that justified the elimination of the sick.

While German academic biologists were not a force for the pseudo-science of racial hygiene (research dedicated to this end was mainly carried out by anthropologists and jurists), there were several significant biologists who attempted to bring the 'discipline' in line with Nazi policy and ideology. Ludwig von Bertalanffy promoted the idea of drawing an equivalence between biological organisms and the state, for 'the hope that the atomizing conception of state and society will be followed by a biological one that recognizes the holistic nature of life and of the *Volk*' had now been realized.[12]

Then there was Ernst Lehmann, Professor of Botany at Tübingen, who was to praise Hitler for perceiving the dangers of racial degeneration, urging that it should be 'the task of biology to promote the spread of this knowledge and to forge new weapons

for the struggle to come'. Lehmann's mission statement involved
the subservience of the individual to the commonality of the *Volk*.
Individualism was a tumour, he declared, that needs to be cut out.
Lehmann founded a new monthly periodical, *Die Biologie*, and a
society known as the Association for German Biologists. One of
Die Biologie's leading contributors was the Austrian Konrad Lorenz,
a behaviourist zoologist, who inveighed against the softening of the
race, a process he compared with the domestication of pet animals.
Lorenz would eventually become a member of the Nazi Party's
Office of Race Policy, and call for the 'elimination' of supposedly
'genetically inferior' people.

Lorenz was also responsible for attempting to give academic
respectability to a view that would form direct connections with
Nazi eugenicist policies. It was the duty of racial hygiene to elimin-
ate morally inferior human beings. In prehistory, he argued, selection favoured endurance, heroism and social usefulness by
environmental factors. That process must now be assumed by
human organization to prevent the degenerative phenomena that
go with domestication. The view provided a 'scientific' basis for
Nazi sterilizations, which would be compulsory for those who
suffered from hereditary conditions such as schizophrenia, blindness
and other weaknesses, a theme picked up in the next chapter.

Lamarckianism and Weismannism

Ernst Haeckel, Darwin's friend, had worked closely with the
German zoologist August Weismann of Freiburg in promoting
Darwinism in Germany. Weismann proposed the theory of germ
plasm, a hereditary material carried in the parents' germ cells which
accounted for every aspect of their potential offspring's characteristics – moral, aesthetic and intellectual as well as physical. His
theory, which sought to settle contemporaneous questions about
nature and nurture, carried important social and political implications.

Throughout the nineteenth century the notion of inheritance
of acquired characteristics, propounded by the French biologist

Jean-Baptiste Lamarck, had been widely accepted, while the disciplines of biology awaited an explanation for transmission of heredity. Such a theory had been established by the Austrian monk and plant botanist Gregor Mendel, studying in the seclusion of his monastery at Brünn in what is now the Czech Republic. Working on the colour and shape of several generations of pea seeds he demonstrated that two features – yellow versus green; round versus wrinkled – were transmitted independently between generations according to definite quantifiable rules. His experiments involved the breeding of some 28,000 plants, fertilized by hand, while employing a number of varieties as control groups. Mendel, unlike his predecessors and most of his contemporaries, demonstrated the importance of quantification, of counting. But his research, published in an obscure journal in 1865, lay dormant until 1900 when three scientists – Carl Correns, Hugo de Vries and Erich von Tschermak – resurrected his theory independently of each other.

In the meantime Weismann had attempted to resolve the issue with a model that distinguished two kinds of cells within an organism: somatic (body) cells and germ (reproductive) cells. Somatic cells are involved in the development of the organism but are not passed on to the offspring. Germ cells, on the other hand, do not undergo development but form the genetic material of heredity. In this way he sought to explain how organisms were stable despite nurture and environmental factors.

By the early twentieth century the contentions over nature and nurture, heredity and environment, had acquired a strong political dimension. The proponents of racial hygiene rejected Lamarck and supported the teaching of Weismann with its insistence that germ plasm was passed down through the generations unaffected by the influences and vicissitudes of the somatic cells of the organism. For the racial hygienists the germ cells accounted for the nature of human characteristics. By the same token, they held that social influences had no power to shape the human condition or destiny. The polarities of the argument – Mendelian (or Weismannian) versus Lamarckian – were later exploited by Nazi ideologues as a demonstration of the contrast between race and the influence of class and economics.

In the Soviet Union, where Lamarck's views were favoured as a mechanism of radical societal change, the germ-line cell theory was interpreted as an unproven, reactionary, pseudo-scientific proposal to combat calls for radical socialist and communist reforms. A dramatic manifestation of the lethal mix of ideology and biological science occurred in agricultural practice, leading to disaster. Following the rise of Stalin after Lenin's death, and the collectivization of farming in the Soviet Union in the late 1920s, an obscure autodidact agronomist called Trofim Denisovich Lysenko leaped to prominence as a result of an article in *Pravda*. A prophet of environmentalism, Lysenko would attack 'bourgeois genetics' of the kind taught in Germany as fascist science, aimed at undermining socialist Lamarckian principles. Stalin would favour Lysenko's Marxist biology even when his attempt to revolutionize grain yields by 'vernalizing' winter wheat, chilling the seeds so as to plant them early, ended in disaster (after World War II Lysenko managed to persuade Stalin that it was not the theory that was wrong but sabotage on the part of reactionary biologists, 3,000 of whom were consequently sacked).

The historian Robert Proctor notes that a lesson to be drawn from the disputes between Nazi and Soviet ideologues over Lamarck versus Mendel is 'that political motivations can be as important in justifying correct views in science as they are in justifying false views'.[13] As it happened, the biologizing of race and an overemphasis on heritable characteristics formed connections in Germany with an enthusiasm for eugenics, which in different ways and with different emphases had become popular in the United States and other countries in Europe. In Germany, however, the symbiosis between racial hygiene and eugenics would become a leading feature of National Socialist ideology.

6. Eugenics and Psychiatry

Parallel with the growth of racial hygiene in Germany was the rise of 'eugenics', the science of encouraging 'good' offspring. Eugenics is associated with Charles Darwin's cousin, Sir Francis Galton, who had a mania for measuring everything, from intelligence to physical beauty. His book *Hereditary Genius*, first published in 1869, explored the heritability of intelligence, with scant interest in environmental factors, emphasizing the advantages of selection of highly intelligent members of the population for ideal breeding. Galton was captivated by the power of the Gaussian normal distribution graph, known popularly as the 'bell curve'. Since the distribution of physical norms and exceptions – height, weight and so forth – could be usefully described in a population, Galton reasoned that intelligence could also be quantified, and that different levels of intelligence could be accounted for by the levels of intelligence in parents. During a period of rising crime and poverty in the second half of the nineteenth century in great cities like London and New York, Galton believed that a more healthy, more intelligent, better-behaved population could be encouraged by appropriate planning on the part of governments: he called his scheme 'eugenics'. One of his proposals was to offer cost-free marriages at Westminster Abbey for unions that met certain 'eugenic' criteria.

One of Galton's most ardent disciples in England was Karl Pearson, an austere, hollow-cheeked individual, a Quaker of socialist leanings, who studied mathematics at Cambridge, then in London, before studying biological sciences in Germany. He became the founder of modern statistics, his original contributions including the introduction of the chi-squared test of statistical significance and the concept of standard deviation.

Dedicated to the task of quantifying biological information

mathematically, he eventually settled at University College London, the founding home of utilitarian philosophy. He attracted funds for 'biometrics', the quantifiable aspects of biology, which helped establish a biometric laboratory and a periodical, *Biometrika*. Following Galton's death in 1911, Pearson became the founding professor of the Galton Chair of Eugenics and headed the Galton Laboratory for National Eugenics, where facts and figures were gathered on a wide range of characteristics deemed to be inherited. These included various diseases, alcoholism and intelligence, although the principal measure of the latter, prior to intelligence tests, was the judgment of a child's class teacher.

Observing the correlation between 'intelligence' of parents and family size, Pearson concluded that the British population was headed for imminent degeneration. Pearson espoused the notion that war had positive eugenic effects as consequences of selection through struggle. 'This dependence of progress on the survival of the fittest race,' he wrote, 'terribly black as it may seem to some of you, gives the struggle for existence its redeeming features; it is the fiery crucible out of which comes the finer metal.'[1]

Enthusiasm for eugenics and the benefits of war was not confined to Britain. In the United States Roland Campbell Macfie claimed that war created a shortage of men, and, as a result, a more 'careful weeding of women'. War did not simply mean a 'martial selection' of men by chance destruction so much as a 'deliberate stringent matrimonial selection of women by the critical eyes of men'. War would mean an improvement in the 'health and beauty of the combatant races'. As Stefan Kühl comments, the view 'fitted well into the militarist, imperialist thinking of the period and linked eugenicists with nationalist movements within the different countries'.[2]

Karl Pearson's counterpart in the field of eugenics in the United States was Charles Davenport, who similarly had combined biology and mathematics to reach eugenic conclusions about moral behaviour and intelligence. He was persuaded that certain ethnic groups suffered stereotypical moral failings; that criminality and prostitution were heritable. He established a Eugenics Record

Office, where he mounted a scheme to process data on the characteristics of the populations of the United States. Davenport worked closely with Henry Goddard, who along with the psychologist Lewis Terman introduced the French 'Binet' intelligence test into the United States, where it became known as the 'Stanford-Binet' test. Convinced that there was a single gene for intelligence, they were also responsible for introducing 'Intelligence Quotient', or IQ, into the language. The application of the test to immigrants led to the rejection of certain ethnic groups. Goddard advocated a simplistic Mendelian view of heredity and intelligence: so two heterozygous parents would each possess one 'intelligence' gene and one 'moron' gene; assuming no dominance, the expected ratio of types in their offspring would be one moron, to one intelligent offspring, to two intermediates. Such views encouraged eugenic policies that would ban the reproductive union of people with genetic illness, syphilis and anti-social behaviour, including even habitual masturbation. Some states, like Indiana, introduced laws for compulsory sterilizations conducted by castration and irradiation.

Meanwhile the development of eugenics in Germany had, like racial hygiene, been increasingly shaped by pseudo-Darwinism as well as Aryan supremacist mythology. Among the leading lights were Philalethes Kuhn, who held the chair of clinical hygiene at the *Technische Hochschule* of Dresden from 1920, Hans Reiter, who taught hygiene with a racist dimension at Rostock from 1919, and Fritz Lenz. Lenz held the first chair in race hygiene at Munich University from 1923; Hitler read and admired his views on race and medicine when he was in Landsberg prison in 1924. Lenz returned the compliment by becoming an enthusiastic member of the Nazi party.

Despite the optimism expressed in the first decade of the century for the eugenic beneficence of war, World War I had radically altered attitudes. Edward Poulton and Leonard Darwin agreed in the pages of the British *Eugenics Review* that 'war unquestionably killed off the better types, and was therefore highly dysgenic'. Following World War I members of the International Organization of Eugenicists united to denounce war for this reason. Ploetz led German eugenicists in advocating a 'eugenic peace order', and the Nazi race politicians would endorse his arguments. Walter Gross, head of the Racial Political Office of the party, declared: 'Because Nazi Germany thinks racially, it wants peace.'[3]

While laws supporting positive eugenics were to find favour in Scandinavia and Germany in the period between the wars, such measures were eventually resisted in Britain despite the continued advocacy of a small elite. When Professor E. W. MacBride wrote to the science weekly *Nature* in 1936 urging a government-sponsored scheme for eugenics, he drew a combative riposte from the redoubtable Joseph Needham, biochemist and expert on China: 'It is difficult to express the dismay experienced in seeing these doctrines, so dangerous to humanity, receiving the imprimatur of what is perhaps the most famous scientific weekly in the world.'[4] Another influential British anti-eugenicist in the 1930s was Lionel Penrose, father of the mathematician Roger Penrose. Penrose had succeeded in Pearson's chair at University College, London, and had 'eugenics' removed from the title. He was adamant that intelligence was owed to a variety of genetic factors as well as to environmental circumstances.

Psychiatry and 'Euthanasia'

Another far-reaching effect of World War I had been a general decline in the practice and reputation of psychiatry as a scientifically respectable discipline, paralleled by the scandal of widespread neglect and mistreatment of mental patients in institutions. The gap between Friedrich Schallmayer's bold proactive recommendations and the reality could not have been more stark. Malnutrition, disease and neglect reigned in Germany's asylums, culminating in a massive 30 per cent of fatalities in the inmate population during the war years.[5] At the same time, soldiers suffering from shell-shock and other effects of trench warfare were subjected to abusive 'psychiatric' treatments, including the application of punitive electro-convulsive 'therapy'. Matters did not improve in Germany's mental hospitals after the war as the economic and political crises deprived mental institutions of funding for the most basic needs.

The historian Michael Burleigh notes that, despite a temporarily active psychiatric reform movement calling for patients' rights and checks on committal procedures, there were indignant voices complaining of the waste of resources on the mentally ill. Leading advocates for a solution to the expanding armies of 'mental defectives' were the lawyer Karl Binding and the psychiatrist Alfred Hoche, who popularized the chilling phrase *lebensunwertes Leben* (life unworthy of life) in their key tract *Die Freigabe der Vernichtung lebensunwerten Lebens* (Lifting Controls on the Destruction of Life Unworthy of Life), published in Leipzig in 1920. Binding recommended allowing doctors to terminate a painful existence on their own initiative; by the same token, he urged that society should terminate a life that is a burden to society. Hoche, taking a profoundly utilitarian view, cited the resources squandered on maintaining defective lives. Idiots had no right to life, he argued, because they lack the essential qualities that make life sacred. To terminate the life of an idiot, he declared, is not to kill in the commonly accepted legal understanding of the act.[6] This text, arguably more than any other, made available to the Nazi regime an 'ethical' rationale for 'euthanasia'. Although in the early days of the regime the public discussion would focus on the prevention of offspring with hereditary disease, hence sterilization, the destruction of life unworthy of life would spread as an unspoken principle, encouraged by relatives who subscribed to the solution.

Meanwhile many leading academic psychiatric researchers in Germany leaned towards National Socialism. The most prestigious psychiatric research centre was the Kaiser Wilhelm Institute of Psychiatry in Munich, founded in 1917 with money from the Rockefeller Foundation and the private benefaction of James Loeb, an American-born Jewish philanthropist. Its director, Professor Emil Kraepelin, brought together a distinguished group of clinicians and researchers, including the neurologist Alois Alzheimer, who discovered the disease of that name. Ernst Rüdin led a research programme on genealogy and demographics through the 1920s that became a classic in the field of genetics and schizophrenia.

As Hitler's party made its bids for power, Rüdin became an enthusiastic exponent of eugenic and racial hygiene policies. His high reputation lent respectability to the Nazi policies of enforced eugenic sterilization. Under his aegis schizophrenia and manic depression were judged categories suitable for sterilization. As the policy expanded, the Kaiser Wilhelm Institute of Psychiatry became a source of expert opinion on eugenic questions and individual cases.

The 'medical' and 'scientific' basis for the notion of eliminating the mentally ill was thus laid in the years following World War I, preparing Germany for the propaganda that would lament the cost of maintaining the 'ballast' of the mentally ill and those with congenital diseases.

PART TWO

The New Physics 1918–1933

7. Physics after the First War

World War I was known to succeeding generations of scientists as 'the chemists' war', a sobriquet amply merited by the work of Fritz Haber and his poison gas colleagues. Physics nevertheless played a varied role in a war in which guns were still the main weapons on land and at sea, and the principal problem was how to aim projectiles accurately, regardless of weather, smoke, distance, darkness and the pitch and roll of heavy seas. To calculate the position of an enemy gun, and other targets emitting observable signals, warriors of the First War had employed the physics of acoustics, optics, electromagnetism and seismometrics.[1] Scientists involved in basic research on both sides, the Allies and the Central powers, threw themselves into work on military technology, showing how temporary and fragile is the division between pure and applied science in wartime. Max Born of Göttingen researched the influence of wind and altitude on sound. Max Wien of Jena worked on wirelesses in aircraft. Arnold Sommerfeld of Munich, albeit a noted theoretical physicist, studied ways of listening in on enemy telephone conversations; his electron tube-amplified bugging devices could detect signals more than half a mile from a phone line.

British scientists too old to fight were also drawn into war work, while many of the younger ones went enthusiastically to the trenches, where all too many died. Most notably, the brilliant protégé of Rutherford, H. J. G. Moseley, was killed by a sniper's bullet in the Dardanelles.

British physicists, as we have seen in the case of Arthur Pollen, had long been grappling with the problem of gunfire control (range-finding equipment) on warships, although hampered by Admiralty prejudices. William Henry Bragg, father of Lawrence, and joint founder with his son of X-ray crystallography, worked on sound-ranging to locate enemy guns; while Ralph Fowler (who became Rutherford's son-in-law) studied at the British Munitions Inventions Department the use of optical instruments to improve gunfire control on anti-aircraft guns. Frederick Lindemann (the future Lord Cherwell, scientific adviser to Winston Churchill) worked at Farnborough on infra-red heat detection as a possible early warning system for approaching aircraft. Meanwhile Ernest Rutherford, at Manchester, was researching means of locating U-boats.

British scientists had realized the importance of science and technology for warfare at an early stage in World War I. The editor of *Nature* urged the government in its 17 June issue of 1915 to make better use of the 'hundreds of men of science in the country whose energies and expert knowledge are not being effectively used . . . The organization of the scientific intellect of the country is essential, yet almost nothing has been done towards its accomplishment.' Like their German counterparts, some of the great British men of science had committed themselves to war work not necessarily compatible with their gifts and expertise. Rutherford, famous by World War I for attempts to understand the atom, believed that acoustics were the only way to detect submarines. He and his team concentrated on the development of 'hydrophones' and other sorts of listening devices, including the use of piezoelectric quartz (crystals which generate electronic pulses when stressed or compressed by sound waves) to locate high-frequency sound. This technology anticipated the detection technique known as sonar. Rutherford, at least, had an inkling of the future possibilities of atomic physics by the last year of the war. There is a story that when he was reproached for his failure to attend a meeting of experts on anti-submarine technology, Rutherford, known for his very loud voice, responded: 'Talk softly, please. I have been engaged in experiments which suggest that the atom can be artificially disintegrated. If it is true, it is of far greater importance than a war.'[2] He declared, in time, intriguingly, that atomic science would never release energy in large quantities. Rutherford, Bragg and J. J. Thomson (known as JJ, the Cambridge physicist who discovered the electron) were members of an Admiralty Board of Invention and Research, but the navy was reluctant to discuss tactics and strategy with civilians, and many opportunities were lost.

By the end of the Great War, few in government or the military needed convincing that physics was essential for weapons and defence technology and that substantial government funding was required if science was to play a major part in future planning. Military physics had become well established by 1918, but few could have guessed the extent to which the discipline would dominate weapons and defence systems innovations in World War II, especially in the detection of aircraft at night. The predicament is captured by the writer Russell McCormmach in his novel *Night Thoughts of a Classical Physicist*. Reflecting on these matters in the depths of the First War, Professor Victor Jacob, the protagonist of the narrative, comes to the conclusion: 'If, God forbid, the world should go to war again, it will use modern physics. But *this* war was fought with *classical* physics, with every branch of it.'[3]

So what was 'modern' physics?

Modern Physics

The first year of the new century, as revealed by the impressive exhibits at the Paris International Exhibition, saw the celebration of the triumph of Germany's scientific attainments, the expression of confidence in the Fatherland's technological future which now, it was widely accepted, rested on secure foundations. But as the exhibition visitors flocked to Paris to admire the impressive artefacts and new chemical and industrial processes, there were developments in progress in the laboratories and seminar rooms of Europe that were soon to transform the way nature was observed and understood, making way for technologies undreamed of.

Until the turn of the new century, scientists had viewed nature and the universe as more or less predictable, determinist and mechanistic, with material effects following material causes. Promising young physicists were being advised to move on into other disciplines, since physics had more or less completed its project, or so it was thought. The prevailing, complacent view was that knowledge of energy, motion, radiation and electromagnetism was now more or less complete. Scientists worked on the assumption that the general structure of the laws of nature was known, that the business of science for the future was to fill in the details; matter and energy were perceived in terms of common-sense mechanical models, corresponding to human intuition. The British scientist William Thomson, Baron Kelvin, had declared that he never satisfied himself until he could make a mechanical model of a thing. All that was about to change, and Germany was to play a major role in the transformation.

'There are times,' writes Eric Hobsbawm in his *Age of Empire 1875–1914*, 'when man's entire way of apprehending and structuring the universe is transformed in a fairly brief period of time, and the decades which preceded the First World War were one of these.'[4] Hobsbawm is referring to turning points that occurred in science at the end of the nineteenth century and the beginning of the twentieth in psychology, biology, physics, innovations known to, and appreciated, at first, by only a small minority of the scientific and intellectual elite of the United States and Europe. Among these innovations were the publication of Freud's *Intepretation of Dreams*, the resurrection of Mendel's experiments in genetics and the work of the German scientist Max Planck, who held the chair of theoretical physics at Berlin University.

Planck, a notoriously reticent and undemonstrative individual (one of his students described him as a 'spare figure in the dark suit, starched white shirt and black bow tie . . . like the typical Prussian civil servant'),[5] had been working on the nature of electromagnetic radiation. In 1900 he proposed, in a profound break with ordinary physical ideas, that radiation exists only in integral multiples of tiny amounts of energy, called 'quanta'. In other words, only by assuming that light is emitted not in waves of arbitrary energy, as was generally thought, but in tiny energy parcels, could his calculations make sense. When he announced this quantum model at a meeting of the Berlin Academy of Sciences, neither he nor his audience realized the implications of the quantum revolution in physics that was about to break over their heads.

Planck's hypothesis had been, in fact, one step in a series of extraordinary developments in physics at the end of the nineteenth century. Five years earlier another German, Wilhelm Röntgen, had discovered the presence of X-rays, a new kind of radiation that undermined accepted assumptions about the physics of the spectrum and methods of experiment. Two years later the Cambridge physicist J. J. Thomson showed that cathode rays were made up of discrete particles two thousand times lighter than an atom of hydrogen. Meanwhile in France, the physicist Antoine-Henri Becquerel discovered that uranium created a cloud on unexposed photographic plates. It was Marie Curie, the Franco-Polish researcher, who first employed the term radioactivity, while coincidentally conducting a series of experiments which revealed that some heavy elements emitted different kinds of radiation. At the turn of the new century researchers had noticed the phenomenon of radioactive decay in uranium. X-rays, radioactivity, beta decay, and spectra were now raising questions about the uncontested primacy of classical physics, the comfortable relationship between scientific explanation and common-sense intuition. Just as Darwin's theory had proposed the evolution of species and the breakdown of fixed natural kinds, so atomic science was about to alter the understanding of the domain of the physical world.

In the following months and years, although many measurements confirmed Planck's formula for the spectrum of emitted radiation, based on the quantum hypothesis, he and many colleagues continued to be puzzled by what this meant. One scientist wrote that Planck had explained an incomprehensible fact by the even more incomprehensible assumption that light waves occurred in jerks.

In 1905, the year he published his epoch-making special relativity theory, Einstein wrote a paper revealing that Planck's quanta accounted for experiments that had been conducted by the German physicist Philipp Lenard (a scientist destined to be identified with

'Nazi' science). Einstein employed Planck's quantum idea to interpret the photoelectric effect. The maximum energy of an emitted
electron depended only on the wavelength of light and not on its
intensity, as was predicted by classical physics. Light, he revealed,
was behaving in the manner of a swarm of particles or *photons*.

Here was a remarkable paradox. The outstanding achievement
of nineteenth-century physics had been the demonstration by James
Clerk Maxwell that light behaves as a wave of an electromagnetic
field; he had set down mathematical equations which described
this. But how could light be both a particle and a wave? Counterintuitive, based on probabilities, non-local qualities and difficult
mathematics, quantum theory undermines the deterministic trajec-
tories of Newtonian physics and brings into our view of nature an
elusive fitfulness. It is the subject of many unresolved debates to
this day, not least its disjointedness with classical physics. Despite
its daunting conceptual difficulties, and its accessibility to only a
small elite of scientists and mathematicians, quantum theory was to
give rise to a stunningly successful account of the atomic and
sub-atomic structure of the world, an intellectual achievement of
great beauty and imagination. It would exert a powerful new
influence in the disciplines of physics and chemistry, and it would
enjoy a remarkable agreement between observation, experiment
and theory. It was to lead to fission and to chain reactions; it would
make available new technologies, from transistors to nuclear
energy, creating the world-transforming technologies of the computer age, and at the same time the mushroom-shaped cloud of the
atomic bomb.

Who should take the accolade, or the blame, for the unfolding
discoveries that flowed from Planck's and Einstein's work in the
first decade of the new century? One thing is certain: while German
talent made a signal contribution to the development of quantum
theory, and while Göttingen, Munich and Berlin would become
major centres of endeavour, its nurture, its source of imagination,
its intellectual stamina and genius originated in cities as diverse as
Cambridge, Manchester, Copenhagen, Vienna, Moscow, Paris and
Zurich. Quantum physics thrived on freely shared information and

wide-ranging collaboration between scientists of different nationalities and cultures.

A key event illustrating international collaboration in the early story of quantum theory was the First International Solvay Congress in Physics held in Brussels at the Hotel Metropole at the end of October 1911. Ernest Solvay was a wealthy Belgian chemist who had made a fortune out of a process for industrial production of sodium carbonate. His industrial empire spanned the United States, Europe and Russia. A thinker and philanthropist, he was keen to promote the sociology and politics of science. It was arguably the first instance of an international meeting focusing on a specific agenda in contemporary physics. Many more, under the auspices of Solvay, would follow down the decades. The media overlooked the significance of the event, although the rumoured elopement from the meeting of two participating scientists – Madame Curie and Paul Langevin – was a subject of widespread gossip in scientific circles.

Ernest Rutherford's contributions at the Solvay Conference stimulated the early quantum work of a young Dane called Niels Bohr in Copenhagen. Henri Poincaré, the French mathematical physicist, influenced another participant, the Englishman James Jeans; then Poincaré returned to Paris to spread the message in a series of papers that caught the attention of his French colleagues.

Convened by the physical chemist Walther Nernst, the Solvay participants included Planck, Einstein, Sommerfeld, Kamerlingh-Onnes, Rutherford, Langevin and Curie. One of the young secretaries of the meeting was the 25-year-old Frederick Lindemann, who would become Winston Churchill's scientific adviser two decades later. Lindemann left a personal account of the meeting. The discussions were 'interesting', he informed his father in Devon, 'but the result is that we seem to be getting deeper into the mire than ever. On every side there seem to be contradictions.' He liked Einstein, and had observed that he had 'a Jewish nose'. Einstein apparently asked the young Lindemann to come and stay with him, 'and I nearly asked him to come and see us [in Devon] . . . However, he does not care much for appearances and goes to dinner in a frock coat. He says he knows very little mathematics and can only set up general considerations, but he seems to have a great success with them.'[6]

In the final session of the conference Einstein declared: 'We all agree that the so-called quantum theory of today, although a useful device, is not a theory in the usual sense of the word, in any case not a theory that can be developed coherently at present. On the other hand, it has been shown that classical mechanics . . . cannot be considered a generally useful scheme for the theoretical representation of all physical phenomena.'[7]

Whatever his reservations, the congress had put Einstein firmly at the centre of the new physics community. Einstein showed at all times his respect for H. A. Lorentz, the chairman of the congress, and Lorentz subsequently tried to tempt Einstein into succeeding him in the chair of physics at Leiden. Einstein turned it down in a letter redolent with regret and deep respect.

Relativity and the Anti-relativists

Albert Einstein, born in Ulm, Germany, in 1879, had transcended his German background and roots at an early stage to such an extent that Germany may hardly claim him. As a boy he had attended the Luitpold Gymnasium in Munich, a highly disciplined, militaristic institution that had not appealed to him. His teacher encouraged him to leave at sixteen, claiming that his indolence affected the diligence of his classmates and that he would never make a success of anything. He was a natural outsider and resented regimentation; it is now believed that he may have been dyslexic. But he had shown an early love of music and began to study the violin. When he began to recognize the underlying mathematical structures, his aptitude and passion for music expanded. In later years he wrote of his lack of practical skills and 'imagination', emphasizing his insights that were 'too deep for words'.

After a brief spell in Milan, where his businessman father had attempted to redeem the ailing family fortunes, Albert settled in Zurich to attend a four-year course in mathematics and physics at the Federal Polytechnic School. He had told his father that he had decided to renounce his German citizenship and his Jewish faith.[8] Authors Michael White and John Gribbin comment that the decisions 'were of course part and parcel of his avowed intention to maintain an isolation from the world and to be master of his destiny'.[9] In January 1896, on payment of three marks, he was officially relieved of his German citizenship and remained stateless until five years later, when he acquired Swiss citizenship. In 1911, on taking up a post at the German University in Prague, he was obliged to accept Austro-Hungarian citizenship; but he retained his Swiss papers.[10] Zurich, a city of 153,000 inhabitants at the turn of the century, was a hospitable metropolis for free and revolutionary spirits. The 'Athens on the Limmat', as it was known, shut no one from its frontiers or its schools. In its crowded cosmopolitan cafés could be found the analysts Adler, Carl Gustav Jung and Freud, as well as Lenin. Excluded, nevertheless, from employment as a schoolteacher, Einstein went to Bern, where he had gained a post as examiner in the Swiss patent office. In 1901 his Serbian lover, Mileva Maric, gave birth to a girl out of wedlock. The child was adopted and died aged two. In 1903 he and Mileva married, moving into a one-bedroom apartment. Continuing his studies, he received a doctorate in physics from the University of Zurich in 1905, and Maric gave birth to a son, Hans Albert.

In that same year Einstein published his special theory of relativity, which became an essential basis for understanding atomic and sub-atomic particles. In many respects it is counterintuitive in terms of our understanding of space and time. Einstein based the theory on the observation that the speed of light is not affected by the velocity of the source that produces it, or the velocity of the observer who receives it. This led him to a basic postulate that the laws of physics must be the same in all reference frames moving relative to one another with a constant velocity. Consequently observers in different uniformly moving frames will find that the passage of time occurs at different rates as seen in different moving frames. According to Einstein, then, in special relativity there is no such thing as absolute space and time. All observations – the timing of an event, the length of a piece of string, or the weight of an object – are relative, depending on the speed of the observer. At the same time he posited the formula $E=mc^2$, relating mass (m) with energy (E) (which in classical physics are distinct) with a proportionality constant involving the speed of light. Einstein later generalized this approach in his General Theory of Relativity, published in 1915, which relates gravity and accelerated frames of motion. He showed that the presence of mass alters the structures of space and time to produce an apparent force which causes accelerated motion of bodies in space anticipating explanations for important cosmological phenomena.

Although general relativity in its broad outlines came to be known at an early stage, the theory was not widely publicized until after the First War. The event that prompted its recognition was an eclipse expedition. Einstein had predicted that light from a star passing close to the sun would be apparently bent by the sun's mass. He also calculated the degree of curvature. In 1919, under the auspices of Arthur Eddington, the Cambridge astrophysicist, expeditions were mounted to Brazil and to the Principe Island off the west coast of Africa. As a result Einstein became a worldwide celebrity. *The Times* of London ran the headline: 'Revolution in Science: New Theory of the Universe, Newtonian Ideas Overthrown.'[11]

Paul Dirac, the Cambridge mathematical physicist who achieved an early link between aspects of quantum theory and Einstein's special theory of relativity, wrote that after the First War, 'relativity came along as a wonderful idea leading to a new domain of thought. It was an escape from the war . . . Relativity was a subject that everybody felt himself competent to write about in a general philosophical way.'[12] Einstein himself disowned the notion that his theories were 'revolutionary', preferring the term 'evolutionary' and stressing that his work was the result of the foundations laid down by Newton and James Clerk Maxwell, whose laws of electrodynamics unified electricity and magnetism as a single measurable phenomenon.

Relativity soon acquired fashionable applications within literature, aesthetics, economics and philosophy, far from its original scientific authenticity. José Ortega y Gasset, for example, saw relativity as 'a marvellous proof of the harmonious multiplicity of all possible points of view'.[13]

The reception among scientists varied from country to country. In France it was regarded with suspicion as a 'German' theory; in the United States there was a tendency to deem it contrary to common sense. A professor of physics at Princeton condemned it, since fundamental physical theories 'must be intelligible to everybody, to the common man as well as to the trained scholar'.[14] In Britain, where the mistaken theory of ether still held sway, scientists were slow to accept it. According to the wave theory of light, the hypothesis of 'ether', an all-pervasive cosmic medium, supposedly explained the presence and movement of light waves. Waves, it was hypothesized, had to propagate in some medium, like waves in water. A wave in a vacuum seemed meaningless.

In Germany, however, relativity prompted lively and informed debates. And it was here that the anti-relativists emerged, a group of physicists as right-wing in politics as they were conservative in science. The leading anti-relativist figures, Philipp Lenard and Johannes Stark, both Nobel laureates, form direct connections with the antagonism towards theoretical physics and quantum physics, anti-Semitism and the phenomenon of Aryan physics under Hitler more than a decade later. They condemned Einstein's theory as laughable, degenerate and unproven. Four-dimensional space-time, curved space and the twin paradox, they declared, lacked physical meaning. Relativity had nothing to do with science, they insisted; the theory was 'ideological' and linked with the internal politics of power in German science. Einstein was quick to point out that the anti-relativist campaign was in fact about anti-Semitism.

Philipp Lenard

A striking feature of the rise of National Socialism in the 1920s, in parallel with the growth of the new physics, was the conviction of a link between German racial purity and 'authentic' science; a link between Jewishness and 'bogus' science. What is remarkable about the racialization of science in 1920s Germany was that its two leading proponents, Philipp Lenard and Johannes Stark, were talented researchers who had made significant contributions to science long before their ideological notions had taken root. Individual brilliance and early personal success in science are evidently no guarantee of rational, dispassionate, universalist principles. Personality and emotions, it seems, are inseparable from the ways in which scientists promote themselves and their science. The existence of a scientific milieu, well funded, independent and ideal for groundbreaking discovery, is no guarantee for the shaping of mature political consciousness.

Philipp Lenard, with sunken eyes, bushy brows, wild hair and an angry beard, was a man of vitriolic and fanatical zeal whose seething envy had many targets. He suffered for many years from a disease of the lymph nodes that caused one ear to press down against his shoulder. It was eventually relieved by radical surgery. Born the son of a wine merchant in the Hungarian city of Poszony (later Bratislava), he came to Germany, where he worked under Heinrich Hertz, who in 1888 performed a series of experiments which produced electromagnetic waves (he had a unit of frequency named after him – 'the hertz'). Lenard subsequently became Hertz's assistant in the University of Bonn before being called to a chair in physics at the University of Heidelberg. Lenard won the Nobel prize in 1905, aged forty-three, for his work on cathode rays. The 'Lenard window', one of his discoveries, involved a method whereby the beam in the cathode-ray tube could be directed on outside targets. He revealed that cathode rays can pass through atoms, and he explained how much of an atom was empty space where positive and negative charges are separate.

Despite his brilliance and success as an experimental physicist he was a great hater of certain of his peer group scientists, his enmities anticipating his xenophobic and racist rage in the early years of the Third Reich. He detested Wilhelm Konrad Röntgen, the disoverer of X-rays, an achievement he thought more properly his own. Lenard had helped Röntgen obtain the highly sensitive tube necessary to generate X-rays, but the older man failed, in Lenard's view, to give him appropriate credit. He also loathed his English former collaborator J. J. Thomson. He accused Thomson of exploiting his research on the photoelectric effect without acknowledgement and consequently nourished a deep contempt for English science as a whole, which he thought to be sloppy. He denounced Thomson to this effect when delivering his Nobel award acceptance speech. Severely practical, his peculiar bent meant that he fell behind in mathematical expertise and theory. He developed a distaste for mathematics and mathematical physics at a point when these disciplines were increasingly crucial. It has been said that his contempt for J. J. Thomson at Cambridge was a token of Thomson's love for mathematics.[15] By the beginning of the First War, his hatred of the English had culminated in the conviction that the island race were a collection of self-seeking, duplicitous tradesmen, whereas the Germans were a nation of heroes. The continent of Europe, he believed, should impose an intellectual blockade on England. His belief that the English had suffered a fall from intellectual grace was confirmed by reading the racist meanderings of Houston Stewart Chamberlain, the Englishman who, as we have seen, had ingratiated himself with the Wagner family and adopted Germany as his homeland, endorsing the myth of Teutonic racial supremacy.

Lenard had earlier expressed respect for Einstein, and as late as 1913 considered inviting him to a chair in theoretical physics at Heidelberg. But he seems to have revised his attitude towards Einstein in the course of the war, and Einstein began to associate Lenard with the anti-relativity campaign that was building in Germany after the war. In 1920, a series of popular lectures attacking relativity was delivered in Berlin. Einstein responded in the *Berliner Tageblatt*,[16] showing that he was capable of doughty polemic. 'A motley group has joined together,' Einstein began, 'to form a company under the pretentious name "Syndicate of German Scientists" currently with the single purpose of denigrating the theory of relativity and me as its author in the eyes of non-physicists.' One of the principal speakers, he noted, a Mr Weyland, 'does not seem to be a specialist at all (is he a doctor? engineer? politician? I could not find this out), presented nothing of pertinence. He broke out in coarse abuse and base accusations.' In the course of the article Einstein claimed that Lenard's objections to relativity 'are already invalidated on the basis of my general proof that on first approximation the statements of the general theory of relativity conform with those of classical mechanics'.

The attack was enough to inflame Lenard's undying wrath. But other factors may have fuelled his anger. His only son Werner died in February 1922 as a result of malnutrition suffered during the war. Lenard, moreover, was one of the investors who exchanged gold for bonds, which became worthless in the inflationary Weimar years. He was convinced that he had been swindled of his fortune by Jewish politicians. Above all he was made the target of a demonstration at Heidelberg in June 1922, when he failed to fly the flag of his institute at half-mast to honour the death of Walther Rathenau (a politician, industrialist and Jew, who Lenard believed had caused irreparable harm to Germany). A band of students demonstrated outside the building and were hosed down with water from an upper storey, doubtless directed by Lenard; whereupon they broke in and frog-marched Lenard to an auditorium, where they subjected him to a torrent of political invective. Lenard was subsequently disciplined by the university for failing to honour the official day of mourning. Thereafter he entered the ranks of the National Socialists.

From this time Lenard's anti-Semitism became a feature of his view of science and scientists. He was the first German scientist of note to condemn 'Jewish physics' – namely, theoretical physics – and to insist on a return to 'Aryan' or experimental physics. In 1922 he charged his colleagues with having betrayed their race, proclaiming that Jews were well known for twisting objective arguments into personal conflicts.[17]

Lenard's first influential text was his *Grosse Naturforscher* (*Great Natural Scientists*), published in 1929, a pseudo-anthropological biographical encyclopedia of sixty-five scientific heroes of history, marked by their Aryan-Germanic heredity, from the time of Hipparchus of Nicea in the first century BC to Heinrich Hertz, his erstwhile mentor. Hertz, ironically, was half Jewish, but Lenard entirely ascribes his hero's experimental aptitude to an Aryan mother, and his unfortunate lapses into theoretical physics to his father. The explanation of Jewish talent by reference to the presence of Aryan blood in their make-up became a familiar argument in his rantings. Another foible was his habit of diverting attribution. For example, in the case of Einstein's irrefutable equation $E=mc^2$, the relation of energy to mass, Lenard claimed that the discovery was owed to a little-known Austrian, and purely Aryan physicist, called Hasenohrl, and that the equation was stolen by Einstein.[18]

The historian Alan D. Beyerchen argues that Lenard's career reveals a background fairly typical for early Nazi activists. 'His upbringing in a German borderland, his romantic yearning for great figures to lead the way, and his frustrated need to feel genuine human contact and belonging were three of the most common characteristics of converts to the Hitler movement.'[19] Beyerchen adds that Lenard's envy and frustration over the praise accorded Einstein explains the growth of his anti-Semitism and that his alienation from the physics community was a major factor in his resort to extremism.

Nazi Physicist Johannes Stark

Johannes Stark, son of a landowner and native of Bavaria, was Lenard's partner in the campaign for Aryan physics. He enjoyed a highly successful early career, discovering in his early thirties the Doppler effect in 'canal rays' (a form of radiation emitted from the holes in positive electrodes) and the influence of electric fields on spectral lines, the 'Stark effect', for which he won a Nobel award in 1919. In these early days he too was impressed by Einstein and invited him to write a review article in a science journal he had founded and edited. For some two years Einstein and Stark corresponded. Stark was one of the first physicists to employ the concept of light quanta in his own work and he later wrote a paper describing an experiment that verified Einstein's light quanta theory.

Stark was a prolific author of scientific papers, but he was what the Nobel laureate James Franck has described as 'a pain in the neck' over intellectual property rights. Franck conceded that Stark had good ideas, and that he had them 'early'. 'He had the idea that photo-chemistry would be a quantum process. Not as clear as Einstein, but he had it.'[20]

Before World War I, however, Stark had been angered and humiliated by a clash with Arnold Sommerfeld over a university appointment. He had high hopes of being called to a chair at Göttingen, which went instead to Peter Debye, one of Sommerfeld's favourite students. Stark's rationale, following his disappointment, foreshadowed later racist prejudices. The fixed appointment was the work of a 'Jewish and pro-Semitic circle' and its 'enterprising business manager [Sommerfeld]'.[21] After the war, stern of eye and with flourishing walrus moustache, Stark worked in Greifswald, a small university town, where he had become active in nationalist conservative politics in reaction to the threat of communist revolution. From there he went to the University of Würzburg, where he began to involve himself in the politics of physics. In opposition to the generally liberal German Physical Society, Stark set up an alternative reactionary organization called the German Professional Community of University Physicists, in the hope of influencing appointments and funding. Thwarted at every stage in his ambitions to control the physics community, Stark became ever more vitriolic and intemperate. When the thesis of one of his protégés, Ludwig Glaser, was mocked by colleagues as excessively applied (on the optical properties of porcelain), Stark concluded that he was surrounded by Einstein lovers and resigned his chair. He had aspired to the presidency of the Reichs Physical-Technical Institute, which would have given him a new and prestigious base in the physics community; but he was passed over and became even more embittered. He was not to get another academic post until Hitler came to power in 1933. Then came Einstein's Nobel prize (for 1921), awarded in 1922.

Stark now unleashed his venom against Einstein and relativity in his book *The Contemporary Crisis in German Physics*. He stopped short of outright anti-Semitism, but he characterized the foundations of the theory and method of publication as anti-German. Stark charged that Einstein and his followers had promoted the 'revolutionary' nature of the theory around the world, drawing parallels with social and political revolution in Germany at the time. Einstein's method of self-publicity, according to Stark, was a betrayal of Germany and German science.

Stark's book, however, was dismissively reviewed by the distinguished physicist Max von Laue, who won a Nobel prize in 1914 for his discovery of the diffraction of X-rays by crystals. This drew, as historian Mark Walker has put it,

the battle lines for the subsequent struggle over Einstein's science: on one side, scientific support of the theory of relativity and opposition to the racist, political, and ideological attacks against its creator; on the other side, escalating personal attacks on Einstein and his work which had less and less to do with science and more and more to do with the National Socialist movement.[22]

Laue closed his review: 'All in all, we would have wished that this book had remained unwritten, in the interest of science in general, of German science in particular, and not least of all in the interest of the author himself.'[23]

Two years later, six months after Hitler had led the march from a beer hall in Munich in an attempt to topple the city government, Lenard and Stark jointly published an article entitled 'The Hitler Spirit and Science' in the *Grossdeutsche Zeitung* for 8 May 1924. The authors compared Hitler with the giants of science, celebrated Aryan genius and condemned the corrupting influence of Judaism down the ages. Eulogizing the 'spirit of total clarity, of honesty towards the outer world', the authors linked these virtues in science with the spirit of the great scientists of the past: 'Galileo, Kepler, Newton, and Faraday'. That same spirit, they continued, 'we admire in Hitler, Ludendorff, Pohner, and their comrades . . . consider what it means to be privileged to have this kind of genius living among us in the flesh'.[24] Yet their readers must not be deceived: 'the Aryan-Germanic blood, the carrier of this unique spirit, is already in rapid decline,' while that 'alien spirit' flourishes, the same that 'brought Christ to the cross and Jordanus Brunus to the stake'. Then came the rallying cry: 'We now place the highest emphasis on protecting that which is inherited in our blood, because we have learned to recognize it as the blessing of all mankind. But we will not simply protect our own intellectual identity, we will first start digging it out again from under the alien-spirited rubble within ourselves – primarily in order to find ourselves again. We need lucid minds not only as scientists – no, people should not be segregated according to their occupation (a frequently used Semitic deception!).'

The phenomenon of *Deutsche Physik* (German Physics) was now revealing itself in Germany in several disparate but converging influences. There were those scientists who were intellectually conservative; who were opposed to relativity and quantum physics, since these insights threatened to undermine intuitive, mechanical models of the world. There were those who envied and despised Einstein because they were convinced that he headed a disruptive Jewish conspiracy within the world of physics. Then there were conservative nationalists who despised Einstein's internationalist, pacifist stand during World War I, as well as his support of the democratic spirit of the Weimar Republic.

Lenard and Stark embodied all these convictions, ten years before Hitler's accession to power, thus anticipating in the domain of physics an enthusiasm for National Socialism and the potential for a Nazi science.

8. German Science Survives

The quantum revolution, or evolution, as Einstein preferred, was taking place against the background of Germany's recent defeat in war, poverty and political upheaval and fragmentation. In the immediate aftermath of the Great War Germany had been plunged into crisis, its economy close to collapse, its alliances in shreds, its military might defeated, its society vulnerable to revolution and civil war. Shamed, oppressed by harsh peace terms at Versailles, impoverished, under-funded, Germany's scientific community was in hard straits. In 1919 the victorious Allied powers had founded the International Research Council; but Germany and Austria, as well as countries that had remained neutral during the conflict, were barred. It was not until after the Locarno Pact in 1925 that a more benign spirit prevailed between the former belligerents and Germany was accepted into the council. But when the official invitation came, it was turned down by the German and Austrian scientists themselves.

German scientists, moreover, were boycotted by many of their foreign peer group, their applications for visiting professorships rejected, their access to foreign-based symposia denied. Between 1919 and 1928 German scientists were banned from many international science conferences. Out of some 275 international meetings between 1919 and 1926, 165 were without German representation – notably, the Solvay Congresses of 1921 and 1922 on atoms and electrons, and electrical conductivity of metals. The quality of these meetings understandably suffered from the absence of the Germans.

Einstein did not approve of such bans. When he was finally invited, alone, to the next Solvay meeting as an honorary non-German, he refused as a gesture of solidarity with his German colleagues. 'If I took part in the Congress I would by implication

become an accomplice to an action which I consider most strongly
to be distressingly unjust.'[1]

Einstein, despite the enmities of Lenard and Stark, had become
a celebrated international figure during these troubled times, and
there were those who feared he might be lost to Germany.
A diplomat attached to the German embassy in London wrote
to the Foreign Ministry in Berlin in September 1920: 'Professor
Einstein counts as a cultural factor of the first rank for Germany,
since his name is known far and wide. We should not drive such
a man out of Germany; we could use him for effective *Kultur*
propaganda.'[2]

The anxiety over Einstein revealed the importance of science in
a country in which most features of national pride had collapsed.
'There is one thing which no foreign or domestic enemy has yet
taken from us,' Max Planck told the Prussian Academy of Sciences
in November 1918, 'that is the position which German science
occupies in the world.'[3] Science deserved the support of the German
people, Planck argued, not only because it would lead to economic
recovery, but because it was Germany's chief cultural resource.

Despite the privations and upheavals during the 1920s, public
understanding of science was given a boost in Germany by the
construction of a new planetarium by the Zeiss lens corporation in
Jena.[4] Enthusiasm for popular astronomy had been generated by
media coverage of a new theory of origins of the universe in a
primal explosion, a notion prompted by the theories of Einstein
(although he himself was sceptical about the idea), and the American
astronomer Edwin Hubble's observations of receding galaxies.
Despite economic crises, the planetaria prospered through public
support and newspaper sponsorship. The design of the modern
Zeiss planetarium was originally conceived in 1919 by Walther
Bauersfeld, who abandoned earlier models of a large rotatable
hollow sphere, opting instead for a collection of projectors creating
luminous images of the stars on to a stationary hemispherical dome
which served as a screen. Guided by electronic motors, the projectors created patterns illustrating the motion of the heavenly bodies
in conformity with nature, although vastly speeded up, of course.

The first planetarium was completed by a team of scientists, engineers and mechanics by March 1923 at the Zeiss factory. By October the model was transported to the Deutsches Museum at Munich, where it was viewed by thousands of people. It was replaced by a larger, second-generation planetarium in May 1925 (which survived the war). The design of the first planetaria provided for a perspective from a single geographical latitude. The model continued in development through the 1920s, until the astronomer Walter Villiger designed an instrument that could demonstrate the motions for any desired geographical latitude. Between 1923 and 1930 no less than seventeen state-of-the-art planetaria were built in major cities of Germany, including Munich, Jena, Düsseldorf, Berlin, Mannheim and Nuremberg. (Seven of them were destroyed during the war.) Strangely, in view of Hitler's stated enthusiasm for astronomy, none were built during the period of the Third Reich, although the Zeiss corporation continued to take orders for planetaria overseas, five of them in the United States.

Physics and Weimar

By the mid-1920s physicists were beginning to collaborate once again with their overseas colleagues. Science funding, hard to come by in Germany and Austria up to 1923, now began to benefit from a new system which eked out scarce resources with additional money from industry and some notable sources abroad. A new agency was established in Weimar Germany called Notgemeinschaft der deutschen Wissenschaft (Emergency Society for German Science). The Notgemeinschaft represented the Kaiser Wilhelm Society and its institutes, the universities and technical universities and the academies of science. It provided money not only for institutions but for research programmes and individual scientists. Its funds came from the Reich government in Berlin, but it also attracted funding from General Electric and the Rockefeller Foundation in America. German industry tended to support the rival Helmholtz Society and applied scientific projects.

In the tense post-war period the physics community in Germany was fragmented politically and intellectually, but generally united in its detestation of the Weimar Republic. The ultra-right, represented by the fascist-minded Lenard and Stark, were, as we have seen, antagonistic towards theoretical physics. Yet there were brilliant right-wingers, like Pascual Jordan, also destined to become a Nazi, who were making major contributions to quantum theory. There was, besides, a rump of generally conservative physicists, like all scientists in Germany politically constrained by their traditional role as civil servants, who were against the Weimar government, under which they had lost prestige and taken pay cuts. Even civilized physicists, like Planck and Sommerfeld, tended to be against Weimar and the fragmentation it represented, as they saw it, of culture and good order. Anti-Semitism was rife among the conservatives, although this was no different from attitudes within scientific communities in other countries, including the United States.

A token of the anti-Semitism that reigned in many universities and research institutions during the 1920s is the story of Richard Willstätter, the chemist and Nobel laureate (he won the prize in 1920 for plant pigments, and chlorophyll in particular). During the war Willstätter, who was Jewish, invented a triple-layered gas-mask, for which he earned an iron cross. A close friend of Haber in the pre-war and wartime days, Willstätter was appointed in 1916 to a chair at Munich. He reports in his memoirs that when the King of Bavaria, Ludwig II, came to sign the document of his appointment as professor, his majesty murmured to a minister present: 'That's the last time I sign in a Jew for you.'[5] Willstätter found Munich so intensely anti-Semitic that he resigned his post in 1924. His main grievance was discrimination against Jews being considered for appointments and promotion, unless they were outstandingly brilliant.[6] But there were other menaces. Willstätter wrote that Munich had never settled down after the Hitler-putsch and that large numbers of patriotic youths were driven to despair because of inflation, the injustice of the Versailles Treaty and the hatred of France towards Germany. He literally saw the writing on the wall when Nazi posters went up around the city with the slogan: 'No German youth may in future sit at the feet of a Jewish teacher.'[7]

Quantum Mechanics

The Danish physicist Niels Bohr, one of the youngest of the early quantum physicists of the pre-First War era, was responsible for taking quantum theory forward in the 1920s, beyond the original ideas of Planck and Einstein. His love of scientific collaboration, and his hopes for an international community of physicists operating as an ideal society, stands in contrast to the enmities and conflicts that reigned in the milieu dominated by Stark and Lenard. His relationship with Werner Heisenberg, one of his German protégés, was to lead to a historic meeting in the depths of the war in occupied Denmark.

Bohr, a fleshy man with a large head, was described by many who knew him as eccentric in speech and mannerisms. The recollection of his appearance by the physicist James Franck is typical: 'sitting there almost like an idiot . . . face empty, limbs were hanging down'. Then suddenly a 'glow went up in him' and he would say: 'Now I know.'[8] Bohr's biographer Abraham Pais attempted to explain Bohr's foibles as a matter of 'deep thought as he spoke'. He remembered how one day Bohr was finishing part of an argument, then said: 'And . . . and . . .', then was silent for a moment, then said, 'But . . .', and continued. Between the 'and' and the 'but', Pais relates, 'the next point had gone through his mind . . . he simply forgot to say it out loud and went on somewhere further down the road'.[9]

Born in 1885, the son of a professor of physiology in Copenhagen, Bohr had been working in the first decade of the century on a model of the atom originally suggested by Ernest Rutherford at his Manchester laboratory in England. Rutherford understood atoms to be much smaller than previously envisaged. He discovered that the preponderance of an atom (99.9 per cent of its weight) is in the central nucleus, while the rest is space patrolled by electrons, like planets orbiting a sun. The problem with Rutherford's model was that it did not account for the stability of the atom, which is one of its most important features. Niels Bohr applied Planck's quantum hypothesis to the hydrogen atom, the lightest of all atoms, which contains only one electron. According to this model, electrons remained in their orbits because their angular momentum came in discrete units: in other words, they were 'quantized'. Bohr won the Nobel prize for physics in 1922 for his work on the atom.

But his model only raised more questions. Otto Frisch, who was to take part in the quantum revolution as a young physicist in Germany in the 1920s, puts it this way:

Why should the electron be confined to specified circular orbits around the nucleus, as Bohr claimed? Was the hydrogen nucleus surrounded by a set of concentric rails on which the electron had to travel? And how did the electron manage, on jumping from a bigger to a smaller circle, to push its surplus energy out as light quantum? All this was totally alien to traditional physics.[10]

The questions were even more vexing when it came to predicting the structures. Do two electrons in an atom of helium follow each other around the same circle, with the nucleus between them? Or do they go in concentric circles? Or intersecting circles, somehow avoiding collisions? What were the underlying laws? The answers to some of these questions were provided in the early 1920s by Arnold Sommerfeld in Munich, and by Wolfgang Pauli from Vienna. Sommerfeld had classified possible orbits of electrons including elliptical orbits of electrons, while Pauli formulated in 1924 his 'exclusion principle', whereby more than one electron cannot exist in the same quantum state. Pauli won the Nobel award for his discovery.

Given the problems of direct observation at the level of the sub-atomic, quantum mechanics was increasingly about what happens when atoms are subjected to specific influences – light, heat, magnetic and electric fields. The new physics would be driven both by experiment and by sophisticated mathematical models, the stuff of the theoretical work of Werner Heisenberg, the head of the Nazi atomic programme in years to come.

Heisenberg first joined the quantum physicists when he accompanied Sommerfeld at a 'Bohr Festival' in 1922 at Göttingen, where the celebrated Danish physicist was giving a series of lectures on quantum mechanics. Heisenberg found 'Göttingen, in gorgeous summer weather . . . resplendent with gardens and flowers, and . . . the excitement of the students who filled most of the auditorium' He enjoyed the experience to the full despite the fact that it was 'often on an empty stomach, as was normal in those days for a student in the fourth semester'.

Heisenberg stands out among his cosmopolitan physicist colleagues as a stereotypical German youth: blond, blue-eyed, given to dressing in *Lederhosen*. He was a member of the 'Pathfinders', which emerged in 1919 from the German Youth Movement (Jugendbewegung) founded in 1913, an organization that promoted patriotism and outdoor life. The Pathfinders liked to trek in the hills and mountains and sit around campfires. They sought to oust stilted old-world manners and charm with 'inner truthfulness'. They took an oath which proclaimed: 'We want to be ready with advice and practical assistance whenever there is a chance of helping a good and just cause. We want to follow our leaders.'[11]

Heisenberg shared a love of music with many of his colleagues, although it was said that his playing was technically correct but without emotion. The son of a professor of Greek, Heisenberg had been taught by Max Born at Göttingen, one of the great mathematical centres of the world. It was Max Born, in fact, who first coined the term 'quantum mechanics' (*Quantenmechanik*) in 1924. By 1921 Born had become director of the Institute for Theoretical Physics at Göttingen; the following year, after a visit by Niels Bohr, he had called for new rigour in the discipline:

The time is perhaps past when the imagination of the investigator was given free rein to devise atomic molecular models at will. Rather, we are now in a position to construct models with a certain, although still by no means complete certainty, through the application of quantum rules.[12]

The following year he took on Heisenberg as his assistant and the triangle of quantum research and discovery began.

In a memoir of those early beginnings, Werner Heisenberg recollected the three intellectual aspects of the new discipline: 'Sommerfeld's school in Munich with the phenomenological approach, the Göttingen centre with the mathematical, and the Copenhagen group with the philosophical tendency, though the transitions are naturally fluid'.[13]

The sense of impermanence, of excitement, of scientists running off with a new idea after each revelatory seminar or symposium, gave the new physics the semblance of a mad hatter's tea party. Arnold Sommerfeld in Munich told his students that before they entered into the house of physics they should take note of the warning sign outside: 'Caution! Dangerous structure! Temporarily closed for complete reconstruction!'[14]

By the summer of 1925 Heisenberg, while recuperating from an attack of asthma on the island of Heligoland, produced 'matrix mechanics', which proposed a formula relating different quantum states of electrons. It was under the aegis of Bohr in Copenhagen, however, that Heisenberg later formulated his famous 'uncertainty principle'. The nub of the principle is that the velocity and position of a fundamental particle cannot both simultaneously be determined with precision. The 'uncertainty principle' was published in Heisenberg's paper 'On the Intuitive Content of Quantum Kinematics and Mechanics' in 1927.

Meanwhile, Erwin Schrödinger, an Austrian-born mathematician and physicist with an aptitude for philosophy, demonstrated a mathematically equivalent equation, which he called 'wave mechanics'. The implications of Heisenberg's and Schrödinger's equations confirmed what Born had already stated, and what had now become widely established, that particles – less than one-billionth of an inch – are not easily visualizable and that measuring their properties went beyond the limits of intuition and perception. Otto Frisch, breaking the caveats of the physicists, nevertheless hazarded an image, following Born's visualization of the new atomic model: 'Here was a new model for the atom,' he wrote, 'though not so easy to visualize; the electron now looked more like a pulsating cloud than an orbiting little planet. But for people who like to think in models it is still the best one we can offer.'

By 1927 the Cambridge mathematical physicist Paul Dirac brought Schrödinger's equation into line with Heisenberg's matrix mechanics, then in a stunning intellectual leap created a quantum mechanical theory, in line with special relativity, that described the behaviour of the electron. Dirac, an angular young man, once described as 'shy as a gazelle, and modest as a Victorian maid', nevertheless stole the show at the Fifth Solvay Conference in Brussels (1927), which was devoted to the new physics. His contributions stretched the brilliant audience to such an extent that Schrödinger said to Bohr: 'He has no idea how *difficult* his papers are for the normal human being.' The comment evinces the inaccessibility of the new physics even to its practitioners.

While quantum theory, by the late 1920s, worked well for descriptions of slow-moving electrons, it broke down when describing electrons moving at or near the speed of light, or when light bounces off an object. Dirac employed Einstein's theory of special relativity to solve these problems. Then he went on to describe, elegantly and profoundly, the behaviour of the motion of electrons, in what came to be known as the 'Dirac equation'. Meanwhile Max Born had ventured that the 'waves' in quantum mechanics were probability amplitude waves. An electron did not have a specific location; an electron orbiting a hydrogen nucleus had a 'certain probability', as John Polkinghorne, physicist and popular expositor of quantum mechanics, puts it, 'of being found here and a certain probability of being found there'.[15]

German Physics Before the Storm

By the late 1920s, despite the manifest economic problems of the day, and general dissatisfaction with the Weimar Republic, German physics was once more taking a major role in the world. There was a period of uneasy calm before the onslaught of Nazism. One has an impression of open-ended seminars, of heated arguments on long country walks; and always music: concert music, and music making in the home.

There was Max Born, who 'kept somewhat aloof from his students and had a rather formal approach to physics'. Victor Weisskopf complains that Born, who had earlier suffered a stroke, and was reclusive, tended to express everything in complex mathematical terms. His presence in the opinion of many was awesome. Weisskopf once went to see Born to express his concern that a life in science would isolate him from the concerns of humanity. Born told him: 'Stay in physics. You will see how deeply the new physics is involved with human affairs.'[16] Naturally Born could not have foreseen at this point the eventual development of the atomic bomb.

A leading professor in experimental physics was James Franck, who had worked with Haber and Hahn on poison gas during the war. Franck surrounded himself with students and lively assistants. He had an intuitive understanding of science, and seemed able to predict with great accuracy the outcome of an experiment or the result of a calculation even when he was not acquainted with the mathematical methods. His students said that he 'had a direct wire to God'.[17] The leading mathematician was Richard Courant, who introduced to Göttingen the 'mathematical practicum', involving study groups of two or three students who worked together on assigned problems. Courant, according to Weisskopf, invited his students home to play and listen to music.

Also at Göttingen was the mathematical legend David Hilbert. Hilbert was a man of severe logic and occasional oddity. One of his students committed suicide, ostensibly because he failed to solve a mathematical problem. Hilbert was asked to talk at his funeral, so before the student's relatives at the graveside he explained that the maths problem that had defeated the unfortunate youth and caused his death was in fact fairly simple. The student, he told them, had simply looked at it from the wrong premise. When he lectured, Hilbert's assistant would write on the blackboard for him. Frustrated with unsatisfactory comments in the classroom he would mutter, 'Crazy, crazy, crazy.'

In Berlin during this period lived and worked Lise Meitner, one of the leading woman physicists in Germany. She was to become a crucial figure in the discovery of atomic fission in Germany. Born into a Jewish legal family in Vienna in 1878, Meitner had struggled to gain a scientific education through private tuition, since schooling in Austria for girls ended at the age of fourteen. Her mentor in the University of Vienna had been Ludwig Boltzmann, who refined the concept of entropy and formulated the statistical description of energy states.

Meitner began her research under Boltzmann on alpha rays, one of the types of radiation (along with beta and gamma rays) emitted by radium. Alpha rays, it was known, shot out of radium atoms at 9,000 miles per second (about 5 per cent of the velocity of light). After Boltzmann's death, she moved to Berlin, where she met Otto Hahn, and formed a partnership with him to study radioactivity at the Kaiser Wilhelm Institute for Chemistry. Meitner and Hahn complemented each other perfectly: he the chemist, she the physicist. Their relations were always proper for the era: never walking out together, or eating together. It took them sixteen years to drop the formal mode of address: Fräulein and Herr.

The circumstances of her research were eloquent testimony to the difficulties confronting women who had decided on a life in science in Germany. Hahn and Meitner applied to Emil Fischer, director of the institute, for bench space; but he would not hear of a woman working within his laboratory. Eventually he permitted her to set up her experiments in a maintenance area in the basement so long as she was not seen elsewhere in the building. She was obliged to use the toilet facilities of a neighbouring café.

Berlin

Otto Frisch, Meitner's nephew, describes in his memoir the personalities and the milieu in Berlin during the 1920s: the drama of discovery, the routines of research, the easy association between scientists, many of them Jewish, some of them world famous at the time, others to become so.[18] Aged just twenty, Frisch came to

Berlin to work as a researcher in the optics division of the PTR (Physikalisch-Technische Reischsanstalt), similar to the National Physical Laboratory in Britain, or the Bureau of Standards in the United States. Meitner helped Frisch find digs in Dahlem, close to her institute but a long bus ride away from his laboratories. She soon introduced him to Otto Hahn, who, according to Frisch, liked to tease people by whistling a jazzed-up version of Beethoven's violin concerto, then asking: 'Is that how it goes?'

Frisch was working on a new unit of brightness to replace the candle power measurement then in universal use. The PTR was involved at that time in work on atomic spectroscopy. 'I stayed long hours,' comments Frisch, 'trying out half-baked ideas of my own . . . Most of my ideas didn't work, but that's how one learns.' Scientists would meet in the lab canteen, but Frisch regretted that a dozen women known as the 'fever girls', whose job it was to calibrate thousands of clinical thermometers, never came in to eat. Like Otto Hahn, it seems, young Frisch was in the habit of whistling. He liked to give a rendering of Bach's Brandenburg Concertos as he walked down a corridor in the vicinity of Walther Bothe's lab (another future Nobel laureate working on radio-activity). Bothe would consequently get distracted while trying to count alpha particles, this being the days before calculators, so he sent out a message ordering the young whistler to desist once and for all.

Frisch's mother was a concert performer and had taught him piano from the age of five. In Berlin, he wrote, he played the piano 'with more panache than skill' and loved to show off with Chopin scherzos. Lise Meitner would play duets with him, especially a reduction of Beethoven's septet. Together they went to concerts; accompanied by his aunt, he first heard the symphonies of Brahms and much classical chamber music.

Every week there was a colloquium at the university. The front bench was occupied mainly by Nobel prize winners. Max Planck was invariably in attendance, as were Haber, Einstein and Walther Nernst. Frisch describes how Nernst, a short man now in his seventies, would often get to his feet after a lecture, waving his hands around and crying out, 'But that is just what I said forty years ago!'

Pranks and lugubrious practical jokes sometimes punctuated the routine of research at the Kaiser Wilhelm Institute for Chemistry. Frisch recounts how Gustav Hertz came to visit his aunt's laboratory. Hertz refused the cup of tea that had been offered and asked for something stronger, instructing a student to fetch down a bottle of pure alcohol from the laboratory shelf. Meitner was appalled and attempted to remonstrate with him, but he insisted on drinking a whole glass of the stuff without any apparent effect. Then it turned out that he had got the student to fill the bottle with water beforehand.

Work, however, was everything. Frisch gives us an insight into his working habits.

I regularly came home, had dinner at seven, had a quarter of an hour's nap after dinner, and then I sat down happily with a sheet of paper and a reading lamp and worked until about one o'clock at night – until I began to have hallucinations ... I began to see queer animals against the background of my room, and then I thought, 'Oh, well, better go to bed.'

Frisch comments that this was for him an ideal life: 'I'd never had such a pleasant life, ever – this concentrated five hours work every night.'[19]

One day Meitner introduced Frisch to Einstein as they walked through the entrance hall of the university. 'I hastily transferred a pile of books from my right arm to my left and dragged off my glove while Einstein patiently waited with his hand held out, typically informal, seemingly quite relaxed.' Frisch adds a comment, indicating that despite the everyday calm of the period, there were dangerous undercurrents threatening to disrupt this contented science community. 'Idolized by millions who did not understand his theories, he was also under vicious attack from some of his colleagues, for precisely the same reason, and with an increasingly anti-Semitic flavour.'[20]

PART THREE

Nazi Enthusiasm, Compliance and Oppression 1933–1939

9. The Dismissals

When Jewish scientists were dismissed en masse from their jobs in the spring of 1933, following the rise of Hitler, their jobs were promptly taken by their non-Jewish junior colleagues. This so puzzled British scientist George Barger, professor of chemistry at the University of Edinburgh, that he wrote to Karl Freudenberg, a well-known non-Jewish professor of organic chemistry at Heidelberg, asking how scientists in Germany could collaborate with the purge by taking these jobs. Freudenberg replied as follows:

There are orders which you simply have to comply with. It is my firm conviction that a cure of the body of the German people was necessary, something which probably only very few will deny. The way it has been carried out cannot be subject to lengthy considerations in this country, simply because there are orders, and it does not matter at all, what the viewpoint of the individual is.[1]

The dismissal of Jewish researchers and teachers from their posts constituted one of the first state applications of racial hygiene in Hitler's Germany, as if to purify the body of science and technology of a dangerous virus. In terms of the everyday social practice of science and its bonds of mutual respect, acceptance of the dismissals without protest, and readiness to benefit from the departures, revealed a profound moral lapse within Germany's communities of science.

Hitler had been sworn in as Chancellor on 30 January 1933, along with Hermann Goering, who doubled as Minister of Aviation and Prussian Minister of the Interior. Goering now controlled the police in Prussia and had wide-ranging powers of coercion to purge the party's opponents. The new Minister of Defence was General Werner von Blomberg, a Nazi sympathizer who had become captivated by Hitler's charisma. Alfred Hugenberg, the leader of the ultra-conservative German National People's Party, took the dual role of Minister of Economics and Minister of Agriculture.

Hitler, however, was not to be tamed into any kind of power-sharing. Immediately he called new elections for 5 March, and set about using his chancellorship to control the media, to oppress the opposition democratic parties and to launch attacks on Jews and socialists. In the March election Hitler still failed to obtain an absolute majority for the National Socialists. But with temporary allies in the form of Hugenberg's right-wing nationalists, he scraped together a majority of 52 per cent, securing 340 out of 647 seats in the Reichstag. In a turnout of 88.7 per cent, the Nazis obtained more than 17 million votes. The socialists dropped to 18.3 per cent of the vote, and the Centre Party, which had conducted a courageous campaign in the face of widespread Nazi intimidation, remained impressively solid at 13.9 per cent. The opening of the newly constituted Reichstag was celebrated on 21 March in the Old Garrison Church in Potsdam, the burial place of Frederick the Great, in the presence of Hitler and Hindenburg. Even as Hitler spoke moderately to the nation on the radio, his brownshirts were rounding up the 'enemies' of the regime and transporting them to concentration camps at Dachau and Oranienburg. Within two days parliamentary democracy came to an end as Hitler won a majority in the Kroll Opera House (the Reichstag being in ruins after an arson attack, probably perpetrated by the Nazis themselves) in favour of the Enabling Act that gave him dictatorial powers under a cloak of spurious legality.

In the last week of March thirty brownshirts broke into Jewish homes in a small town in south-west Germany, herded the occupants into the town hall and beat them up. The attack was repeated in a neighbouring town, resulting in the deaths of two men. A week later, on 1 April, the Nazis launched the *Judenboykott*, the boycott of Jewish businesses across the entire country.

An early indication of the reshaping of academic, intellectual and scientific life in Germany was the closure of publishing houses, or their Nazification, and the establishment of the Reich Chamber of

Culture in pursuit of *Gleichschaltung*. Bernhard Rust, aged fifty, was appointed Prussian Minister for Culture in February 1933. His responsibilities included science and education. He declared his aims:

Everybody must recognize that what we are experiencing is not a change of direction but the fundamental fact that the German people is waking up to itself. This movement is progressing relentlessly until one day the whole German people will have been won over to the new cause and will have created for itself its own organizations for politics, the economy, and culture . . . Our *Gleichschaltung* means that the new German ideology as the only valid one will assume the supreme position over all the others.[2]

Rust had since 1930 been an unemployed provincial school-master, relieved of his duties that year by the local authority in Hanover for suspected mental instability; he had been accused of abusing a schoolgirl. In the First War he had received a head wound (and a subsequent Iron Cross), which, in the view of some, was associated with his mental and behavioural difficulties. The accusation did not affect his political career. The following year he was elected to the Reichstag, and rose rapidly through the SA to the rank of *Obergruppenführer*. His early boast was to have succeeded overnight in 'liquidating the school as an institution of intellectual acrobatics'.[3] The following year he was appointed Reich Minister for Science, Education and Popular Culture.

Thomas Mann, already a reluctant émigré, noted in his diary for 3 April 1933 that Germany, which now 'lacked any sense of evil', had 'relapsed into darkest barbarism'; he would 'never yield, but sooner die than come to terms with it'.[4] Another who was not prepared to walk in step was Albert Einstein, who had declared before leaving the country for the United States on 10 March that 'civil liberty, tolerance, and equality of all citizens before the law' had vanished in Germany, and that justice was now in the power of 'a raw and rabid mob of Nazi militia'.[5] On 28 March he resigned his membership of the Prussian Academy of Sciences, pre-empting the decision of Bernhard Rust to expel him. The academy members, including Max von Laue and Max Planck, made no move to protest Einstein's expulsion. Einstein made his feelings explicit: 'The conduct of German intellectuals – as a group – was no better than the rabble.'[6]

At the end of the first week of April 1933 the new regime passed the Law for the Restoration of the Career Civil Service (Gesetz zur Wiederherstellung des Berufbeamtentums), throwing socialists and non-Aryans out of their civil service jobs. A non-Aryan was defined as someone who had one parent or grandparent who was non-Aryan – so as to include even those with 'a quarter' Jewish blood. Since academics and members of the Kaiser Wilhelm institutes were civil servants, about a thousand university teachers, many of them scientists, including 313 professors, were dismissed. Hitler refused to be deterred by the evident damage that would be done to German science. He was adamant.

At first, applications for exemption were allowed in the case of Jews who had been employed before World War I, or who had fought in the war, or lost a father or son in the war. The situation of those entitled to apply was invidious. Nobel prize winner James Franck, a First War veteran (one of Haber's poison gas pioneers) who headed the Second Physics Institute in Göttingen, decided that to apply was to collude with the regime: 'We Germans of Jewish descent,' he wrote in his letter of resignation, 'are being treated as aliens and enemies of the Fatherland . . . Whoever was in the war is supposed to receive permission to serve the state further. I refuse to make use of this privilege.'[7] Some forty-two of Franck's colleagues denounced him for stirring up anti-German propaganda by his resignation.

Max Born, director of the Institute for Theoretical Physics in Göttingen, also refused to apply. Einstein wrote to him in May: 'You know, I think, that I have never had a particularly favourable opinion of the Germans (morally and politically speaking). But I must confess that the degree of their brutality and cowardice came as something of a surprise to me.'[8]

The young biochemist Hans Krebs (who was to win the Nobel prize for physiology or medicine in 1952 for his discovery of the citric acid cycle) was at that time working in the laboratories of a hospital in Freiburg. He has written an impression of the immediate impact of the anti-Semitic campaign: 'Within a few days of Hitler's assumption of power, Nazi uniforms appeared everywhere. Colleagues in the hospital who had admitted at most only mild sympathies with Hitler were suddenly to be seen in the uniforms of Nazi organizations.'[9] Krebs writes that four months earlier the dean of the faculty, the professor of surgery E. Rehn, had signed a letter which had warmly recommended his appointment as a teacher in the Medical Faculty. On 12 April he received a letter from this same Rehn, officially informing him that at the request of the Minister of Education he was to consider himself on immediate leave of absence: 'By order of the Office of the Academic Rector I hereby inform you, with reference to the Ministerial Order A No. 7642, that you have been placed on leave until further notice.' The difference between the two documents, Krebs comments, 'demonstrates the breakdown of social tenets under ruthless political pressure – threats of being deprived of one's livelihood, of being discredited by public defamation, of being thrown into jail or concentration camp, or of being murdered'.[10]

Meanwhile German schools were also being brought within the ambit of National Socialism and made subject to racial discrimination. On 25 April 1933, the Nazi government passed with much trumpeting its Law Against Overcrowding of German Schools and Universities, aimed a reducing the number of places available for Jewish students. The act laid down a strict quota (1.5 per cent of school and college enrolments) deemed appropriate for the size of the non-Aryan or Jewish population at large.

Textbooks were rewritten and new curricula devised to inculcate National Socialist ideology with an emphasis on racial doctrines. Teachers were required to join the National Socialist Teachers' League, and as members of the civil service were subject to the anti-Semitic provisions of the Restoration of the Civil Service Act. Teachers critical of the party were dismissed; those who remained were required to be trained in National Socialist principles. Was it possible for a teacher to retain a sense of independence and work against the system? Was it right for teachers to remain within such a system? There were teachers, as there were scientists, who agonized over the predicament.

A geography teacher, writing in 1938, commented on the circumstances of the profession, his reflections connecting with the rationale proffered by others, including distinguished scientists like Werner Heisenberg, who chose to stay rather than leave Germany.

I am trying through the teaching of geography to do everything in my power to give the boys knowledge and I hope later on, judgment, so that when, as they grow older, the Nazi fever dies down and it again becomes possible to offer some opposition they may be prepared . . . if I went to America and left others to do it, would that be honest, or are the only honest people those in prison cells?"[11]

Before Hitler's rise to power schools had come under the auspices of the *Länder* (the provincial states). Now they came under the auspices of Berhard Rust's Reich Ministry. His office had also assumed responsibility for appointing rectors and deans of the universities, once the prerogative of local faculties. Rust even took to appointing the leaders of students' unions who would act as watchdogs over academics.

Book-burning

On 10 May the night sky over Berlin lit up as students and teachers took part in the burning of books in the square before the Opera. The list of banned books ran to some 10,000 titles and included Kafka, Marx, Heine, the brothers Mann and, of course, Einstein. The act was organized by the Deutscher Studentenbund (German Students' League) with the full support of Goebbels and Rosenberg. The watching crowds cheered as the flames rose higher. This holocaust of the books was repeated across Germany that month in other university towns. And as if to emphasize the quasi-religious parallel, a Nazi commentator described the cultural task ahead:

To operate within the cultural professions separating the tares from the wheat, and to decide between the fit and unfit . . . to divide by blood and spirit German from alien . . . What within the new structures will be created is a tremendous leader corps, made up out of all who participate in any way in the process of forming the national will, from the greatest spiritual genius to the most insignificant helper, from the man who does the creative work to the last retailer who hawks literature and journals in the streets and at the railway stations.[12]

Before the end of the month, Martin Heidegger, the philosopher and recently appointed rector of the University of Freiburg, delivered to the students and faculty his inaugural address, entitled *The Self-Assertion of the German University*. He proclaimed that the new spirit abroad in Germany signified the resurrection of true philosophical inquiry in its original pre-Socratic spirit. Nazism, in this view, involved a large element of risk-taking, an adventure of the mind and spirit into the unknown. Given his reputation as a leading philosopher, Heidegger's enthusiasm lent substantial credibility to the regime. Some commentators have seized upon Heidegger's omission of Hitler's name, or even explicit mention of the National Socialist Party, in mitigation of his collaboration with Nazism: he resigned his post in 1934.

The opposite case, however, might be argued: the fellow traveller who took benefits from the regime (Heidegger joined the party in April) and failed to criticize it, while appearing to remain aloof, exerted considerable influence over the undecided, the young especially. Heidegger's mistake was to believe that the new regime would enable his university to be self-determining, self-asserting. He soon discovered that National Socialism meant subjection of the university to the oppression of Nazi thugs. In contrast, the rector of Frankfurt University, Ernst Krieck, who had no such illusions, announced that as 'Führer of the university' he would cooperate with the National Socialists and that 'as a matter of principle the university will in future be an organ and element of the state' which 'under authoritative leadership will have to carry out its educational work within the context of a racial–political basis and goals'.[13]

If there was a new spirit in the air it was that articulated by the ideologue Alfred Rosenberg, who had called in his prolific papers, books and speeches for a 'new belief' in the *Mythus*, trust in an inchoate, mystical 'searching for truth', a 'complete unfolding of self' from the 'experienced Mythus of the Nordic racial soul'. This Mythus, Rosenberg insisted, was utterly opposed to the 'scholastic–logical–mechanical' path to knowledge.[14] It rejected 'mathematical schematicism' and a Hobbesian 'mechanical atomism' for an 'organicism' in which the 'blending together of the individual and the general consummated itself in the form of individuality'.[15] It was not a *Weltanschauung* that lent itself easily to the spirit of the natural sciences and mathematics, nor to the spirit of a technologically advanced nation state.

Meanwhile the dismissals went ahead without protest from the majority of students and colleagues, revealing the extent to which the universities and institutes had been hospitable to right–wing ideology, and anti–Semitism – from professors down to junior students. Those Jews who attempted to claim exemption were soon disabused as to their chances of survival as academics and scientists. They faced heckling in their lectures and treachery among their colleagues.

Even those who argued the special case for isolated exemptions betrayed a degree of collusion. Heisenberg, pleading with Max Born to seek exemption, wrote:

Since . . . only the very least are affected by the law – you and Franck certainly not, nor Courant – the political revolution could take place without any damage to Göttingen physics . . . Certainly in the course of time the splendid things will separate from the hateful.[16]

The dismissal of Jews was met with complacency on the one hand and acquiescent conformism on the other. Readiness to take the places of the dismissed also revealed a strong component of opportunism as well as a form of warped chauvinism. Historian and biologist Ute Deichmann cites the case of Paul Harteck, a young physical chemist who was in correspondence with Karl-Friedrich Bonhoeffer, a young professor of physical chemistry (and eldest brother of Dietrich Bonhoeffer, the theologian), on the question of the dismissals and the opportunities that thus arose. Harteck, who had studied in Cambridge in 1932, wrote to Bonhoeffer in April 1933 warning him not to be taken in by the complaints of dismissed Jews. In May he wrote: 'In London the Jews, the half and quarter Jews of Germany are gathering. If these people have ever had a liking for Germany, it can only have been a very superficial one, because now you really don't notice anything of it.'[17] In June Bonhoeffer could write to Harteck telling him of 'many openings occurring simultaneously', and that 'you should actually get one of them'.[18] Harteck, who will return to our narrative as one of the team destined to work on the Nazi atom bomb, eventually took the chair made vacant by the Jewish physical chemist Otto Stern at Hamburg University in 1934.

Demonstrations of solidarity with dismissed Jews were virtually unknown. A rare example of a refusal to take the place of a dismissed Jew, however, occurred in the University of Düsseldorf.[19] Otto Krayer, a biologist, rejected the offer of a chair which fell vacant as a result of a Jewish colleague being sacked and proceeded to complain to the government about the dismissals. Krayer was himself dismissed and departed for the United States. Another was Kurt Kosswig, a geneticist at the *Technische Hochschule* of Brunswick. Kosswig refused to apply the racist doctrines of the regime and denounced the treatment of Jews. He managed to escape the country before being arrested.[20]

An argument used by those who were unhappy about the dismissals, but who stayed on and kept silent – for example, Planck, Laue and Heisenberg – was the need to preserve German science for better days. Laue, usually praised for his resistance to Nazism, even went so far as to rebuke Einstein for his public resignation from the Prussian Academy of Sciences ahead of the Restoration Act in March 1933. Einstein remained intransigent in his view that scientists should not support the Third Reich. 'From the situation in Germany you can see just where such self-restraint leads. It means surrendering leadership to the blind and irresponsible. Does not lack of sense of responsibility lie behind this?'[21]

There were timid moves on the part of some Aryan scientists to collect signatures of colleagues to protest against the dismissals. But Planck, according to Otto Hahn, said: 'If today thirty professors get up and protest against the government by tomorrow there will be also one hundred and fifty individuals declaring their solidarity with Hitler, simply because they're after the jobs.'[22]

Civil Servants and the Nazi Salute

An incident, small in itself, but significant in terms of the acquiescence of the German great and good in science, occurred when Max Planck opened the new Kaiser Wilhelm Institute of Metals in Stuttgart, a year after Hitler came to power. As Planck came to the rostrum to deliver an address, the audience watched expectantly to see whether he would start with the prescribed 'Heil Hitler'. According to an onlooker, Planck 'lifted his hand half high, and let it sink again. He did it a second time. Then finally the hand came up, and he said, "Heil Hitler."' The writer, the prominent physicist P. P. Ewald, commented: 'Looking back it was the only thing you could do if you didn't want to jeopardize the whole Kaiser Wilhelm Gesellschaft.'[23] According to an anecdote circulated among physicists to this day, Laue was in the habit of carrying a parcel in each arm everywhere he went so as to preclude the necessity of raising his arm in a salute when accosted by Nazis.

Planck opened subsequent meetings of the Kaiser Wilhelm Society with the Nazi salute, apparently without a qualm, and signed official letters with Heil Hitler. He encouraged colleagues not to emigrate and to stay in Germany to await better times, and his speeches indicated a desire to remain on good terms with the regime.[24]

Understanding Planck's Hitler salute, and the acquiescence of those scientists who stayed and stepped willingly into the shoes of the dismissed non-Aryans, involves not only a recognition of the views expressed by the geography teacher above but an appreciation of the role of the civil servant in modern Germany. Whether they taught in universities or the Kaiser Wilhelm Institutes, scientists were servants of the state alongside school teachers, postal workers and tens of thousands of other public employees. The traditional stance of the civil servant was to subscribe to forms of apolitical nationalism. For decades academics in the higher echelons had enjoyed freedom of research and teaching, and privileged rank guaranteed by their loyalty to the state and their apolitical status.[25] Subjected to inflation and reductions in pay under the Weimar Republic, especially the 'Hunger Chancellor' Brüning's austerity measures of July 1930, as well as disruptions in grading, the educated civil service did not regret the fall of the Weimar republic. Loyalty to the state, and to the administration in power, however, was traditionally a lifetime commitment that went far beyond consideration of salary. Hitler's promise of a return to patriotism and 'clarity' found a ready echo in a disaffected civil service. Nazi propaganda, moreover, promised restitution for loss of earnings, along with the restoration of the 'loyalties' that could not include Jews and subversives. The notion of 'restoration', however, was no more than a thin disguise for anti-Semitism.

When Max Planck, as president of the Kaiser Wilhelm Society, called on Hitler to put in a good word for Fritz Haber, whose resignation as president of the KWI for Physical Chemistry had been accepted, Planck evidently attempted to persuade Hitler that certain Jewish scientists were worth nurturing for the benefit they might bring to the state. Planck's own transcript of the meeting, written in 1947, claims that he told Hitler 'that there were different sorts of Jews, some valuable and some valueless for mankind, among the former old families of exemplary German culture, and that one had to make distinctions in such matters'.[26] Hitler said: 'That's not right. A Jew is a Jew; all Jews cling together like burrs. Wherever one Jew is, other Jews of all types immediately gather.'[27]

Planck then countered that the importance to Germany of the work of Jewish scientists meant that it would be no less than an act of self-mutilation to force valuable Jews to emigrate, and that

foreign countries would be the first to reap the benefits. Planck
wrote that Hitler did not respond to this point but began to rant,
speaking faster and faster. Slapping his knee with a powerful blow,
Hitler concluded: 'People say that I sometimes suffer from weak
nerves. That's slander. I have nerves of steel.' Hitler was in such a
rage that Planck fell silent and took his leave.

Haber's End

Haber's enthusiastic work for the state, his encouragement of scientists to devote themselves to the German war machine, his public
rejection of the religion of his forebears for Christianity, went for
nothing. As far as Hitler's regime was concerned he was just a Jew,
and an enemy. As if to eradicate the least gesture of honour accorded
Haber for a lifetime's devotion to German science, a ceremonial
tree planted in the grounds of his institute to celebrate his sixtieth
birthday was torn up.

At his departure from the Kaiser Wilhelm Institute, Haber wrote
a dignified letter of resignation, stating that he had always felt
himself to be a good German, despite the fact that he was Jewish,
and that he had always put his country first. Since he was no longer
able to do that, he had no choice but to leave the Fatherland. Haber
wrote to Willstätter: 'I am bitter as never before, I was German to
an extent that I feel fully only now and I find it odious in the
extreme that I can no longer work enough to begin confidently a
new post in a different country.'[28]

Einstein wrote to him from the United States: 'I can conceive
of your inner conflicts. It is somewhat like having to give up a
theory on which one has worked one's whole life. It is not the
same for me because I never believed in it in the least.'[29]

Suffering from heart disease, Haber now devoted himself to
finding work for his Jewish staff abroad. Eventually he departed for
Cambridge, England, where he was greeted hospitably enough by
the professor of chemistry, William Pope, another poison gas
expert; but Haber was understandably given the cold shoulder by
British lab technicians who had fought in the trenches during the

war. He was invited to join the new Daniel Sieff Institute (today the Weizmann Institute) in what was then Palestine and he accepted. Einstein wrote to Haber warning him against working with the Hebrew University; yet Einstein could at least be 'pleased that . . . your former love for the blond beast has cooled off a bit'.[30] Haber never made it to the Middle East. He died at Basel in Switzerland while visiting his family in January 1934 at the age of sixty-five. Amidst protests by the Nazis, Max Planck organized a memorial service to celebrate Haber's life a year later. It was done in defiance of the Nazis' explicit veto. Max Planck and Otto Hahn gave short addresses. It was considered an act of bravery when Professor Max Bodenstein read a commemorative speech on Haber at the Preussische Akademie der Wissenschaften on 28 June 1934. He said:

Haber's significance for science and its nurturing is not remotely exhausted by his own scientific works and with those of his school. He was not merely a man of science and technology; he was also highly competent in matters of organization and financial management . . . Haber was a Jew, as were the overwhelming majority of his colleagues. This made conflict with the National Socialist state inevitable.[31]

Haber's Institute for Physical Chemistry in Berlin now bears his name, a fact that attracts controversy to this day. As Fritz Stern comments in his elegant essay on Haber and Einstein, 'The memory lives on – dimly in distorted controversy.'[32]

The Exodus

The exodus of Jewish scientists was devastating in its consequences for Germany.[33] Some 25 per cent of the pre-1933 physics community was lost to the country, including Einstein, Franck, Gustav Hertz, Schrödinger, Hess and Debye – all Nobel laureates. Other lost laureates included Stern, Bloch, Born, Wigner, Bethe, Gabor, Hevesy and Herzberg, as well as mathematicians Richard Courant, Hermann Weyl and Emmy Noether. Most of the lost physicists were scientists of high originality and unique experience; they were irreplaceable. Almost half of Germany's theoretical physicists went, and many of its top experts in quantum mechanics and nuclear physics.

The loss to Germany was a huge gain to Britain and the United States. In Britain, William Beveridge (who later headed the London School of Economics and inspired the formation of the welfare state), A. V. Hill and the biochemist and Nobel laureate Frederick Gowland Hopkins set up the Academic Assistance Council to provide jobs for displaced academics. Those who came to Britain were to have a profound and lasting effect on the country's culture across a broad span of disciplines and cultural influences. Leo Szilard, the tireless physicist and polymath, worked in collaboration with Beveridge to establish dismissed scientists from Germany before moving to the United States in 1938 where, as we shall see, he became active in nuclear fission research and the politics of the atomic bomb.

In terms of overall numbers through the 1930s, however, the German physics community did not shrink in absolute numbers because of an increase in applied physicists in the universities; but the quality of the scientists declined, and basic research stagnated. Of all the German universities, with the exception of Berlin, Göttingen, a world centre of mathematical physics, was hurt the most. Lost to this birthplace of quantum mechanics were Max Born, James Franck, Walter Heitler, Heinrich Kuhn, Lothar Nordheim, Eugene Rabinowich and Hertha Sponer.

In addition to the loss of many leading researchers German science became increasingly isolated, as foreign scientists avoided travelling to Germany, ceased collaborating in research programmes with Germans and cancelled membership of scientific societies and subscriptions to journals. At the same time, the regime was making it difficult for academics to travel outside of the country and placed restrictions on membership of organizations deemed inimical to National Socialism.

In the heightened atmosphere of National Socialist 'revolution', the spectre of scientific secrecy loomed, along with the breakdown of foreign peer group discussion and collaboration. Germany was to make progress in many areas of applied physics, especially in processes and technology with military applications; but academically it would begin to fall behind. The director of research at AEG, one of Germany's largest companies, noted that according to records American citations (in Germany's leading physics review) had risen between 1913 and 1938 from 3 per cent to 15 per cent, while German citations (in the leading international physics review) during the same period had dropped from 30 per cent to 16 per cent.

In a move that revealed the depths of race hatred within the government education bureaucracy, Rust ordered on 15 April 1937 that 'Jews who are of German nationality, are no longer to be admitted to examination for a doctor's degree. In addition, renewal of doctor's diplomas held by the same is to be discontinued.'[34] The following week, indicating the strangeness of the times, saw the opening of the new Kaiser Wilhelm Institute of Physics in Dahlem, close to the other institutes. The Institute of Physics had originally been established as a scientific fund towards the end of World War I. A proper home for laboratories and offices was made possible in 1937 by a late Rockefeller Foundation grant, as well as Reich funding, revealing that even after the rise of Hitler major institutions in the United States were not ashamed to support a regime that racially abused its scientists.

10. Engineers and Rocketeers

After the First War, Germany, one of the great armaments producers of the world, had ceased to forge guns, armour-plating and ordnance as a result of the terms of the Versailles Treaty. Germany's ingenuity in aeronautical and maritime technology, born out of close synergies between science and industry, had been drastically and humiliatingly curtailed. According to the treaty, military aircraft were banned, and the German navy, formerly the third largest in the world, had been ignominiously scuttled in the cold grey waters of Britain's home fleet harbour, Scapa Flow. All Germany could now boast was a pathetic flotilla of obsolete coastal defence vessels under 10,000 tons. Meanwhile the Reichswehr, the remaining Germany army, was not allowed by the terms of the treaty to exceed 100,000 men; tanks were banned, and guns were not allowed to exceed 105 mm. The tasks of the army were mainly border-control and internal policing. By another provision of Versailles the German Military and Technical Academy (Militärtechnische Akademie) in Potsdam closed its doors.

During the Weimar period, however, the government and German industrialists managed to continue weapons development under cover: by pushing forward dual-use technology, by conducting military research and development in secret, and by placing research programmes outside the country.

Article 191 of Versailles had expressly forbidden the building of submarines by Germany. Nevertheless the German navy set up two secret funds which paid for research and development in thirty different firms, some of them foreign, for example, a torpedo factory in Cadiz, some of them clandestine, such as a Berlin 'bacon' company. Meanwhile three German industrial firms, Krupp, A. G. Weser and Vulkanwerft, set up a joint stock company in Holland in 1922 to build eight U-boats with German technology for the

Kawaski shipyard in Kobe, Japan. They also built U-boats for Finland, Spain, Sweden and Turkey.[1] In 1925, moreover, sums of money from the prodigious $195 million Dawes Plan rescue package were diverted to AEG, Krupp, Simens-Schuckert and Thyssen for rearmament projects such as a special bureau for 'antisubmarine warfare questions', which was a clandestine U-boat development scheme. Under the retired U-boat specialist Admiral Arno Spindler, the navy embarked on a determined and complex clandestine development scheme. The plan included submarine R and D conducted by foreign-based German companies; industrial espionage aimed at collecting information on foreign submarine developments; and media propaganda to expose the unfairness of the Versailles ban on Germany's submarine technology.[2]

The Versailles restrictions against the continued development of German air power became the focus of anger and injured pride during the Weimar years. There was enormous enthusiasm for flying among German people, and gliding became a national passion with military overtones. At Wasserkuppe, where a war memorial dedicated to flyers dominated the surrounding countryside, thousands of spectators gathered to watch gliders swarming in the skies. An inscription in stone read:

We dead fliers
Remain victors
By our own efforts
Volk fly again
And you will become a victor
By your own effort.[3]

German aeronautical research had developed and expanded despite the constraints of Versailles. At the level of basic research, for example, Ludwig Prandtl and the Göttingen airfoils assured the future of German design of lift surfaces for aircraft. And by placing R and D outside the country, in Sweden and Spain for example, Germany managed to evade Allied scrutiny of aircraft design that might have obvious military applications. At the same time, planes like the high-speed Heinkel transporters and the Dornier flying boat, and the diesel-engined Junkers passenger liner, continued to stretch expertise in aeronautical technology.[4]

Nazi Engineers

The advent of Hitler, however, found the German engineering communities in a state of widespread demoralization, and not a little puzzlement. The Nazi ideologues had been sending mixed messages about the status of technology in the Third Reich. National Socialism in its early manifestation had involved a hybrid of technocracy and return to nature. Some senior Nazis, however, were outspokenly hostile towards technology. Wilhelm Stapel, editor of *Deutsches Volkstum*, for example, declared that 'the words engineer and organizer should be turned into the nation's terms of abuse'. Alfred Rosenberg, rabid anti-Semite and editor-in-chief of the Nazi daily paper *Völkischer Beobachter* (*People's Observer*), attempted to lay the confusion to rest when he wrote in his notorious *Myth of the Twentieth Century* that those who condemn technology 'forget that it is based on an everlasting German drive which would disappear together with the downfall of technology'.[5] Despite his addiction to vague mystical notions such as the *Mythus* of blood and his rejection of systematic science, Rosenberg insisted that loss of German technological edge would 'deliver us to the barbarians and leave us in the condition to which the Mediterranean civilizations once declined'. It is not technology that kills everything vital, he added, 'rather human beings who have degenerated'.

Hitler's dictatorship, meanwhile, announced a rapid expansion of military technology and new prospects undreamed of.[6] After years of inactivity and hopelessness, a period in which some 5,000 frustrated engineers and technicians had taken jobs in the Soviet Union, Hitler was offering opportunities for their skills.[7]

Being civil servants, most German engineers were apolitical and owed no allegiance as a group to the Nazi party when Hitler came to power. By the beginning of the summer of 1933, however, wooed by Nazi propaganda that sought to boost their status within the new Germany, the largest engineering associations joined the Nazi-controlled Reich Society for Technological and Scientific Work (Reichsgemeinschaft technisch-wissenschaftlicher Arbeit). The leading engineering periodical *German Technology* (*Deutsche Technik*) extolled technicians and engineers as a 'crucial force' in the new aims of the nation state, while denouncing the new physics and reminding its readers of the special technical giftedness of the German race.

In the early days the fundamental struggle for the minds and hearts of engineers involved, in the view of the Nazi ideologues, the rescuing of engineers from the clutches of capitalism and their integration into the national community. But this depended on engineers adopting the *Führerprinzip*, creating a fusion of technical ability and ideological commitment. Fritz Todt, Nazi Armaments Minister and builder of the autobahns, put it thus in a 1936 issue of the periodical *Deutsche Technik*:

It is so obvious as to be hardly worth mentioning that just as technical ability on its own is not enough for the engineer in the Third Reich, nor is an ambitious will to leadership [*Führerwille*] of a purely political person. One-sidedness of the latter variety may produce a worthwhile speaker or political administrator, but not a National Socialist engineer.[8]

The Rocket Engineers

Meanwhile, in greatest secrecy, a remarkable rocket development project had been taking shape at secret testing sites near Berlin and on the Baltic coast under the auspices of the army. The ambitions would in time emerge as a combination of advanced technology and fantasies of phallic terror. One day, before the war's end, the fruit of the engineers' labour and ingenuity would rise with thunderous noise and fire from the secluded forests of northern Germany.

As with other military services which conducted their clandestine projects despite the rules of the armistice, German Army Ordnance had been active during the Weimar years. The head of the army's Ballistics and Munitions Branch, Lieutenant Colonel Karl Emil Becker, had gathered about him a small team of qualified engineers, including Walter Dornberger and Leo Zanssen. One of the technologies that lay outside the provisions of the Versailles Treaty was rocketry, on which Becker had kept a watchful eye through the 1920s. As a result he had encountered the young rocket enthusiast Wernher von Braun, a name that would one day be inextricably linked with Hitler's V1s and V2s, which rained down on London during the latter part of the war.

The Germans had employed rockets for limited purposes in the trenches in World War I, specifically to fire grappling hooks to remove barbed wire, while the French used them against observation balloons and Zeppelins. But the testing conditions of trench warfare exposed the weakness of rocket technology at that time. The gunpowder, or black powder, used as a propellant proved to be extremely hygroscopic, or prone to damp, and tended to burn unevenly.

An important aspect of R and D for the post-war enthusiasts was aerodynamic shape. As Peter Wegener, a Nazi rocket engineer, put it: 'A rocket must have a low air drag, it must be stable, and it must fly on a controllable predetermined course . . . stability of flight in the denser layers of the atmosphere is provided by fixed external wing-like surfaces and movable rudders.'[9] Then there was the fuel or propellant. Should it be solid or liquid? The idea to use liquid, as opposed to the more conventional solid fuel, had been pressed by three prominent enthusiasts. The American Robert Goddard Clark of the University of Massachusetts and the Russian Konstatin Tsiolokovsky had recommended liquid as a means of gaining massively improved thrust, weight for weight. But it was Hermann Oberth, a German-speaker from Transylvania in Romania, who achieved what the historian Michael J. Neufeld has termed a 'seminal' influence on rocketry in his book *Die Rakete zu den Planetenraumen* (*The Rocket into Interplanetary Space*), published in Germany in 1923. Oberth discussed how the problems of space flight might be solved by the use of alcohol/liquid-oxygen and liquid-hydrogen/ liquid-oxygen multi-stage vehicles. His book was a prelude to a mania for imaginary rocket scenarios after the First War in Germany.

Enthused by the promotion of a freelance writer called Max Valier, the motor manufacturing heir Fritz von Opel backed a series of rocket-car stunts, which led to a craze for rocket spectacles involving all manner of vehicles, including railway carriages and sleds. In October 1929 Fritz Lang's silent movie *Frau im Mond* (*Woman in the Moon*) featuring a montage of impressions of space flight was released, creating mass interest in space flight. In an attempt to promote the film Oberth had funded a self-taught engineer and conman, Rudolf Nebel, to build a liquid-fuel rocket. The scheme did not succeed. But that same year Johannes Winkler, first president of the Verein für Raumschiffahrt (VfR – Society for Space Travel), an assorted group of rocket enthusiasts, conducted private and company research at Junkers aircraft company in Dessau on the potential use of liquid rocket fuel for aircraft propulsion. Less auspicious was the 'research' of Max Valier, the writer, and his two assistants, Walter Riedel and Arthur Rudolph, who were trying to build a rocket-car engine fuelled by paraffin. On 17 May Valier was killed in a lab explosion. As a result of that accident, and a second rocket accident involving a teenager, the Reichstag considered passing an act to ban all such experiments. The banning initiative did not succeed, largely, it is thought, because Oberth, in advance of the vote, had managed without accident to achieve a smooth ninety-second test of a rocket motor at the Chemisch-Technische Reichsanstalt.

Meanwhile, in September 1930, Rudolph Nebel (who continued solitary research on a small rocket called MIRAK ['Minimum Rakete'] on a farm in Saxony) set up a research team for the VfR on a former military facility north of Berlin. They called the base the *Raketenflugplatz* (the Rocketport). Among these rocket aficionados was Wernher von Braun, an eighteen-year-old mechanical engineering student at the *Technische Hochschule* in Berlin. Born in Wirtsitz, in Germany, in 1912, von Braun came from a

Prussian Junker family; his father was a former Minister of Agriculture. Rejoicing in a baronial title he disdained to use, von Braun had graduated aged twenty from the Charlottenburg Institute of Technology, a college that insisted on practical apprenticeships for its engineering students. Von Braun had worked in a machine shop at the Borsig Works factory, as well as in a foundry and on an assembly line.

Among those early enthusiasts was a young historian of rocketry, Willy Ley, author of *Rockets, Missiles, and Space Travel*. An unashamed romantic, he described the flight of a rocket from a swampy district of wasteland near Berlin on 10 May 1931 thus:

Our rocket testing-ground was covered with the young green of pine shoots and new birch leaves, the depressions between the hills were full of young willows. Crickets sang in the high grass and frogs croaked somewhere in the distance . . . But the beast flew! Went up like an elevator, very slowly, to twenty meters. Then it fell down and broke a leg.[10]

Von Braun Joins the Army

Despite his youth, von Braun was the following year recruited into the army's secret rocket research programme by Becker and Dornberger. At the same time he was enrolled into a doctoral programme in applied physics at Berlin University. His thesis, which the army insisted should be kept secret, would describe the rocket development conducted under his direction at the army facility at Kummersdorf.

When Hitler came to power in 1933, the records of the VfR were seized by the state. A member of the VfR overheard a company manager at Siemens informing an official in one of Hitler's ministries: 'Now I've all the rocket people safely on ice around here and can watch what they are doing.'[11]

Von Braun's activities were now entirely under the direction of his paymasters, the army. In later years he attempted to give the impression that he had exploited the army, rather than the other way about:

Our feelings toward the Army resembled those of the early aviation pioneers, who, in most countries, tried to milk the military purse for their own ends and who felt little moral scruples as to the possible future use of their own brainchild. The issue in these discussions was merely how the gold cow could be milked most successfully.[12]

The sentiment echoes the dependency factor in many realms of science and technology, the need to follow the funding, the supply of equipment, laboratory space: the everyday realities of researchers. A combination of alibi and understandable pressure to conform, von Braun's comment on his relationship with the regime would be repeated by a generation of German scientists who put their talents and knowledge at the service of Hitler's war aims.

In December 1934, just two years after he had joined the army rocket research team, two years in which three ill-fated early rockets, known as A1, had blown up on take-off at Kummersdorf, Wernher von Braun and his technicians received shipment of two rockets at a launch site on the lonely windswept island of Borkum in the North Sea. The missiles, known as Max and Moritz after the German version of the comic characters 'The Katzenjammer Kids', were also known officially as A2. The moment of truth had come for the latest model of Germany's secret missile programme.

The A2 was of a revolutionary design, the precursor of the guided missiles of the future. It had separate fuel and liquid-oxygen tanks to prevent explosions caused by leaking. There was a gyroscope between the tanks near the rocket's centre of gravity which maintained good control in the first stages of flight. The combustion chamber had been lengthened to give the fuel more time to burn.

Max was launched on 19 December. The flight was perfect. The rocket reached an altitude of 1.4 miles before falling by parachute to earth half a mile from the launch pad. The following day Moritz performed equally well.

Military officers present asked von Braun whether he thought such a rocket could be used as a weapon. He was equivocal. Yes, it could be used as a weapon, he conceded: but what was the point? Its range was less than that of conventional artillery, nor could it deliver a better payload. Von Braun's reaction, at this early stage, revealed that the young engineer could compartmentalize his interests in the purposes of missile technology, which had started not with military aims but romantic dreams of the conquest of space.

Three months later Hitler repudiated the Versailles Treaty and its military provisions, giving the green light to massive rearmament. As a result, the rocket development project at Kummersdorf, to be known as 'Experimental Station West', qualified for increased funding and expansion. The Gestapo counter-intelligence, as if it were necessary, banned all overseas collaboration and transfer of information. While the ultimate shape of the programme was as yet uncertain, the object of the group was to maintain Germany's lead in rocket technology. The most important aspect of their work for the future, according to Leo Zanssen of Army Ordnance, was going to be 'the element of surprise'.[13]

The rocket programme would be 'in-house' at every level but outside procurement contracts would be necessary, albeit circumscribed by heavy regulations. The crucial guidance systems technology – involving sophisticated gyroscopic principles – was placed with Kreiselgerate GmbH, which itself was a high-tech company secretly owned by the German navy. Inter-service collaboration was another matter. By 1935 the Luftwaffe under Goering forged an alliance with the rocket programme, bringing in massive new funding and mutual technology gains from work on projects such as rocket-assisted aircraft take-off.

Given the rapid expansion and tumescent ambitions of the army rocketeers, the need for a permanent site on the Baltic now became urgent. Eventually Peenemünde was chosen, a district along a lonely stretch of the Baltic coast with capacity for testing sites, factories, laboratories and impenetrable security. By this time Becker had been promoted to general and head of the Testing Division; and he was determined to match Luftwaffe funding with army funding. By April 1937 most of the team at Kummersdorf had moved to the new site.

The German rocket project now became one of the first examples of state mobilization of major engineering and scientific resources for the enforced realization of new military technology. The project would be matched, and eventually exceeded, in the course of the war only by the Manhattan Project for the building of the atomic bomb in the United States.

11. Medicine under Hitler

Surgeon and part-time author Dr Curt Emmrich records that, as Hitler took up the reigns of unchallenged dictatorial power in 1933, he noticed that colleagues began to turn up in the common room of his Hamburg hospital, 'somewhat sheepish at first', in brown or black uniforms. Another doctor noticed that when attending a reunion of naval physicians in the same year, only a minority were not members of the National Socialist Party.

Physicians, as a group, outnumbered other professionals in enthusiasm for Nazi Party membership. Within the Reich Physicians' Chamber Nazi membership peaked at 44.8 per cent. Lawyers formed the next largest Nazi membership, but never exceeded 25 per cent.[1] Those physicians licensed to practice between 1925 and 1932 showed the largest proportion of membership, at 53.1 per cent, while the smallest proportion, 39.1 per cent, was among those registered between 1878 and 1918. The historian Michael Kater deduces from this that the most enthusiastic Nazi physicians were those who had suffered professionally during the Great Depression. 'Up to and including 1933,' according to Kater, 'they approached the Nazi camp out of resentment for past and present conditions and in hope of a brighter future at the hands of the predestined leaders.'[2] Till Bastian, the historian of medicine in the Third Reich, notes that the profession had been notably conservative and chauvinist from the outbreak of World War I.

After Hitler's rise to power early in 1933 the majority of German physicians welcomed the new regime, hoping for a redress of anomalies in health care, administration, pay and working conditions under the Weimar Republic. Junior doctors had found it difficult under Weimar to set up in practice because of a system that allowed only one doctor per 600 patients. The salaries of doctors in Germany during the 1920s and 1930s far outstripped those of comparable professionals, including lawyers. Yet doctors' pay had sunk to an all-time low in the years before Hitler came to power. The austerity measures of Chancellor Heinrich Brüning's administration, and the huge unemployment figures, had driven patients away from the surgeries. As the depression lifted and the numbers of doctors dwindled, due to the disappearance of Jewish physicians, doctors' pay in 1934 began to rise above the worst figures of 1932. By 1937 the profession was enjoying salaries comparable to those in 1928.

Racism and anti-Semitism were rife in the profession by the early 1930s. The publisher Julius Friedrich Lehmann, as we have seen, had been instrumental in promoting the racist ideologies of Ploetz and Günther during the 1920s: Lehmann used his publishing outreach and personal wealth to spread the Nazi message among medical students and practising physicians. His lavishly coloured textbooks were best-sellers within the profession, and he owned a weekly medical paper in Munich. Exploiting the popularity of his medical titles, he published many *völkisch* and racist tracts. Among his initiatives, he funded Ludwig Schemann, the editor of the translated works of the racist Arthur Comte de Gobineau; in 1926 he published the highly popular *The Physician and his Mission* by the Danzig surgeon Erwin Liek: by the end of the decade 30,000 copies of the title had sold. As an esteemed medical publisher his promotion of racist ideology lent respectability to the Nazi movement among doctors. Lehmann died aged seventy a year after Hitler came to power, but he had already managed to publish leading Nazi figures such as Richard Walther Darré and the Nazi ideologue Alfred Rosenberg. Hitler praised Lehmann for his tenacity in combating the tide of opposition to National Socialism; only subsequent generations, Hitler declared, would honour Lehmann for what he had done for Germany.[3]

Lehmann's medical propaganda discouraged specialization in medicine and promoted general practice, a tendency associated with the party's holistic view of medicine and health. 'In principle,' writes Kater, this holistic view did not allow for the singling out of any of the body's organs for closer attention, but because it was biologically slanted, a certain positive bias was attributed to the mechanics of reproduction and a negative one to matters of the brain. This dependency was in perfect harmony with the Nazis' constant emphasis on things physical as well as their overall anti-intellectuality.[4]

At the same time, under the influence of National Socialist thinking within the profession, doctors began to expound the importance of the health of the entire nation as opposed to the health of the individual.[5] In the teaching hospitals and universities academic medicine was generally criticized by Nazi doctors for its 'negative features', which were described as 'liberalism, individualism, mechanistic-materialist thinking, Jewish-communist human ideology, lack of respect for the blood and soil, neglect of race and heredity, emphasis on individual organs and the undervaluing of soul and constitution'.[6]

Jewish Doctors

A week after Hitler's rise to power in January 1933 the co-founder of the Nazi Physicians' League, Dr Kurt Klare, wrote to a colleague that 'Jews and philosemites ought to note that the Germans are masters of their own house once more and will control their own destiny'.[7] The dismissal of Jewish doctors, who numbered some 16 per cent of the medical profession,[8] began in March with the removal of Jewish officers from national medical associations and local groups. This was followed by the Jewish boycott orchestrated by Goebbels on 1 April, when Jewish-run business and professional services, including doctors' clinics, were attacked or put out of bounds. In Nuremberg it was announced that social insurance vouchers issued by Jewish doctors would not be honoured, and several Jewish doctors were taken to a park near the Lehrter Bahnhof and shot in the legs. A Jewish physician in Munich was arrested on charges of performing illegal abortions and kept in custody for a week. Another welfare physician in Berlin was imprisoned for nine months. While the Jewish members of the medical profession braced themselves for anti-Semitic legislation, contracts were broken, Jewish doctors were forbidden to deputize for non-Jewish colleagues, payments were suspended and in the province of Bavaria Jewish doctors were forbidden to work for the public school system. In the province of Baden membership of Jewish physicians on recognized panels was restricted to their percentage in the population as a whole – deemed to be 1.5.[9]

On the basis of questionnaires, from April of 1933 the government aimed to bar all doctors who had set up in practice after 1 August 1914, or who had not seen front-line duty, or had not lost a father or son in the war. Those who thought themselves within the law were obliged to apply for exemption. By early 1939 some 2,600 physicians had been dismissed, their places taken by Aryans. Casual torture, terror and murder of Jewish doctors continued through the year: trumped-up charges included the issuing of incorrect prescriptions, sexual attacks on women patients, illegal abortions and embezzlement. Suicide among Jewish doctors became common.[10] But still Jewish physicians continued to practise, partly under the exemptions, partly because patients still valued their professional services. Nazi activists vented their rage against the patients of Jewish doctors, accusing them of treachery.

A Jewish Doctor's Story

The diary of Doctor Hertha Nathorff, a gynaecologist raised as a Jew within a respectable family in Germany, provides a moving impression of the everyday predicament of many Jewish doctors who stayed on in Germany under Hitler.[11] Nathorff, whose maiden name was Einstein (she was distantly related to the scientist), trained in medicine in Heidelberg during World War I, and worked for the Red Cross, where she became chief doctor in a maternity unit. Aged thirty-eight in 1933, she was running her own gynaecology clinic for 150 patients in Berlin. She had a son aged eight, and had been married to a senior Jewish hospital doctor since 1923.

In January of 1933 we find Nathorff wanting to believe in Hitler following his first successful election, but by the following month she had noticed a marked increase in anti-Semitic remarks by her patients in the clinic. Two weeks after the Jewish boycott of April 1933 she reported Jewish surgeons being thrown out of operating theatres and Jewish doctors being forbidden entry to their hospitals. Her husband was sacked by his hospital and he began to work privately. Jewish doctors were put on to the backs of open trucks and paraded around the streets while pedestrians jeered. The following week she was sacked from her part-time role as a pregnancy counsellor, although she continued to run her clinic.

In May a patient broke down and wept during a consultation; the woman had been on the point of gassing herself because she believed that, since she once had sexual relations with a Jew, she would not be able to conceive a pure Aryan child. A week later one of Nathorff's medical Jewish colleagues committed suicide, unable to cope with the altered circumstances of his life. Nathorff visited the mental asylum to see a woman Jewish physician who had gone mad after being sacked from her job and jilted by her Aryan lover.

One of her patients, unaware that Nathorff was Jewish, told her 'everything will be fine when all the filthy Jews have been thrown out of Germany'. Nathorff informed the patient that she herself was Jewish. The patient sent flowers and a note of apology.

Nathorff was no longer allowed to treat national insurance patients; many brought her flowers at their last visit: she felt that she was attending her own funeral. A gynaecology professor to whom she regularly referred patients told Nathorff that he had joined the Party in order to control the excesses from within. She responded angrily that when she wanted to prevent folly she combats it rather than collaborate with it.

The oppression continued unabated into the following year. A Jewish female colleague broke down in Nathorff's house early one morning: her house had been searched during the night by the Gestapo looking for subversive literature. Her midwife assistant at the clinic of many years' standing came to inform her that she had been advised that she should join the Reich Midwife Association, which precluded her working any longer for a Jew. A patient came in to say that she could no longer come to consultations because her teenage sons had threatened to denounce her. Former patients cut her dead in the street, including her former secretary whom she had helped through a financial crisis. A patient came asking for poison to end her life: after discovering she had a Jewish grandmother, she had lost her job and her boyfriend.

The tales of routine persecution, told in her diary, continued, year after year into the late 1930s. Things came to a head on the *Kristallnacht*, Night of Broken Glass, 9–10 November 1938. That night, according to the diary, Nathorff's husband and other Jewish colleagues were out all night in Berlin treating the injured and the sick; many Jews suffered heart attacks. She stayed at home taking telephone calls and coordinating her husband's visits. The police came to the house the next morning looking for her husband. They threatened to shoot her and her son in the head. Her husband arrived, exhausted from his night's work, and was arrested at once and taken away.

On 12 November she stood in line at the American consulate hoping for a visa, only to be told to come back another day. Two days later she learned that her husband was alive, but under arrest for 'racial defilement'. She was visited by a man who asked for money in return for a withdrawal of the charge. When he threatened her with a gun, she handed over a large sum of money and then collapsed. Soon after, her husband was released from detention.

The following spring they successfully sent their son to England, where they planned to join him before proceeding as a family to New York. After obtaining permission to travel, they were obliged to hand over their silver cutlery and jewellery. They were allowed to take only the barest essentials and limited funds in cash.

The Nathorff family eventually settled in New York, where they started a new life. After several years of hardship, during which Nathorff worked as a cleaner while her husband learned English in order to practise medicine again, they began to prosper once more.

Her husband died in 1954; Nathorff published her diary in German in 1987, but she never returned to Germany.

The Fate of Psychoanalysis

Sigmund Freud and Albert Einstein, the world's two most famous living Jews – many would have regarded them as the two greatest living scientists – were distantly known to each other. They had both put their signatures in 1912 to the foundation of a new scientific association 'quite indifferent to metaphysical speculation and so-called critical transcendental doctrine', a scheme which foundered on the outbreak of world war. They had met briefly in the 1920s in Berlin, respecting each other across an appreciable gulf of mutual ignorance of each other's disciplines. Einstein was not inclined at that stage to give credit to Freud's theory of repression (although he would later, in 1936, concede its validity on the occasion of the psychoanalyst's eightieth birthday), whereas Freud is recorded as saying of Einstein: 'he understands as much about psychology as I do about physics,' a basis on which, Freud remarked, they were able 'to talk pleasantly to each other'.

As the Nazi Party gathered strength in Germany during the summer of 1932, Freud had agreed to a public exchange of views with Einstein which would be promoted by the International Institute of Intellectual Cooperation (founded by the League of Nations in 1926). Einstein had asked, rhetorically: 'Is there any way of delivering mankind from the menace of war?' Einstein had responded to his own question optimistically. He argued for an international authority militarily strong enough to enforce peace, while conceding that human beings were filled with a 'lust for hatred and destruction'. But that was precisely where there was a role, as Einstein now proposed, for psychoanalysis in the life of individuals and nations in the pursuance of peace.

Freud, however, was not inclined to optimism on the efficacy of psychoanalysis as a cure-all for the aggression of human beings. In September of 1932, he wrote to Einstein:

There is no likelihood of our being able to suppress humanity's aggressive tendencies. In some happy corners of the earth, they say, where nature brings forth abundantly whatever man desires, there flourish races whose lives go gently by, unknowing of aggression or constraint. This I can hardly credit . . . The Bolshevists, too, aspire to do away with human aggressiveness by ensuring the satisfaction of material needs and enforcing equality between man and man . . . Meanwhile they busily perfect their armaments, and their hatred of outsiders is not the least of the factors of cohesion amongst themselves.[12]

When Hitler came to power the following year, the published Einstein-Freud text, entitled *Why War?*, was banned in the Third Reich. In May Freud's works were thrown on the Berlin bonfire of books along with Einstein's. Freud later remarked that at least he was burning 'in the best of company,' a comment replete with painful irony, since four of his five sisters were to die in the death camps by the end of World War II.

Freud seemed aware of the danger of Hitlerism spreading to Vienna, but he was resigned. 'If they kill me – good. It is one kind of death like another.'[13] He admitted in 1934, however, that he would leave Austria if a puppet of Hitler's led the country. 'The world is turning,' he remarked, 'into an enormous prison, and Germany [is] its worst cell.'[14]

Already, in Berlin, the German Psychoanalytical Society had suffered the same Jewish dismissals as the medical profession. Max Eitingon, head of the society, was a Russian Jew: he fled to Palestine and was succeeded by his number two, Felix Boehm. But German psychoanalysis was soon under the effective control of Professor M. H. Goering, cousin of Hermann, the Air Minister. Professor Goering let it be known that Hitler's *Mein Kampf* would be the basic textbook of psychoanalysts in the Third Reich. In 1936 the institute's library was confiscated as well as the assets of the *Internationaler Psychoanalytischer Verlag*, the institution's publishing house in Leipzig.

Meanwhile the German Society for Psychotherapy, sister organization to the psychoanalysts' society, had been reorganized under

firm Nazi control as the newly named International General Medical Society for Psychotherapy. The president, Ernst Kretschmer,
had resigned and Freud's one-time colleague and rival, Carl Gustav
Jung, took over, also assuming the editorship of the psychotherapists' periodical *Zentralblatt für Psychotherapie*. Jung was to defend
himself against charges of Nazi collaboration by insisting that he
had intervened with the regime on behalf of dismissed or persecuted
psychotherapists, many of whom were Jewish. Freud, however,
had maintained ever since 1912 that Jung was guilty of anti-
Semitism. Since Jung expatiated in 1934 on Freud's deleterious
influence on psychoanalysis, Freud felt vindicated in his long-term
verdict. Certainly Jung now made Freud a personal focus of his
critique of psychoanalysis as unremittingly Jewish. Jung argued that
he had merely resisted Freud's 'soulless materialism' and charged
that Freud encouraged anti-Semitism by his readiness to 'scent out
anti-Semitism everywhere'. In a typical disquisition on the issue in
his own periodical, Jung wrote in 1934:

It has been a great mistake of all previous medical psychology to apply
Jewish categories which are not even binding for all Jews, indiscriminately
to Christians, Germans or Slavs. In so doing, medical psychology has
declared that most precious secret of the Germanic people – the creatively
prophetic depth of the soul – to be a childishly banal morass, while for
decades my warning voice has been suspected of anti-Semitism. The
source of this suspicion is Freud. He did not know the Germanic soul any
more than did all his Germanic imitators . . . Has the mighty apparition of
National Socialism which the whole world watches with astonished eyes
taught them something better?[15]

Glowing encomiums of the Germans' 'depth of soul' a year into
Hitler's regime were not likely to encourage the view among
practitioners, many of whom were Jewish, that psychoanalysis and
psychotherapy were in good hands. Little wonder that Wilhelm
Stekel, one of Freud's earliest and most loyal supporters, wrote to
Chaim Weizmann urging that the University of Jerusalem should

now become the centre for Germany's 'prohibited branch of psychoanalysis'.

Analysis continued in a clandestine fashion in clinics for nervous disorders, the analysands being virtually smuggled into such institutions. Sessions in the analysts' private homes came to an end from 1938, when Hitler's police state made such practices too dangerous. The Freudian terminology of analysis was masked with alternative phraseology. For example, the word analysis itself was substituted by 'real deep-psychological treatment of long duration'.[16]

In the meantime, although psychoanalysis did not die out altogether in Germany, the centre of gravity of the practice shifted to the United States, where many of the home-grown analysts were German trained.

The migration to the United States of practitioners, many of whom had been mentors of the American community of psychoanalysts, coincided with the worst years of the Depression. Ernest Jones, analyst and Freud's biographer, wrote in April 1933:

Some seventy thousand Jews got away from Germany as the blow was falling. Among them are most of the German psychoanalytical society; I only know of four or five members still left in Berlin. The persecution has been much worse than you seem to think and has really quite lived up to the Middle Ages in reputation. It is a very hunnish affair.[17]

Jones was busy raising funds to receive the influx of psychoanalysts.

The American community of analysts did not take to the influx with equanimity. Their arrival had exacerbated an ongoing set of schisms within psychoanalysis in America, characterized by Ives Hendrick, at the time a youthful radical, as a conflict of 'Jew against gentile, American against European, older against younger generations, city against city, personal rivalries of intensely narcissistic people'.[18] Nevertheless, as the social anthropologist the late Ernst Gellner has commented, psychoanalysis and its various manifestations could not prosper in authoritarian regimes such as Nazi

Germany. In America it could offer 'pastoral care' for relationships, the source of most anxiety in modern life, and a realistically tragic view of humankind in a violent era. In the post-war era, moreover, psychoanalysis would form an important element of the procurable self-improvement market.

Freud, who was suffering from cancer, remained in Vienna until the *Anschluss*, when the German race laws were applied to Austrian citizens. After many tortuous negotiations in Britain, Berlin and Vienna, Freud was eventually allowed to leave Austria for Britain, where he took up residence in London. Before he was finally allowed to set off for the station the Gestapo arrived carrying a document, stating that he had been properly treated, which they ordered Freud to sign. Freud wrote his signature, adding: 'I can heartily recommend the Gestapo to anyone.'[19]

Constraints on Psychology

The fate of schools of academic or scientific psychology in Germany followed the pattern of the predicament of most other communities of scientists in Germany. Once purged of its Jewish and left-wing psychologists (six of fifteen full professors teaching psychology in German universities lost their posts), psychology survived after a fashion, subject to oppression from within and without.

The German Society for Psychology was purged of its Jewish board members in April 1933, while individual academics vied with each other to demonstrate the Nazi-oriented credentials of their theories and research programmes. Hitler was extolled as 'a great psychologist', and was characterized in the keynote speech of the 1933 meeting of the German Society for Psychology as a 'bold and emotionally deep Chancellor'. The themes of upcoming conferences and congresses now focused on racial issues and Nazi communitarianism. Schools of psychological theory were twisted or hijacked to fit Nazi ideology. Gestalt psychology was adapted at Jena University to notions of the 'German soul', while other schools sought to expound the racial basis of psychological types. Erich Rudolf Jaensch of Marburg University and Felix Krueger of Leipzig attacked the Berlin University Gestalt school under Wolfgang Köhler as excessively materialistic. Köhler asserted a structural correspondence between the external world and the mind, whereas the Nazi perspective advocated a holism that called for a return to feeling, manliness and heroism. Krueger enthusiastically employed his vague notions of psychological holism to promote the need for Germans to lose their individuality in group consciousness, and even to support geopolitical expansionism and racism.

The fate of the Psychological Institute in Berlin illustrates the attempts of individual academics to comply with the regime in order to preserve a valued school of research in the face of pressure for conformity from below as well as the vindictiveness of Nazi academics in neighbouring faculties. The Psychological Institute was headed by Köhler, who was not a Jew. The drift of his psychology (which had no relationship to Gestalt therapy in America of the 1950s, nor the Gestalt theory that was to oppose behaviourism) was concerned with the quest for a theory of subjective impressions of form and their relationship to objective physical phenomena, especially in the processes of vision and sound. Apart from Jaensch's self-serving critiques, Nazi academics were quick to point out that one of the eminent cofounders of Gestalt psychology, Max Wertheimer, was a Jew.

Köhler's reaction to the first wave of dismissals appeared ambivalent. He attempted to retain the services of his distinguished Jewish colleague, Kurt Lewin, while seeming to endorse in public a qualified acceptance of denominational quotas in the professions. In an interview with the *New York Times*, 7 July 1933, entitled 'Köhler Foresees a Liberal Germany', he declared that the National Socialist government was to be commended for apportioning top jobs in the universities and the civil service according to religious affiliation. 'This,' he said, 'would have a most tranquilizing effect on the outside world, and would win over to our new government the adherence of a multitude of German citizens, including some of the most valuable.' The reflection paralleled Max Planck's observation to Hitler (as we saw earlier) that there were different sorts of Jews, worthy and worthless, and distinctions must be made between them. Given the menace and uncertainties of the time, however, Köhler was evidently engaged in diplomatic tactics for survival. His Gestalt cofounder, Kurt Koffka, wrote at about this time: 'Köhler is a very courageous man, much more than one could know.'[20] But like many scientists and academics, Köhler attempted to survive by separating his public role as a scientist civil servant from his private status as a 'free' individual.

On the first day of the new academic year, addressing the students and faculty, he gave a weak Nazi salute, at the same time announcing that he did not see eye to eye with the Nazi regime but that he was obeying the requirements of the university administration. That month an article appeared in the students' newsletter of the Philosophical Faculty, of which Ludwig Bieberbach, noted mathematician and recent convert to Nazism, was dean, asking whether the Psychological Institute had submitted itself to *Gleichschaltung*. The author was Hans Preuss, student leader of the university's Science Office. Preuss charged that the Psychological Institute was a stronghold of communists and Jews and that German students felt isolated there. Köhler's students rallied around their professor, countercharging publicly that it could not be denied that students were selected, but that this was strictly on the basis of their scientific abilities.

The following month members of the Nazi League broke into Köhler's offices to search for evidence of treason. Nothing incriminating was found. A second raid occurred in April of the following year, 1934, instigated directly by Ludwig Bieberbach, with the same result. Early in the summer the institute students called for a torchlight procession in honour of Köhler. As one of the students remarked later: 'We were naïve enough to believe that we could build a wall around ourselves and do science inside, while the Nazis busied themselves with their Volk.'[21] The students lost their bid to parade for their professor, and Köhler was destined to lose his battle to build up a special position in which he could give the Nazis their margin while remaining aloof as an individual and as a scientist.

In June of 1934 the students of the Psychological Institute were subjected to interrogation by members of the German Students'

League led by Berlin student leader Richter. Students were asked, for example, how they could support a professor who was associated with the Jew Wertheimer. Under interrogation, some of these students appeared to adopt Köhler's position on Jewish quotas: 'I know that Professor Köhler,' answered one of them, 'in no way denies the Jewish problem.'[22]

Köhler, who had by now offered his resignation, took on the Nazi students to the extent of arranging a meeting between a high-ranking official in the Prussian Ministry of Science and Education and a psychology student and his assistant, Hedwig von Restorff, representing the institute. Their arguments included a plea for the orderly operation of the *Führerprinzip*, which, they said, was threatened by the unruly conduct of students who were attempting to run things from the bottom up. As a result the Ministry later confirmed their trust in the professor, and issued an order for the Nazi students to be disciplined. A year later the Nazi German Students' League was neutralized.

But Köhler's tribulations were far from over. In June of 1935 the Gestapo instructed the university to fire two of his assistants, Karl Duncker and Otto von Lauenstein, for anti-Nazi activities. Köhler's position was now untenable. He again offered his resignation, which was accepted on 6 September, and he departed for the United States to take up a post as visiting professor at Swarthmore College supported by a Rockefeller Foundation grant.

Bieberbach the mathematician and Johann Baptist Rieffert, a philosopher and psychologist, now made their move to take over psychology in the university. Both enthusiastic Nazis, they had plotted with student accomplices to denounce the institute's assistants as a prelude to unseating Köhler. Rieffert, who had worked in the army psychological service, joined Berlin University in 1931, offering a course in 'character study'. From this vantage point he joined the Nazi Party in March 1933 and the SA in July. In July of 1935 Bieberbach urged the appointment of Rieffert as director of the Psychological Institute in the event of Köhler's departure. He repeated the request in September, asking that Rieffert should convert the work of the institute to 'race psychology'. In the event Rieffert was appointed temporary director, but his reign was short-lived. By the following year it was discovered that he had been a member of the Social Democratic Party. He was sacked from his university post and dropped from the Nazi Party.

By the late 1930s the psychology courses offered in Berlin University were a mix of experimental and developmental psychology and 'race psychology' devised by the Philosophical Faculty under Hans Günther, as well as a course titled 'racial soul studies' (*Rassenseelenkunde*). An American visitor, Barbara Burks, reported the 'utter barrenness' of the intellectual climate of what had been the former stronghold of the Gestalt school.[23]

12. The Cancer Campaign

While some scientific disciplines in Germany degenerated under the influence of Nazi ideology and oppression, some, like basic biology, merely stagnated, and others even appeared to flourish.[1] The historian Joseph Needham has commented that the tree of an evil regime does not, and can not, bring forth the good fruit of science. But scholars have been revising the bad-tree-bad-fruit view, taking into account what appeared to be the progressive Nazi campaign against cancer. 'How many of us know?' asks the historian Robert Proctor,

that Nazi health activists launched the world's most powerful anti-smoking campaign? How many of us know that the Nazi war on cancer was the most aggressive in the world, encompassing restrictions on the use of asbestos, bans on tobacco, and bans on carcinogenic pesticides and food dyes?[2]

The phenomenon is of great interest, since it prompts searching questions about science itself: that it so easily adapted to Nazi politics. Even at its most scientific and apparently progressive, from the point of view of public health policy, the anti-cancer campaign in Hitler's Germany was inextricably linked with racism and anti-Semitism.

Having led the world in dyestuffs technology and industrial chemistry, the Germans were the first to discover in the latter decades of the nineteenth century links between coal tar derivatives, aniline dyes, chromates and various forms of cancer. In 1902 German medical scientists were the first to detect a link between cancer, and especially leukaemia, and X-rays. The first state-supported anti-cancer agency was founded in Germany in 1900 as a result of the perception of rising incidence of cancers.

From the outset the agency decided on prevention as a key, indicating a conviction that the causes were environmental. By the 1920s national campaigns under the Weimar Republic expanded to include early detection and health and safety measures in the work place. Under the Nazis these initiatives were taken to remarkable new lengths.

After Hitler came to power campaigns were launched in the press and on radio advocating regular examinations. Poster campaigns advised men to have their colons checked for cancer as frequently as they serviced their motor cars, and for smokers to desist. Mass X-rays were carried out in schools, the army, in factories and offices. Since the Nazi war on cancer was conducted on the level of prevention rather than cure, the determination to succeed switched from biomedical research to public health campaigns, the gathering and analysis of statistics, or epidemiology, and surveillance for early detection.[3]

Combating asbestosis was a striking example of an enlightened campaign against occupational disease. Early in the regime, an anti-dust campaign had targeted asbestos fibres as a serious hazard, resulting in mandatory rules for the construction of ventilators and measures to reduce exposure. By 1938 German and Austrian researchers had found compelling evidence for twelve different types of cancer associated with asbestos; their findings became the basis for ever-widening research including experiments showing the growth of tumours in the lungs of exposed mice. By 1942 a German review concluding that asbestos was dangerous contrasted with the American and British view that working with the mineral was harmless. The following year the Nazi government accepted the principle that workers affected by asbestosis should receive compensation.[4]

Anti-Semitism in the anti-cancer campaign soon emerged in the metaphorical exploitation of cancer by the Nazi ideologues. Richard Doll, a distinguished cancer epidemiologist in Britain after World War II, remembers that while studying in Germany he saw a film slide in which Jews were depicted as cancer cells, while curative X-rays targeting these tumours were depicted as Nazi storm troopers.[5]

Despite the tendency to see many forms of behaviour and diseases as heritable, the conviction that cancers were caused by environment, occupation, lifestyle and culture, or 'civilization', was endorsed under the Nazis, although it was suggested that people of weak genetic make-up might be more susceptible than others. Arthur Hintze, a leading radiologist and surgeon in Berlin, believed that there was a link between cancer and diet; he also thought that religious practices might play a role. Otmar von Verschuer, later implicated in the work of Josef Mengele, suggested heredity contributed to only 1 per cent of all cancers. Jews, however, in the view of some 'experts', were carriers of cancer. They had brought tobacco to Germany in the first place and they traded in tobacco globally. But because of their genetic weaknesses, Jews were disproportionately susceptible to cancer.

The enthusiastic use of X-rays for diagnosis had malign implications for the expanding policy of sterilization. X-rays as a diagnostic tool prompted an increase in resulting cancers, especially among medical attendants who worked constantly with X-ray machines. Attendants who fell victim to the disease were known as 'martyrs to science' and celebrated as heroes. Even while one sector of the medical profession was developing the most rigorous protective standards for the use of X-rays as a diagnostic tool, another sector was planning to exploit the technology for the sterilization of 'undesirables'.

Under the Nazis measures to combat a wide variety of contaminants in the workplace put Germany at the forefront of the world in health and safety procedures, although in time of war this became more a matter of rhetoric than reality. Companies like IG Farben, while subscribing to wide-ranging health and safety measures, maintained in wartime a strict limit of 5 per cent on the proportion of workers to be hospitalized as a result of occupational illness. At the same time, the notion of euthanasia as an answer to illness was ever present: the regime wished to prevent illness in the 'master race', but it also believed in a policy of banishing the sick!

In their drive for cancer prevention Nazi nutritionists identified diet as a major contributory factor. Germans were warned against excessive consumption of tinned and smoked meats, fatty foods and sugar, and encouraged to consume cereals, fresh fruit and vegetables. Much was made of the fact that Hitler was a vegetarian, a teetotaller and a non-smoker.

The link between cancer and diet was perhaps a consequence of Swiss and German statistics from the middle of the nineteenth century which revealed that cancer of the stomach accounted for more than a third of all incidence of the disease. The gradual reduction of incidence of stomach cancer, from the high levels of the First War, was most likely a result of an increase in fresh foods, and a decline in adulterated foods, due to modern transportation. Many foodstuffs, moreover, had been subject to harmful dye additives to make them appear fresher than they were.

The drive for better eating was clearly less an encouragement of individual health than the promotion of the well-being of the *Volk*. One Hitler Youth health *Führer* proclaimed: 'Nutrition is not a private matter!'[6] As a propaganda poster put it: 'Dein Körper gehört dem Führer!' (Your body belongs to the Führer). Nazi ideologues contrasted the Nazi notion of health as a communal duty with the socialist conviction that individuals could do as they wished with their own bodies. Another aspect of diet was the Nazi attachment to traditional notions of naturalism: natural foods, such as wholewheat bread, were superior to white bread, which was deemed a 'French revolutionary invention'. But there were economic considerations, besides. Heavy meat eating and white bread eating were deemed a drain on the economy. In his four-year plan, Hermann Goering excoriated the practice of fattening animals on grain that should be used for cereals and bread. It had been pointed out by economists that it took 90,000 calories of grain to produce 9,000 calories of pork.

Suspicions of a link between alcohol and cancer were well entrenched in Germany by the 1930s, and it was widely held that alcohol damaged human genes. Both Hitler and Himmler, being teetotal, put the campaign against alcohol on a level of high priority. Himmler believed alcohol to be a treacherous poison, and Hitler claimed that more lives were lost to alcohol than were lost on the field of battle. Despite the popularity of beer in Germany, not only did the drive against alcohol involve postal campaigns and rationing, but the 1933 Sterilization Law allowed for the vasectomizing of alcoholics, who were incarcerated in concentration camps.[7]

The effort to demonstrate a link between tobacco and cancer, and to inform the German public of the dangers of smoking, was far in advance of the rest of the world. Proctor comments: 'The startling truth is that it was actually in Nazi Germany that the link was originally established. German tobacco epidemiology was, in fact, for a time, the most advanced in the world.'[8] The battle against smoking included a range of measures that were seen only relatively recently in the United States and Britain, and are still lagging in most of Europe. The Nazis banned smoking in many public areas, including offices and waiting rooms. There were bans on advertising cigarettes, with special reference to smokers appearing manly, sporting or sexually attractive, and ubiquitous health notices especially aimed at the young. No-smoking carriages were provided on trains with statutory fines. The president of Jena University, Karl Astel, director of the Institute for Tobacco Hazards Research there, banned smoking on the campus and was known for snatching cigarettes from students' mouths. There were cases, even, of drivers arrested on account of causing an accident while smoking. Tobacco, according to the propaganda, reduced energy for work, it was a cause of impotence in men; it was an 'epidemic', a 'plague', a form of 'lung masturbation'.

It was the German physician Fritz Lickint who in 1929 demonstrated the first statistical evidence linking lung cancer and cigarettes through 'case series' revealing that lung cancer patients were more likely to be smokers. A decade later he was to publish his *Tabak und Organismus* (*Tobacco and the Organism*), a thousand-page compendium revealing the diseases associated with every form of smoking, chewing or sniffing tobacco, as well as cancers of the lung, lip, mouth, jaw, oesophagus and throat. In the meantime two formidable studies were put in hand, eventually published in 1939 and 1943. The first, by Franz Hermann Müller of Cologne, anticipated British and American epidemiological studies of smoking and lung cancer by seven years. His study falls into the category of a 'survey-based retrospective case-control study'.[9] A survey was devised whereby the smoking behaviour of the lung cancer group was compared with the behaviour of a healthy 'control group' of similar age. The survey was then sent to the relatives of the deceased with further questions about the patients' habits and environment. The results revealed that lung cancer victims were more than six times likely to be extremely heavy smokers (ten to fifteen cigars or more than thirty-five cigarettes a day) than less enthusiastic smokers. Among the healthy group, however, there was a much higher proportion of non-smokers than smokers. Müller, who was still under thirty when he published his paper, concluded that tobacco was not simply an important cause of lung cancer, but that 'the extraordinary rise in tobacco use' was 'the single most important cause of the rising incidence of lung cancer' in recent decades.[10] There were no noticeable Nazi or racist dimensions to the study, and three Jewish scientists are mentioned in the text. Before the end of the war Müller disappeared without trace and has been presumed missing in action.

Then at Jena University's Institute for Tobacco Hazards two medical researchers, the pathologist Eberhard Schairer and his young colleague, Erich Schöniger, published a paper in 1943 that surpassed even Müller's study for complexity, comparing lung cancers in men and women, and in rural and urban populations. Their most important conclusion was that smoking was likely to be a cause of lung cancer, but less likely to be the cause of other cancers.

Yet it was one thing to launch campaign after campaign, but quite another to stop people smoking. Tobacco consumption grew in the first six to seven years after Hitler came to power, doubling between 1933 and 1940. Consumption dropped back by virtually a third by 1944, probably a result of rationing. Was the massive increase a form of 'resistance' to Nazi rule? Or an indication of stress, as the German nation faced the prospect of another war? The fact remains that Germany had led the world in a crucial piece of epidemiology, even though the regime, despite its best efforts, did not have the power to discourage people from smoking.

Overall, the Nazi war on cancer showed, as Robert Proctor notes, that 'good' science could be 'pursued in the name of anti-democratic ideals. Public health initiatives were pursued not just *in spite of* fascism, but also *in consequence of* fascism.'[11]

13. *Geopolitik* and *Lebensraum*

As in the medical profession, where Nazism was actively and enthusiastically espoused and promoted by its practitioners, segments of the discipline of geography made contributions to National Socialist ideology. Out of German geography came ideas that justified the call for an expansion of living space, *Lebensraum*, thereby endowing Hitler's aims for conquest with scientific legitimacy.

The 'science' of geopolitics was originally the brainchild of Rudolf Kjellen, a Swedish political scientist, who coined the term in 1905. Kjellen had, in turn, been influenced by the British geographer Sir Halford Mackinder. Mackinder, who became director of the London School of Economics in the first decade of the twentieth century, conjured up the idea of a people's 'heartland, and concept of the nation state as organic in nature, behaving, as it were, like a single organism'. He was convinced of the need, when required, for a country to expand beyond its boundaries.

Mackinder held that different racial types adapted through Darwinian natural selection to their geographical environment. In his view the temperate climes of the British Isles produced the 'John Bull' type, with an aptitude for freedom and civilized values, whereas the Slavic type, reared on the open steppes of Russia, was more adapted and acquiescent to despotic rule. Until complete adaptation to climate and geography had taken place, the peoples of the planet, thought Mackinder, should submit to the guidance of the Anglo-Saxon type; and in the event of their failing to recognize the wisdom of this, Pax Britannica should be imposed for the benefit of the world.[1]

It was Ratzel, however, who coined the term *Lebensraum* in a key essay entitled *Der Lebensraum*, published in 1901, to describe a 'geographical region within which living organisms develop'. The principal scope of the work was the study of plant and zoogeography, but he also applied his sociobiological treatment of *Lebensraum* to warfare between nations.[2]

Ratzel's ideas were exploited in the 1920s by Karl Haushofer, a former general in the Bavarian army, and applied to the process of expelling disruptive minorities in the interests of an expanding nation state: an early version of ethnic cleansing. Haushofer visited Hitler several times in 1924 in the prison at Landsberg where the Nazi leader was detained after the Beer Hall Putsch.

The notion of *Lebensraum*, linked with fear for the survival of *Volk*, now entered the thinking of Hitler. 'Our German nation,' as he put it in his political testament *Mein Kampf*, written during his incarceration,

must find the courage to gather our people and their strength for an advance along the road that will lead this people from its present restricted living space to new land and soil, and hence also free it from the danger of vanishing from the earth or of serving others as a slave nation.[3]

But was this version of geopolitics in any sense a science? As a discipline it owed more to an idiosyncratic version of cultural and political history than to authentic geography. Ratzel and Kjellen believed that a thorough knowledge of a country's geography was essential for shaping national policy on a grand scale. Haushofer declared that Germany's predicament after World War I demanded geopolitical specialists for the resolution of the crisis of defeat, bankruptcy and loss of territory. Just as figures like Max Planck had urged the importance of science as the Fatherland's best asset in the post-war situation, there were those who cited the 'science' of geopolitics as the principal discipline for German recovery.

To promote the new discipline, Haushofer founded and edited in 1924 the periodical *Zeitschrift für Geopolitik*. Packed with inflated claims, the periodical hardly bore the marks of objectivity, rationality and empirical method that are associated with scientific disciplines. He described geopolitics as:

the science of the earth relationships of political processes . . . based on the broad foundations of geography, especially on political geography, which is the science of the political organisms in space and their structure . . . Geopolitics sets out to furnish the tools for political action and the directives for political life as whole. Thus, geopolitics becomes an art, namely, the art of guiding practical politics. Geopolitics is the geographic conscience of the state.[4]

Haushofer laboured the point that Germany's defeat in World War I had stemmed from 'ignorance of the geographical realities' on the part of Germany's leaders.[5] The discipline of geopolitics emerged from an 'elementary craving for better scientific protection of the political unit'.[6] He shared with Oswald Spengler the perception that the West was in decline and argued in the pages of his periodical that it was Germany's destiny to lead Europe back to primacy as the defender of civilized values in the world. Germany was the heart of Europe, and Europe was the heart of the civilized world.

After the harsh measures of the Treaty of Versailles, which threatened to condemn Germany to permanent debt, separate its peoples and deprive the Fatherland of legitimate territory, Haushofer urged geographers to become politicized. Not all geographers agreed with him and the stage was set for tensions between the geographical classicists who argued for an apolitical stance and Haushofer's growing geopolitical group.

During the Weimar period, even as Hitler adapted Haushofer's leading ideas to Nazi ideology, geopolitical ideas began to feature in education. The influence of right-wing politics in geography increasingly adopted a biologically racist tone. Two key concepts associated with *Lebensraum* were promoted by a constituency of geographers who promoted the aspirations of German imperialism: *Drang nach Osten* (Push to the East) and *Volks- und Kulturboden* (Soil of the People and of Culture). The latter notion drew together three interpretations of territory or belonging: the Reich, or state borders; ethnic communities outside those borders; and wider cultural links with Germanhood.[7]

Under the Nazis, from 1933 onwards, geopolitics became a compulsory subject in schools and universities, the newly drafted textbooks quoting extensively from Haushofer and Ratzel in support of Hitler's bellicose *Lebensraum* rhetoric. Schoolchildren were taught to look forward to the formation of Germany's world empire, *Germanisches Weltreich*, a necessary outcome of the Fatherland's just quest for *Lebensraum*.

Students were informed that 'the meaning and the iron will of national socialist statesmanship are the securing of our *Lebensraum* and a sensible organization of space . . . To create a spatial organization suitable to Germany means to undo the damage caused by an unavailing population.'[8]

Haushofer assumed a central role as the father of German geopolitics under Hitler, and he was happy during the 1930s to praise National Socialist foreign policy as a manifestation of his own theories, even though his geocentric ideas for Germany did not imply a racist ideology. He would live, however, to deplore Hitler's invasion of the Soviet Union as the reverse of his belief that Germany, Japan and Russia should make common cause.

The status of geopolitics would continue to rise with Hitler's rapid conquest of territories to the east after the invasion of Poland and of Russia. In the wake of the expansion eastwards, the establishment of German settlements in Poland, the destruction of Jews and transports of slave labourers westwards, geographers would be set to work establishing demographics and assessing the 'Germanic' populations and optimal population densities in Russia and the Ukraine. Anticipating the turn of events after Stalingrad: as the Red Army juggernaut rolled towards Berlin, would sweep away these Nazi geopolitical calculations.

14. Nazi Physics

Unlike medicine, anthropology and geography, which assisted the regime by contributing exploitable ideas with a semblance of scientific respectability to National Socialism, the discipline of physics was less amenable to being hijacked by Nazi ideologies. This did not prevent the Nazi physicists Philipp Lenard and Johannes Stark attempting to dominate the physics community. The story of Nazi physics in the 1930s reveals the extent to which non-Nazi physicists allowed themselves to be pressured, and the extent to which they resisted. They were less prepared to compromise with forces within the discipline than those outside it.

When Hitler came to power in 1933, Lenard and Stark were seventy-one and fifty-nine years of age respectively; Stark was set to become the more active of the two. Lenard went to see Hitler and informed him that science in the universities was in a deplorable state and in need of reform. He offered his services to secure appropriate posts for suitably Germanic physicists. We do not know how Hitler reacted.

Lenard meanwhile embarked on a series of lectures, which he eventually edited into a four-volume work, entitled *Deutsche Physik* (*German Physics*), published in parts through the mid-1930s. Its opening lines set the tone for the racist sub-text of the entire enterprise: ' "German physics?" people will ask. I could have also said Aryan physics or physics of Nordic-natured persons, physics of the reality-founders, of the truth-seekers, physics of those who have founded natural research.'[1]

While Lenard hoped to be a major influence in the wings, Stark had set his heart on controlling, or at least influencing, the key posts in scientific management in Nazi Germany. Stark wanted the presidency of the Kaiser Wilhelm Society for Lenard; for himself he sought the Emergency Foundation with its control of funding and choice of research programmes; and he wanted the presidency of the Physikalisch-Technische Reichsanstalt (the PTR), which he acquired in May of 1933. On his appointment, Lenard wrote a piece in the *Völkischer Beobachter*[2] welcoming Stark's new powers: 'It signifies a definite renunciation of the apparently already inescapable predominance of what might briefly be called Einsteinian thinking in physics; and it is a move towards reaffirming the scientist's old prerogatives: to think independently, guided only by nature.'[3] He added that Einstein's theories were a combination of sound knowledge and 'some arbitrary additions that had been strung together mathematically'. These theories 'are now already falling to pieces, which is the due fate of unnatural products'.

At once Stark sought to apply the fascist 'leadership principle' – the *Führerprinzip* – which advocated strict lines of authority leading up to himself, and he dismissed the 'Jews and leading figures of the previous regime' from the institute's advisory committee, a body which in time disappeared altogether.[4] He hoped for the enlargement of the PTR, to include fifty large institutes, 300 laboratories and thousands of personnel – plans that foundered in the prevailing circumstances of overlap and rivalry. He nevertheless established various research programmes with military applications, working in close collaboration with the Luftwaffe and the army.

By the following year he had become president of the German Research Foundation and could tell Lenard that together they would now control the universities and develop research programmes along 'German' lines, meaning a preference for experimental research over theoretical. But following the appointment of Bernhard Rust to the supreme ministerial post controlling education, science and culture, Stark found himself in conflict with the large power-broking members of the Nazi cartels.

A story that illustrates the fragmentation of the regime in its science policies, funding and appointments, and the ultimate failure of the *Deutsche Physik*, involves the figure of Werner Heisenberg and spans the entire decade to the eve of World War II.

Heisenberg's Chair

In August 1934, a few days after the death of Germany's President, Field Marshal Hindenburg, Hitler announced his intention to combine the offices of the presidency and the chancellorship, thereby aspiring to the role of unchallenged dictator. A referendum was organized for 19 August and a propaganda campaign was launched. Against this background, Johannes Stark attempted to gather the support of German Nobel prize-winners, importuning them to lend their names to a manifesto pledging loyalty to Hitler. Heisenberg, Laue, Planck and Nernst refused, insisting that science and politics had nothing to do with each other.[5] Stark retorted that supporting Hitler was not a political act; it was, he argued, an 'avowal of the German *Volk* to its Führer'. But praising Einstein while refusing to support the Führer, he went on, that indeed was political. The same month, all civil servants, which included scientists working in universities and state-funded institutions like the Kaiser Wilhelm Institutes, were required to take an oath of allegiance to Hitler. Heisenberg notably delayed taking the oath until the following year, and in the meantime further blotted his record by taking a prominent part in a conference in Hanover, in September 1934, 'at which theoretical physics, and Einstein's theories, were enthusiastically endorsed and promoted. Heisenberg seemed bent, like so many others, on attempting to separate science from politics.

Later Heisenberg wrote to his mother explaining how he saw his task as a scientist remaining in Germany and working for the future of science but not the regime. 'I must be satisfied to oversee in the small field of science the values that must become important for the future. That is in this general chaos the only clear thing that is left for me to do. The world out there is really ugly, but the work is beautiful.'[6] Events, and the ugliness of the world 'out there', however, were not to leave him in peace.

Following a Nazi decree in January 1935 forcing professors into retirement at sixty-five, the great Arnold Sommerfeld, aged sixty-six, prepared to leave his post at Munich University.

Sommerfeld had held the appointment for nearly thirty years since taking over in 1906 from Ludwig Boltzmann, the first physicist to employ the theory of probability in investigating energy relations. The university and indeed Sommerfeld himself were determined to appoint Werner Heisenberg, who at that time held a chair at Leipzig University, where he ran the Institute of Theoretical Physics. Heisenberg was still, at thirty-seven, considered young, but he was now the leading theoretical physicist in Germany, and the Munich chair was regarded, in view of Sommerfeld's special contribution to the discipline, an appointment for a theoretician.

But there was now fierce opposition to Heisenberg from the Nazi physicists Lenard and Stark, who were busily rooting out scientists who nourished a 'Jewish' spirit – namely an aptitude for theory over experiment. In December 1935 Johannes Stark delivered a vitriolic speech in Heidelberg, excoriating Einstein and charging that, although he 'has disappeared from Germany . . . unfortunately his German friends and supporters continue to act in his spirit'.[7] Stark mentioned the pillars of the scientific establishment – Planck and Laue – by name, adding that Heisenberg 'the theoretical formalist' was about to be rewarded with a chair. The attack was followed by scathing articles in the Nazi newspaper *Völkischer Beobachter* and Nazi ideologue Alfred Rosenberg's monthly *Nationalsozialistische Monatshefte* (*National Socialist Monthlies*). Heisenberg replied temperately in the *Beobachter*, in February 1936, defending Einstein's theory of relativity, but Stark was allowed to append a rejoinder in which he condemned Heisenberg's work as 'an aberration of the Jewish mind'.[8] The attacks and ripostes continued for more than a year, until an article, partly penned by Stark, characterizing Heisenberg as 'a white Jew', appeared on 15 July 1937 in the SS newspaper *Das Schwarze Korps* (*The Black Corps*, issue 28). The significance of the publication was the apparent auspices of Heinrich Himmler, who had a year earlier been appointed Reichsführer-SS with a place in Hitler's cabinet, and responsibility for special intelligence gathering on social and cultural activities in the Nazi state. The article opened with a general screed:

The victory of racial anti-Semitism is to be considered only a partial war
. . . for it is not a racial Jew in himself who is a threat to us, but rather the
spirit that he spreads. And if the carrier of this spirit is not a Jew but a
German, then he should be considered doubly worthy of being combated
as the racial Jew, who cannot hide the origin of his spirit. Common slang
has coined a phrase for such bacteria carriers, the 'white Jew'.[9]

In the body of the article Einstein, Haber, Sommerfeld and Planck
were accused of excluding real Germans from appointments in
physics, and Heisenberg was characterized as 'that white Jew' and
'representative of the Einstein spirit in the new Germany'.

The sub-text to the attack was reminiscent of the McCarthy
witch-hunts of the 1950s; as historian David Cassidy has pointed
out, 'by exploiting the dominant hate ideology of the day, character
assassination of a key individual could be used to achieve the
influence they craved'.[10] Heisenberg was accused of smuggling an
article defending relativity theory into a Nazi paper, of attempting
to influence the Reich Education Ministry with his treacherous
views, of refusing to endorse Hitler's presidency, of securing his
Leipzig chair by favouritism of the 'white Jewish' establishment, of
harbouring Jews in his department. It was no longer a question of
whether Heisenberg would succeed Sommerfeld at Munich; the
onslaught boded ill for the physicist's future safety. The article
referred to him as the Ossietzki of physics, an allusion to Carl von
Ossietzki, winner of the 1936 Nobel peace prize who was even
then languishing in Dachau, where he would die of torture and
starvation the following year. The editors of the paper commented
that 'white Jews' should be made to 'disappear'.

At this point it must have been abundantly clear to Heisenberg
that he could not bury himself in the 'beauty of the work' and
remain unscathed by the ugliness 'out there'. He decided to fight
back, holding in reserve the ultimate decision to resign.

He was not without supporters in Germany. Planck and many
of Heisenberg's colleagues fired off letters of complaint to university
deans, and local and state education bureaucracies. Heisenberg also
received the backing of his peers by being elected to various bodies,

for example, the Saxon Academy and the Göttingen Academy of Sciences.

The extent of Heisenberg's personal and public dilemma is expressed in a letter he wrote to Sommerfeld:

Now I actually see no other possibility than to ask for my dismissal if the defence of my honour is refused here. However, I would like to ask you for your advice in advance. You know that it would be very painful for me to leave Germany; I do not want to do it unless it must be absolutely so. However, I also have no desire to live here as a second-class person.[11]

The *Schwarze Korps* article was published when Heisenberg had just arrived in Munich with his young wife, Elisabeth, pregnant with their first child. They were on their way to the Alps for a holiday. Heisenberg initially attempted to rebut Stark and the article by writing a letter of complaint to the Reich Ministry of Education, insisting that he would be forced to resign if the ministry stood by the attacks. If on the other hand the ministry disapproved of the allegations, he wished to have the official backing and protection of the state, as would any 'lieutenant in the Wehrmacht' in a similar case. At the same time, Heisenberg took a major gamble by writing in precisely the same terms to the Reichsführer-SS, Heinrich Himmler, with whom his family had a tenuous connection.

Heisenberg's grandfather, a former high school rector, had at one time belonged to the same hiking club as Himmler's father, Joseph Gerhard Himmler, an assistant rector of a gymnasium in Landshut. As a result Heisenberg's mother knew Himmler's mother. Mrs Heisenberg accordingly went to see Mrs Himmler in Munich, where both women lived, at some point in July or August 1937. According to Heisenberg, Mrs Himmler was at first reluctant to involve herself in the affairs of her son, but Mrs Heisenberg broke the ice by saying: 'Oh, you know, Mrs Himmler, we mothers know nothing about politics – neither your son's nor mine. But we know that we have to care for our boys. That is why I have come to you.'[12] Mrs Himmler softened and undertook to pass on Heisenberg's letter to Himmler.

It was not until November that Himmler took up the case and wrote to Heisenberg asking the scientist to respond to the charges. Heisenberg defended himself point by point and Himmler accordingly decided to look into the matter further, initiating an investigation that would take a further eight months, running concurrently with and independently from an investigation on his reputation under way in the Reich Education Ministry. Meanwhile Lenard and Stark continued their vitriolic attacks against 'Jewish physics' with the backing of the Nazi-led Teachers League and Rudolf Hess, Hitler's party deputy, who was intent on exerting his own influence on science. At a meeting of university rectors in December 1937 it was mooted that 'political reliability' should be considered a condition of university appointments.

The SS investigation of Heisenberg had involved the presence of spies in his classrooms and the bugging of his home. Heisenberg's past, moreover, was also being scrutinized for evidence of homosexuality, a crime under Nazi law that could land an individual in a concentration camp. There was a suggestion that he had married hurriedly in order to cover up his guilt. The investigation yielded no evidence.[13] He wrote to his mother in November 1937: 'such a struggle poisons one's entire thoughts, and the hate for these fundamentally sick individuals who torment one eats into one's soul'.[14] During the investigation in the winter of 1937–8 Heisenberg was interrogated in the basement of the SS headquarters in Berlin by investigators who had an academic grounding in physics and mathematics.

It is likely that the regime's decision to clear his name had been more or less made by the spring of 1938, for he was given permission to lecture in England in March of 1938. Himmler sent an official letter to Heisenberg, however, on 21 July 1938, confirming that he, the Reichsführer-SS, did not approve of the attack in *Das Schwarze Korps* and that he had forbidden any further criticisms of that nature. Himmler invited Heisenberg to come to Berlin to talk 'man to man' on the issues, but in the meantime he advised the physicist to separate the personal and political characteristics of scientists from his research: he was referring, obviously, to the work of Albert Einstein. The same day he wrote to Reinhard Heydrich, his secret service chief, requesting that Heisenberg should be allowed to work in peace, and commenting: 'I believe that Heisenberg is decent, and we could not afford to lose or silence this man, who is relatively young and can educate a new generation.'[15]

The results of Himmler's investigation are to be found in a memorandum sent at some point in 1939 to the Reich Education Ministry: Heisenberg was a harmless, apolitical academic for whom 'theoretical physics is merely the working hypothesis with which the experiment inquires of nature in suitable experiments'. He was a man of decent character who had increasingly supported National Socialism, and was positive towards it. Heisenberg's biographer David Cassidy comments, 'Whether completely accurate or not, this remained the Nazi regime's official assessment of Werner Heisenberg until the end.'[16]

So Heisenberg's pact with the regime was an implicit agreement to separate science from the scientist – theoretical physics from Einstein's Jewishness. It was a shameful pact, and it was made under duress at a time when Himmler and his henchman Heydrich were engaged in campaigns of terror, forced expulsions and arrests of Jews.

In November 1938 the infamous *Kristallnacht*, the Night of Broken Glass, would reveal the full fury and violence of the Nazi onslaught against Jews. What induced Heisenberg to stay in Germany despite the high moral price? His predicament reveals a combination of intense pressures, special perhaps to a scientist of world-class distinction in Germany at the time. Heisenberg appears, at this stage, to have drawn an equivalence between his personal survival and the survival of physics in the Fatherland. Later, his compromises would appear less benign.

In the event, Heisenberg was not chosen to succeed Sommerfeld in Munich, despite Himmler's support on several occasions. The job went to the least able of the candidates, Wilhelm Müller, author of a textbook on engineering, and well known for his published support of Aryan physics. In the meantime the chair in physics at

Vienna, which had been another possibility, was also denied to
Heisenberg; his name had been blocked by influential supporters
of Nazi science. As the persecution of Jews, 'undesirables' and
'enemies' of the state increased, he appeared safe, but his rehabili-
tation seemed a strangely qualified victory, his claim to protect
theoretical physics in Germany a weak rationale indeed.

Case of Pascual Jordan

Not all theoretical physicists in the early days of Hitler were apolitical, and not all Nazi physicists followed the lead of Lenard and
Stark. In the ambit of the physics community there emerged the
unusual spectacle of an isolated theoretical physicist who was every
inch a Nazi. Pascual Jordan promoted a wholly contradictory stance
to Aryan physics by proclaiming an equivalence between National
Socialism and a focus of theoretical physics which had been routinely excoriated by Lenard and Stark as 'Jewish physics'. His story
illustrates the pitfalls of drawing parallels between ideology and
science. He was the first, but by no means the last, to produce a
'quantum theory' of society.

Pascual Jordan, rabid Hitler enthusiast, an aficionado of Freud's
unconscious and a brilliant contributor to the new physics, for a time
shared the obloquy poured on Heisenberg (although he defended
himself with even greater passion and vim). One of his harshest
Nazi critics accused him of attempting to sabotage 'logical purity'
and the great Aryan tradition of rigorous science 'with unproven
pseudo-religious fantasies'.[17]

We get a rare physical glimpse of his monomaniacal tendencies
and personality at the Peenemünde rocket plant, where he was sent
to work on the mathematical problems of turbulence. A young
colleague, Peter Wegener, who shared an office with him, remembers how Jordan would neglect his proper research to spend the
day composing a textbook on algebra without ever consulting
notes. Jordan suffered from a marked speech impediment which
meant that he was incapable of speaking in continuous sentences.
'I soon discovered that one had to turn one's back to him while he

sat in his chair, feet on his desk and one hand on his forehead, twisting his nose with his thumb. These were the moments when he spoke rather fluently.'[18]

He was a member of Max Born's young team of quantum physicists in Göttingen through the 1920s, along with Wolfgang Pauli and Friedrich Hund, who took Heisenberg's early ideas and showed how in consequence a systematic theory of atomic behaviour could be developed. He was also from time to time invited to Copenhagen to work with Niels Bohr.

Jordan saw a link between the ideology of National Socialism and the strange counterintuitive dynamics of the Copenhagen interpretation of quantum physics. Just as the new physics was an epochal discovery about the true force of nature, so too, in his view, was National Socialism no mere ideology, political platform or evanescent manifestation of party politics, but a veritable force of nature. Nazism, in his view, was true in the sense that the second law of thermodynamics was true.

Jordan saw the revolution in physics, moreover, as a 'mirror image of the revolutionary transformation of the world'.[19] Describing the early emergence of quantum ideas, Jordan gave an impression of the sense of intellectual excitement they generated:

Everyone was filled with such tension that it almost took their breath away. The ice had been broken . . . It became more and more clear that in this connexion we had stumbled upon a quite unexpected and deeply embedded layer of the secrets of Nature. It was evident that wholly new processes of thought, beyond all previous notions in physics, would be needed to resolve the contradictions – only later recognised as merely apparent – which now came to a head.[20]

National Socialism had the power to liquidate its intellectual enemies, just as National Socialist weapons based on new technologies developed by its scientists had the power to liquidate tanks, planes and armies. As early as 1935 he was predicting that 'a not too distant future may have at its disposal technical energy sources which make a Niagara power station appear trifling and explosive materials in comparison to which all present explosives are harmless toys'.[21]

Quantum physics, according to Jordan, revealed an underlying all-embracing vision of reality, in which conventional cause and effect are overturned and old distinctions between object and subject, the individual and society, are eliminated. Nazism, understood in terms of the new physics, would deliver a death blow to the morally bankrupt influence of the Enlightenment, with its emphasis on the individual and its tendency to objectify nature in terms of mere mechanistic determinism. In quantum physics he saw a direct relationship, and an exemplar, between the behaviour of molecules and sub-atomic particles, and the *Führerprinzip* in National Socialism.

We know that in the body of the bacterium, there are among the enormously many molecules constituting this . . . creature, a very small group of special molecules which are endowed with dictatorial authority of the total organism; they form a . . . [steering centre] of the living cell.[22]

Biology, psychology, consciousness, telepathy, clairvoyance, spirituality, all could now be explained, described and controlled in the light of new quantum insights. Above all, political revolution, involving the redrawing of national boundaries and the thrust towards unified continents, could be seen in terms of the new physics. 'The political transformation already accomplished in so many European states, in the shape of a replacement of the parliamentary forms of government by authoritative and dictatorial methods,' he wrote, 'signifies in no way merely a technical modernization of the apparatus of government; rather it is the eruption of a revolutionary reconstruction of our entire thought, values and action, gradually encompassing all areas of life and culture.'[23] No longer would notions about the unconscious in psychology, or holism in spirituality, inhabit a separate area of thought, as it had done under the influence of the Enlightenment, but quantum mechanics would supply a rigorous scientific underpinning for these hitherto 'fuzzy' modes of thought. Jordan conceded that his ideas would be difficult for ordinary people to grasp; he saw a great gulf between the mind of the 'rare, surpassing researcher', the brain of a 'natural-scientific thinker' and ordinary minds. But he did not despair of providing explanations that could be visualized so that, at least, his ideas would be accessible to an elite audience.

Jordan was not only attacked by Lenard and Stark, but by other Nazi intellectuals such as the eugenicist and idealist philosopher Kurt Hildebrandt for espousing the abstract and relativist ideas that were already corrupting the German people. Jordan responded by counter-charging that Hildebrandt was denying the internationalist science on which the Reich must depend for its technological ability to defend and extend itself: the machine gun, he declared, knew no racial or national boundaries in its effectiveness. His internationalist arguments, moreover, did not mean that he had budged one iota on his anti-Semitism. In his *Physics and the Secret of Organic Life*, published in 1941, he dismissed the personal link between Einstein and the theory of relativity, arguing that someone else would have discovered it anyway. He declared, moreover, that the crucial content of the special theory had been anticipated by the French physicist Henri Poincaré.

With Germany conquering its neighbours irresistibly, Jordan now announced the triumph of technology not as a means of plutocratic plundering in the service of individual capitalists, but as serving of the totality of life of the wider community. 'We are not willing,' he wrote, 'to see any abuse in the coupling of science to military might after military might has proven its compelling *aufbauende* force in the creation of a new Europe.'[24]

Jordan's general ideas, which emerged in a variety of forms and disciplines, brought him no influence in the Third Reich. The interest of his case is the extraordinary adaptability of science, even in a researcher of remarkable talent, to bizarre hybrids of political and scientific myth. Jordan's grasp of modern physics, for all its sophistication, was employed to shape a fascistic view of the world

— indicating how readily in some individuals science can be adapted in the service of totalitarianism, and that great talent in physics is not a necessary, still less a sufficient, basis for moral and political integrity.

15. Himmler's Pseudo-science

In the autumn of 1935 Heinrich Himmler, Hitler's notorious SS Reichsführer SS, penned a letter to Dr Ernst-Robert Grawitz, Reich Physician of the SS and Police:

Following our brief conversation in Gmund about the examination of the question of left-handedness in humans in earlier times, I now reiterate my thoughts in writing and ask that you instruct Professor Janssen to set some doctoral dissertations on this topic. There is much to suggest that in earliest times mankind was left-handed. A large proportion of finds from this period can be cited as evidence for this. It was probably with the introduction of the shield, which protected the left half of the body, in which the heart is located, that mankind switched to right-handedness, albeit without fully neglecting the use of the left hand.[1]

In the lethal mix of power, fear, cruelty and dilettantism, pseudo-science began to flourish virtually unchallenged under the auspices of the SS in Hitler's Germany. An autodidact, lacking in judgment or even the basics of scientific research, Himmler found time to get involved in all manner of research schemes, starting early in the regime with bogus programmes linked with the origins of Aryanism, leading ultimately to the murderous 'medical research' in the death camps. While such activities were marginal to the enormous volume of scientific and technological work being conducted in the Third Reich as a whole, the direction of SS-sponsored 'researches' in time brought appalling suffering and death to their victims and degradation upon all those associated with their conduct and data.

Born in 1900, Himmler was brought up a strict Catholic. He was interested even as a child in animal and plant breeding. He kept a herb garden as a boy, a prelude to his obsession with alternative

medicines and homeopathy, which he thought superior to regular
pharmacological products (in years to come he ordered that herbs
should be grown in the concentration camps). A vegetarian and
teetotaller, on leaving school he studied agriculture in Munich,
graduating aged twenty-two. Shortly afterwards he rejected his
Catholicism to become a devotee of Adolf Hitler, adopting fanatical
beliefs in the superiority of the Germanic people and the need for
Lebensraum.

Within two years of Hitler's assumption of power Himmler
played a key role in establishing the Ahnenerbe (Ancestral Herit-
age), a research society that encouraged a wide circuit of pseudo-
scientific pursuits. He became its curator, and eventually its
president. The Ahnenerbe was principally preoccupied with historical and ethnographical research aimed at discovering the origins
of Aryan superiority. In time it claimed to research the natural
sciences proper and made bids to attract genuine scientists. Himmler
characterized its business in a letter to Goering as the 'implementation of fundamental and most valuable research, which nevertheless
went wholly or partially unrecognized, or was even persecuted by
the official sciences'.[2] Apart from its chronic paranoia and sense of
injured merit, the Ahnenerbe portrayed itself as breaking down
divisions between the natural sciences and the arts, and of encouraging a type of 'holism' that would open up science to the Nazi
Weltanschauung. Its stock in trade was a mishmash of genetics,
geopolitics, philology, anthropology, history and archaeology,
blended with astrology, mythology and the occult. The search for
evidence of origins was unrelenting. There was a decade-long
quest, for example, to recover an early text of Tacitus' *Germania*
from a vault in Italy, for which Hitler himself petitioned Mussolini
unsuccessfully in 1936, and for which SS troops were still searching
in 1943.[3]

Tacitus depicts the German homeland as an unrelieved wilder-
ness, a vast virgin territory of mountains and forests which thrilled
the aficionados of the Ahnenerbe, not least Richard Walther Darré,
who was responsible for the epithet *Blut und Boden* (Blood and Soil)
and who enthusiastically promoted a Nazi policy of *Naturschutz*

(the protection of nature). Others saw references to racial purity in Tacitus' accounts of Germanic mythical hymns in praise of Tuisto, a deity that had risen from the German soil: Tuisto gave birth to the first man, siring three sons who became the ancestors of the German people. Blood and soil were thus inextricably one. As the ancient Roman put it: the people of Germany 'were untainted by intermarriage with other peoples, unique, wholesome, their physiques, despite their vast hordes, remain identical: wild blue eyes, flaming hair, huge, strong' (*Germania*, chapter 4). Sweet music to the racial hygienists.

Himmler was also convinced that the leaders of the Third Reich had descended from the Vikings, and planned an expedition to Iceland to prove the fact. More remotely, as we shall see, he believed that the origins of the Aryan race stemmed from living shoots conserved in crusts of ice in outer space.[4]

Under the special auspices of the Ahnenerbe, and indeed on the particular orders of Himmler, the zoologist Ernst Schäfer mounted an expedition to Tibet in 1938. His aims were ideological rather than scientific. He was convinced that:

we could accomplish more as SS men and do much more for the lack of understanding for the new Germany by being up front about who we are, than by travelling under the disguise of an obscure, if neutral scientific academy; after all we have a clear conscience.[5]

The aim of the expedition, which included the anthropologist Bruno Beger and the entomologist Karl Wienert, was to collect material about the proportion, origins, significance, and development of the Nordic race in the region. This same Schäfer two years later became chief of a section of the Ahnenerbe for Central Asian Studies and Expeditions, which he hoped to expand into a large institute for the 'rebuilding of German science'.

Central to the early philosophy of the Ahnenerbe was an outlandish theosophical theory known as the *Welteislehre*, the doctrine of world ice, with echoes of Nordic 'heroic' mythology featuring apocalyptic struggles between fire and ice. This fantastical notion was revealed in a dream to an eccentric Austrian engineer called Hanns Hörbiger, published in 1913 under the title *Hörbiger's Glacial-Kosmogonie* – '*Hörbiger's Glacial Cosmogony*,' reads the title page,

a new developmental history of the universe and the solar system, based on the opposition of cosmic Neptunism to an equally universal Plutonism, following the most up to date results of all exact branches of research, revised and supported by his own experiences and edited by Ph. Fauth: with 212 diagrams.[6]

All the planets, including the moon, are covered in ice. Our Milky Way is made up of ice crystals. Drawn by the gravitational field of the sun, bits of cosmic ice explode and vapourize. Solar prominences are great geysers of exploding steam erupting out into space where the vapour forms ice-crystal meteors. Ice crystals descend on the planets, forming ice crusts and in the case of the earth rain. When meteors explode on entering our atmosphere they form hailstorms. Our weather, in other words, comes from outer space, and Himmler was keen to exploit the theory to make long-range weather forecasts.

Hörbiger goes on to describe the action of luminiferous ether, a popular notion in late nineteenth-century physics. Ether slows down objects hurtling through space before orbiting with planets, but the fate of all moons is inevitably to spiral down until they crash into the surface of their host planets. Earth once had a second moon which crashed in the ocean, caused Noah's flood and formed Atlantis. The ice ages, and the disappearance of the dinosaurs, the formation of coal and the sinking of Atlantis, were all owed to the descent of earth's second moon.

The *Welteislehre* was a theory of everything, containing all explanations, including the origins of the universe and its inevitable apocalypse. In the fullness of time the earth itself will spiral down into the sun to be destroyed in a final explosion. The theory posited a quite different basis than classical physics for the forces in outer space. It appealed to 'a unified, fantastic and splendid world picture, a scientific foundation for a truly Nordic *Weltanschauung*', which involved a riposte to Einstein's 'dreadful and mistaken theory' and its 'flooding waves of spiritual destruction'.[7]

Himmler was keen to attract regular scientists to his institute to work on the *Welteislehre*. Under the cloak of an 'Institute of Meteorology' (Pflegestätte fur Wetterkunde), he drew together astronomers, geologists and meteorologists to demonstrate the theory's validity. But the Ahnenerbe only succeeded in attracting devastating criticism on this score, in particular from the Nazi physicist and Nobel laureate Philipp Lenard, who lambasted the editor of the *Illustrierter Beobachter* (*Illustrated Observer*), a periodical run by the National Socialists, after its editor announced a long-running series on the *Welteislehre*. The series was pulled, and future research on ice cosmology was conducted more discreetly.

Himmler nevertheless continued to defend the theory vigorously. In June of 1937 we find him excoriating a hapless Dr Otto Wacker of the Ministry for Science, Training and Popular Education for passing on to him a memorandum from Paul Guthnik of the observatory at Neubabelsberg. The scientist had declared that glacial cosmogony was nonsense. Himmler wrote to the Minister:

I state clearly and unambiguously: I am a defender of free research in all its forms, and this includes free research into glacial cosmogony. Moreover I intend to give this free research my warmest support, and in this I find myself in the best possible company, as the Führer and Chancellor of the German Reich, Adolf Hitler, has for many years been a convinced adherent of this theory, which is disdained by the journeymen of science.[8]

The next month Himmler tracked down the culprit responsible for commissioning the memorandum on glacial cosmogony and relieved him of his job, uniform and Party badge. Writing to Reinhard Heydrich, Himmler added that the man 'was constantly attacking glacial cosmogony', . . . and that he 'takes a totally unscientific position' with regard to the phenomenon.[9]

Himmler also set up a research programme within the SD (the Sicherheitsdienst, the SS Security Service) to explore the persecution of witches in Germany during the seventeenth century.

Himmler was keen to demonstrate through witch trial documents that the persecution of witches had in fact been a combined attack by the Catholic Church and the Jews against Healthy Germandom. The research involved the collation of some 33,000 items in a 'witch card index' drawn from some 260 archives and libraries.[10]

Like Hitler, Himmler was preoccupied with health fads. He was fanatical about the banning of artificial ingredients in food, blaming commercial food companies for destroying the natural diet of the German people.

City folk, living through the winter largely on canned food, are already at their mercy, but now they attack the countryside with their refined flour, sugar, and white bread. The war has interrupted these proceedings; after the war we shall take energetic steps to prevent the ruin of our people by the food industries.[11]

Even as war loomed, among his busy initiatives into racial hygiene and aspects of archaic genetics he ordered that SS members must investigate their family trees going back to the end of the Thirty Years' War – if a Jew should appear in a man's family tree after that date, he must leave the SS. Later we find him writing to SS officials:

In Felix Niedner's *Norwegian King's Stories* [translations of old Norse verse] I noticed that in one story a man gives his king two red horses with white manes as a gift. Recently I saw a red-brown horse with a white mane . . . I hereby commission SS-Standartenführer Fegelein to investigate the genealogy of this horse.[12]

The work of the Ahnenerbe, for all its absurd aspects, was to assume a murderous force in the course of the war, when Himmler had the power and scope to combine his unhinged racist theories with practical 'experiments'. At the same time, as the war progressed Himmler would pursue the dream of 'miracle' weapons, pestering scientific institutes to research a catalogue of hare-brained schemes. The phenomenon of Himmler's pseudo-science reveals the prevalence within SS-sponsored science and technology of amateurishness, dilettantism and, in time, sadism and mass murder. Himmler's attempts to create a science special to National Socialism, dedicated to racial origins and genocide, give an indication of what might have become of science had Himmler succeeded Hitler as Führer.

16. *Deutsche Mathematik*

The dismissals policy saw the departure of many distinguished mathematicians to Britain and the United States. In 1934 the Education Minister Bernhard Rust asked the veteran mathematician David Hilbert how Göttingen, formerly one of the world centres of mathematics, had suffered after the removal of Jewish mathematicians. Hilbert replied: 'Suffered? It hasn't suffered, Minister. It doesn't exist any more!'[1]

While some disciplines, like geography, medicine and anthropology, lent themselves easily to Nazification, and even made enthusiastic contributions to National Socialist ideology, the discipline of mathematics appeared a poor prospect for exploitation or collaboration. There was, nevertheless, a mathematician of repute in Germany, Ludwig Bieberbach, who did his level best to bring mathematics in line with National Socialism. We met Ludwig Bieberbach, a lean, dapper individual, as dean of the philosophy faculty at Berlin University, when he encouraged the students' raid on Köhler's Psychological Institute and strove to impose Nazi ideology on faculty and students. His project reveals the extent to which a serious and original thinker could attempt to adapt himself to the prevailing ideology with no conceivable advantage to his discipline.

Bieberbach was a distinguished mathematician remembered for a difficult conundrum known as Bieberbach's Conjecture. Posed in 1916 (and not proved until Louis de Branges's solution in 1985), the conjecture can be applied, for example, to how angles in one spatial domain relate to those in quite different spatial domains; the conjecture is useful in astronomy. Born in 1886, Bieberbach did important work on function theory and had a reputation for being a charismatic but scatty teacher.

Bieberbach became an enthusiastic Nazi very suddenly in the

spring of 1933 as Hitler seized power in the new Reich and anti-Semitism spread in the universities. Bieberbach went over to the Nazis for no very good reason, it seems (other than 'a desire to shine', according to Max Born). Bieberbach's tortuous attempt to put mathematics into the straitjacket of Nazism suggests that he might have been under some kind of personal pressure to conform. This occurred to the British mathematician G. H. Hardy. Writing about Bieberbach's notions in *Nature* in 1934, Hardy made a perceptive comment, which acknowledged the pressures on scientists and mathematicians in times of crisis, while not granting Bieberbach the least benefit of the doubt:

There are many of us, many Englishmen and many Germans, who said things during the War which we scarcely meant and are sorry to remember now. Anxiety for one's own position, dread of falling behind the rising torment of folly, determination at all costs not to be outdone, may be natural if not particularly heroic excuses. Professor Bieberbach's reputation excludes such explanations of his utterances; and I find myself driven to the more uncharitable conclusion that he really believes them true.[2]

Bieberbach joined the SA (storm troopers) in April 1933 and at this point was promoted dean of Berlin's Philosophy Faculty, which incorporated the natural sciences and mathematics. By November he was seen around the university in a Nazi uniform. The following year, after the dismissal of the Göttingen mathematician Edmund Landau, who was Jewish, Bieberbach wrote that his departure signified the important lesson that representatives of different races do not mix. It was doubtless under his influence that a student leader in Bieberbach's circle at Berlin University declared, 'I think it would be better not to teach number theory for ten semesters than to have it taught by a Jew.'[3]

Bieberbach's ideas about German mathematics were based on a theory proposed by the anthropologist Erich Rudolf Jaensch, featuring vague connections between psychology and racial types. Bieberbach thought that Jewish and French mathematicians had a tendency to abstraction rather than to the truth of hard empirical realities and Nordic intuition. Nazi mathematics called for a 'tangible' dimension that could be clearly visualized even when it referred to pure mathematics. In an acidulous essay on 'The Reality of German Mathematics' written in January 1936 a fellow mathematical Nazi spirit, Professor Erhard Tornier, wrote:

Every theory in pure mathematics is justified which is truly capable of answering concrete questions about real objects . . . If not, such theories are . . . a document of Jewish-liberalistic obfuscation, born of the intellect of rootless artistes who juggle with intangible definitions to simulate mathematical creativity to their unthinking captive audience, which is glad if it can slowly learn a few tricks that it can itself use to show off.[4]

Although Bieberbach constantly harped on the theme of the racial expression of mathematics, however, he seemed incapable of articulating a proper description of what this meant, let alone an intelligible basis for the proposition. 'Mathematics,' Bieberbach reported to a meeting of the Prussian Academy of Sciences in 1934, 'may consist of timeless verities, but the ways they are represented, treated and deduced, derive from particular human characteristics.' At the same meeting he opined: 'Mathematical research is a powerful expression of folk characteristics . . . maybe this is less immediately evident than it is in artistic styles as works of mathematics cannot be so readily taken in at a glance.'[5]

Influence of Nazi Mathematics

The attempt to influence mathematics extended right into the schools. Educationist Claus Heinrich Tietjen, in an essay on maths and 'Nordic Man', wrote:

Science teaching therefore brings the school pupil into close contact with two decisive realities of national life: large-scale war and the question of work. These are just the realities which have begotten the National Socialist movement and for which that movement must fight. All the same, little account has been taken of these fundamental issues in the arithmetic books for the Volksschule, and they are also insufficiently brought out in the teaching books for the higher schools. These books are brimful of economic problems and questions about the global economy which breathe much less national spirit. It is both possible and necessary, therefore, to throw out a good deal of ballast which is soon forgotten anyway and is in some cases irrelevant to what is being learnt.[6]

The proponents of *Deutsche Mathematik* were keen to see their discipline taking its place alongside other National Socialist sciences, such as *Deutsche Physik* and racial hygiene: they feared that mathematics would be left behind because of its reputation for being international and value-neutral. Nazi ideology insisted that only those disciplines which directly served the national polity had a right to exist and Philipp Lenard himself had dismissed mathematics as the 'science of counting', declaring that it belonged properly in the humanities. In their enthusiasm to be part of the National Socialist 'revolution', Nazi mathematicians took to writing essays about the link between mathematics and National Socialism, typically emphasizing the importance of devotion to a cause, the need for service, the importance of anti-materialism and order and the rejection of chaos.

The Persecution of Jewish and Dissident Mathematicians

Meanwhile, the fate of Jewish and dissident mathematicians after 1933 was similar to that of other scientists. Ludwig Berwald was deported by the Gestapo to Łódź, where he eventually died in 1942, as did Walter Froehlich; Otto Blumenthal was sent to Theresienstadt, the 'model' camp, much promoted in Nazi propaganda as a humane detention community, where he died in 1944. Paul Epstein, of Frankfurt University, committed suicide after being summoned by the Gestapo in August 1939; Felix Hausdorff of Bonn committed suicide in 1942, when it was no longer possible to avoid being sent to a concentration camp. Robert Remak was arrested on *Kristallnacht* (9–10 November 1938) and detained temporarily in Sachsenhausen concentration camp. In 1942 he was arrested again in Holland and sent to Auschwitz, where he died. Stanislaw Saks and Juliusz Pawel Chauder were killed by the Gestapo in 1942 and 1943. John von Neumann had already departed for the United States in 1930 and settled at Princeton; Richard Courant also crossed the Atlantic, while Max Born went to Edinburgh. Another loss to Hilbert's community of mathematicians at Göttingen was Emmy Noether, probably Germany's most eminent pure mathematician, and a founder with B. L. van der Waerden and others of modern algebra. In 1933 she left Germany for a post at Bryn Mawr in the United States. She died tragically in the course of an operation in the spring of 1935. The *New York Times* published only a brief obituary note on her death, moving Einstein to write an emotional letter to the paper in which he declared: 'In the judgment of most competent living mathematicians, Fräulein Noether was the most significant creative mathematical genius [of the female sex] thus far produced.'[7]

Two extraordinary escapes from Germany were made early in the war by the mathematicians Max Dehn and Kurt Gödel via the Trans-Siberian railway in 1940. Born in 1878 in Hamburg, Dehn was a student of Hilbert who had succeeded to a post held by Ludwig Bieberbach at Frankfurt. He was dismissed from this position in 1935, two years later than many Jewish academics on account of his war service. He was arrested by the Nazis on *Kristallnacht*, but released a day later due to overcrowding in the detention centre, and thence fled to Denmark and so to Norway. When the Nazis invaded Norway in March 1940, Dehn realized that he must escape eastwards, as all routes to the west were closed.

Kurt Gödel was born in Brünn, Moravia, and educated at the University of Vienna, where he took Austrian citizenship. In 1930, at a mathematical conference in Königsberg, he gave the first indication of his famous proof which demonstrated that any system of mathematics will contain propositions which can neither be proved nor disproved, contrary to Hilbert's long-held views. Gödel spent a period at the newly founded Institute of Advanced Study at Princeton during 1933–4, but returned to Austria the following year and fell ill with depression. Finally Gödel and his wife left Austria again for America, crossing Russia on the Trans-Siberian railway in mid-January 1940. Dehn and his wife had made the same journey in late October of the previous year. Gödel took two months to reach San Francisco via Yokohama; the Dehn family travelled to San Francisco via Vladivostok and Japan, taking the same journey time.[8]

PART FOUR

The Science of Destruction and Defence 1933–1943

17. Fission Mania

The science that led to the first nuclear weapons began as an exciting, often hectic drama of basic research. Some of the participants seemed wholly unaware of the implications of their work. Others, against the background of Hitler's regime, seemed painfully conscious of the dangers from the outset.

The story of the discovery of fission and its immediate consequences forms a crucial episode in the chronicle of the successful Anglo-American creation of the first major weapon of mass destruction, and Germany's failure to achieve that dreaded goal. The main elements of the story reveal important aspects of the pressures on the scientists involved: of hubris and competition, eagerness to be first, reluctance to share recognition, evasion of responsibility.

In the spring of 1933, as hundreds of scientists were being dismissed from their jobs in German universities and scientific institutes, Otto Hahn – Lise Meitner's collaborator at the Kaiser Wilhelm Institute for Chemistry – was in the United States lecturing at Cornell and meeting colleagues within his discipline. For the first few months of the Hitler chancellorship, Meitner therefore found herself standing in as director of the institute, assuming responsibility for more than twenty-five scientists, while watching the increased presence of brownshirts in the corridors. She wrote to Hahn a few days after the election of 5 March, indicating cryptically the new state of affairs: 'Today the accounting office ordered us to estimate the cost of our national flag, because it is to be replaced with a black-white-red one which the KWG will pay for.'[1] Two weeks later, she wrote again: 'Everything and everyone is influenced by the political upheavals. Already last week we were notified by the KWG that along with the black-white-red flag, we must also display the swastika . . . It must have been very difficult for Haber to raise the swastika.'[2]

Despite restless nights she kept going on cups of coffee and cigarettes. Among the departures from Germany was her nephew Otto Frisch, her erstwhile piano partner, who had been thrown out of his research job at the University of Hamburg. He was destined for Copenhagen to work in Bohr's physics institute before moving on to Birmingham University in Britain.

Another of her young colleagues, Leo Szilard, of Hungarian-Jewish descent, then aged thirty-five, decided to leave of his own accord, even though there was no pressure upon him as yet to do so. He cancelled a planned lecture series with Meitner and took off for Austria, bound for England, then eventually to the United States, his life's savings hidden inside his shoes. He wrote later:

I left Germany a few days after the Reichstag fire . . . I took the train from Berlin to Vienna on a certain date, close to the 1st of April. The train was empty. The same train on the next day was overcrowded, and was stopped at the frontier. People had to get out, and everybody was interrogated by the Nazis.[3]

Reflecting on this event, he wrote: 'This just goes to show that if you want to succeed in this world you don't have to be much cleverer than other people, you just have to be one day earlier than most people.'[4]

Szilard, a man of remarkable genius and political astuteness, would live out of a suitcase for several years. He immediately set about helping Jewish émigrés, working with Britain's William Beveridge to set up the Academic Assistance Council (later the Society for the Protection of Science and Learning) based in London. In the autumn of 1933, standing at a road intersection near the British Museum, Szilard had an extraordinary scientific epiphany in which he understood the potential for releasing energy from the atom. His insight was to anticipate a technology that would alter the world for ever, endowing the human race with the ability to destroy itself.

Ernest Rutherford, the abrasive New Zealander transplanted to Cambridge via Manchester and Montreal, had proposed an atomic model that resembled a tightly packed nucleus surrounded by orbiting electrons. As his model was increasingly refined by researchers in Europe and the United States, Rutherford sturdily maintained that scientists would not be able to exploit the energy locked within the atom. He was the first scientist, however, to succeed in transmuting elements by bombarding nitrogen with alpha particles (or alpha rays) from radium, which he had 'borrowed' from Austrian friends in Vienna and kept through World War I. Rutherford meanwhile found that when he bombarded nitrogen, its atoms turned into heavier atoms of oxygen and lighter atoms of hydrogen. Alpha particles, being positively charged, could penetrate and transform lightweight atoms which carried nuclei with few positive charges; but they were repelled by heavy nuclei with multiple positive charges in elements like uranium.

Then, the year before Hitler came to power, the English physicist James Chadwick made a crucial contribution to nuclear physics at Rutherford's Cambridge laboratory. In 1932 Chadwick discovered the neutron, an elementary particle that due to its electrically neutral property could penetrate the electric protective barriers of the nucleus of an atom. That same year Fritz Houtermans, noting Chadwick's result, declared in a speech at the Technical Academy in Berlin that the neutron would one day release gigantic forces locked up in matter. But the following year Rutherford told members of the British Association that researchers who expected such energy releases were talking 'moonshine'. One researcher who read Rutherford's assertion, and was instantly challenged by it, was Leo Szilard. That day, 12 September 1933, Szilard was staying at the Imperial Hotel, Bloomsbury, close to the British Museum. He remembered H. G. Wells's *The World Set Free*, published in 1914, in which atomic energy and an atomic bomb are predicted. 'This sort of set me pondering, as I walked the streets of London,' wrote Szilard. 'And I remembered I stopped for a red light at the intersection of Southampton Row . . . I was pondering whether Lord Rutherford might not be proved wrong.' At this point he saw the possibility of a neutron 'chain reaction' that might unleash the energy from within atoms.

As Szilard put it:

It suddenly occurred to me that if we could find an element which is split by neutrons and which would emit _two_ neutrons when it absorbed _one_ neutron, such an element if assembled in sufficiently large mass, could sustain a nuclear chain reaction.[5]

By a 'chain reaction' he meant that if an atom were to be fissioned in a mass of material made of an element that is split by neutrons, there was reason to believe that the two separating parts of the atom would stabilize themselves, and in the process might eject two or more secondary neutrons. Two secondary neutrons might then be prompted to split two more atoms, which in turn would release four additional secondary neutrons, which would then split four additional atoms, which would then release eight secondary neutrons, splitting eight atoms, releasing sixteen neutrons, splitting sixteen atoms, releasing thirty-two neutrons – the process continuing exponentially in a fraction of a second through billions of atoms in the mass of the element: causing a human-devised explosion unprecedented in history.

Szilard spoke to British physicists, but nobody listened. In the end he filed for a patent on the idea, which was granted in Britain on 12 March 1934. Considering the potential for a prodigious weapon of mass destruction, he assigned the patent to the British Admiralty on condition that it should be kept secret.

Szilard, however, was not alone in pondering the possibility of an atomic bomb. Two years later, the French physicist Frédéric Joliot-Curie, when receiving his Nobel award in Stockholm, warned of the potential of nuclear physics for making weapons of mass destruction: 'We are justified in reflecting that scientists who can construct and demolish elements at will may also be capable of causing nuclear transformations of an explosive character.'[6] Ironically, as we shall see, six years later Joliot-Curie would allow his scientific hubris to be first to publish, to get the better of his insights into the danger of an atomic weapon in the hands of Adolf Hitler.

But other developments were in train that were to lead to the experimental preliminary of a viable energy-releasing chain reaction. Through the 1930s Enrico Fermi, the Italian physicist based in Rome, had been systematically researching the properties of neutrons. Fermi was only thirty-two when Hitler came to power, but he had already made a name for himself in the previous decade by developing a statistical method for analysing sub-atomic particles. In consequence he had been nicknamed the Pope of physics, and his laboratory had become a focus of interest for physicists the world over. The young Hans Bethe remarked to his mentor Arnold Sommerfeld that Fermi's lab was of greater importance in the Eternal City than the Colosseum.

Using a glass vial filled with beryllium powder and radon to generate his neutrons, Fermi set about irradiating, one by one, all the chemical elements in the periodic table, transmuting them into radioactive isotopes. Performing experiments in the goldfish pond outside his laboratory, he discovered that the radioactivity of a metal bombarded with neutrons increased a hundredfold when the neutrons had been slowed down by water or paraffin. He also discovered that when he bombarded uranium, the heaviest of all elements, he produced a mixture of radioactive new elements, which he called transuranes, or artificial transuranic elements. In fact, and without knowing it, he had split the uranium atom.

Lise Meitner Stays

Meitner had not followed Szilard out of Germany. Being an Austrian citizen, she was not under duress to leave her post in Berlin, nor did she feel obliged to resign in protest against the persecution of Jewish colleagues. Later, after the war, she regretted that decision: 'I know that it was not only stupid but very wrong that I did not leave at once,' she was to write in 1946.[7]

Ever since Szilard's departure Meitner had been absorbed in a baffling series of experiments. Fermi's claims about transuranes had acted as a stimulus for her, Otto Hahn and their assistant Fritz Strassmann to set about replicating his results in Berlin (Strassmann would later shelter a fugitive from the Nazi regime at risk to his own life). At the same time, two physicists, the married couple Ida and Walter Noddack, working at Freiburg im Breisgau, had come up with a surprising interpretation of their own. Writing in 1934 in the German science journal *Zeitschrift für Angewandte Chemie* (*Journal of Applied Chemistry*), Ida Noddack suggested that when heavy nuclei are bombarded with neutrons, the nuclei in question might break into a number of large pieces. She was speculating, in other words, that neutrons of an energy less than a single electron volt might have split an atomic nucleus capable of repelling bombardments at millions of electron volts. Fermi thought the suggestion nonsensical; Otto Hahn and Lise Meitner tended to agree with him. The scepticism of both parties was a token of the widely held view among the potential atom breakers that special equipment, such as a cyclotron, would be essential for the task. The cyclotron was a device for accelerating charged particles using high voltage difference in a high-intensity magnetic field. The van de Graaff generators and cyclotrons were capable of hurling particles as projectiles at energies of up to 9 million electron volts, but the electric charge of the nucleus had still repelled even these highly charged bombardments. It was like suggesting that a peashooter could destroy a fortification capable of resisting bombardment with explosive shells. Frau Noddack reported some years later that she attempted to persuade Otto Hahn that some reference at least should be made in his lectures and publications to her criticism of Fermi's experiments and her speculation that his neutrons had split the atom. Hahn apparently responded that he did not want to make her look ridiculous.[8]

Meanwhile in their patient and systematic study of Fermi's 'transuranic elements', the Hahn-Meitner-Strassmann team in Berlin found many strange products emerging from uranium. Their results were published in various journals over the next three years. In the spring of 1935, for example, they were reporting substances with three different half lives, and finding it difficult to come up with an explanation that made sense. They were fumbling in the dark.

All this time, as Meitner well knew, Jews in Germany were being subjected to mounting persecution. But she was still thinking in a dissociated fashion of her personal safety and the path of her career in Berlin, rather than the shame and degradation brought on science by the regime. She reasoned that under the protection of Max Planck she could come to no personal harm. She seemed to be distracting herself in a dedicated routine of work.

In 1936 she and Hahn were nominated by Planck for the Nobel prize in chemistry for their work in the field. Later Planck unsuccessfully nominated Meitner alone for a Nobel prize. It seems that Heisenberg, Planck and Laue realized that the prize might offer her special protection in addition to her Austrian nationality. It was Planck's last gesture of support. In 1937, aged seventy-nine, he resigned his presidency of the Kaiser Wilhelm Society, to be replaced by Carl Bosch, Fritz Haber's erstwhile collaborator in the nitrogen-fixing process.

Meitner Flees

In 1938, on 12 March, the German army entered Austria, enforcing the annexation, or *Anschluss*, thus making all Austrians citizens of the Third Reich and subject to the Nazi race laws in Germany. Meitner could no longer depend on her Austrian nationality to protect her. The story of her eventual departure from Germany gives a stark impression of the menacing climate created by the Nazi police state.

She had received invitations from various colleagues to work abroad; but she delayed coming to a decision and continued her highly focused work at the institute despite the mounting dangers. But by now her Austrian passport was invalid and she could not obtain a German one, as her status had been overtaken by a new law which refused permission for technical specialists to leave the Reich. Kurt Hess, a member of the institute charged with

hospitality for visiting scholars, was quick to threaten anyone who
spoke out on behalf of Meitner. Even Hahn became cautious in
her defence,[9] informing her that she should no longer be a member
of the institute but might continue her work there 'unofficially'.
Meitner would never completely forgive Hahn for his pusilla-
nimity.

Carl Bosch eventually came to her aid by applying for a permit
for foreign travel. There had been invitations for her to go to
Holland, to Bohr's institute in Copenhagen, to Liverpool, to Cambridge. Bosch's application dated 22 May noted: 'Miss Meitner is
non-Aryan; nonetheless, with the concurrence of all the official
departments, she has been allowed to retain her position as she is
engaged in important scientific discoveries.'[10] Eventually Bosch's
application was turned down in menacing terms. Meitner copied
the Kafkaesque text of the refusal on the stationery of the Adlon
Hotel, where she had now moved for safety.

There are political objections to the issuing of a passport to Professor
Meitner. It is considered undesirable that renowned Jews should leave
Germany for abroad to act there against the interests of Germany according to their inner persuasion as representatives of German sciences or
even with their reputation and experience.[11]

She could not work in Germany as an active scientist, and yet she
could not leave the country *because* she was a scientist.

On 13 July, after failing to secure funds to establish Meitner
long-term in Holland, the Dutch physicist Dirk Coster of Groningen University 'rescued' her, bringing her to Holland with the
promise of an appointment at his university for at least one year.
On the eve of her departure she worked a normal day so as not to
arouse suspicion. She packed two small cases with her essential
belongings and spent a sleepless night at the home of Otto Hahn.
Germans travelling abroad were not allowed to take more than ten
marks out of the country, so for emergencies Hahn gave her a
diamond ring he had inherited from his mother.

Although Coster was waiting at the station at Berlin, he could not make contact with her for fear of raising the suspicions of watchful Gestapo agents. The train took seven hours to reach the border with Holland. Days before, Coster had been to the border and spoken with immigration officers, showing them Meitner's official entrance permit and asking them to use 'friendly persuasion' with the German guards to let her through. As a result she managed to cross the border safely. When the train at last arrived in Groningen, Meitner and Coster met each other and he took her to his car. Coster, as agreed, sent a telegram to Hahn saying that the 'baby' had arrived.

Coster in the event could not find her a proper appointment, so she proceeded to Stockholm, where she had been offered a place at Manne Siegbahn's new laboratory and where she could work not far from Niels Bohr in neighbouring Denmark. The move probably saved her life, since by 1942 Jews in Holland were being sent in large numbers to the death camps.

Fission Discovered

Before Meitner's departure from Berlin she and Hahn and Strassmann had been intrigued by what looked to be a mysterious new radioactive element which resulted from the irradiation of uranium with neutrons in the Paris laboratory of Irène Curie (Marie's daughter) and Pavel Savitch. A few weeks later, Curie and Savitch published an article claiming that the element behaved as if it were a radioactive isotope of lanthanum, about half the weight of uranium, which suggested that the irradiated uranium atoms had been split.

By the time Meitner had arrived in Sweden, Hahn and Strassmann were attempting to replicate the Curie and Savitch experiments. They irradiated a sample of uranium with neutrons and detected traces of radioactivity which appeared to come from elements chemically similar to radium, but its activity halved in hours. They then dissolved the irradiated uranium in acid and added a salt of barium, a non-radioactive element lighter than radium, and observed that the barium carried the new radioactivity with it, leaving the uranium behind in solution. The significance was that the uranium atom had been split and that one of the pieces was a radioactive isotope of barium. Hahn and Strassmann could not bring themselves to believe what they had evidently achieved, so they continued to test the result in other ways.

Meanwhile, in a letter to Meitner, dated 19 December, and datelined 'Monday evening, in the lab', Hahn informed her of the strange result they had achieved: 'Actually there is something about the "radium isotopes" that is so remarkable that for now we are telling only you.'[12] He wanted her to propose something that perhaps she might publish. It was his way of acknowledging her participation in the momentous discovery that was staring them in the face, at the same time making it clear that a joint paper with a Jewish scientist was out of the question.

Two days later he was writing to Meitner again, asking if she could come up with an explanation, since 'we cannot suppress our results, even if they are perhaps physically absurd'.[13] That same day Meitner was writing to Hahn saying that she could not believe that a complete rupture of the uranium nucleus had occurred. 'But in nuclear physics we have experienced so many surprises, that one cannot unconditionally say: it is impossible.'[14]

Hahn had been completing an article on his experiments for the journal *Die Naturwissenschaften*, which declared:

As chemists the experiments we have briefly described force us to substitute for the [heavy] elements formerly identified as radium, actinium, thorium the [much lighter] elements barium, lanthanum and cerium, but as 'nuclear chemists' close to physics we cannot yet take this leap which is contrary to all experience of nuclear physics.[15]

On receiving Hahn's most recent letter, Meitner went to join her nephew Otto Frisch for a Christmas vacation with friends, the von Bahr-Bergiuses, at Kungalv on the Swedish coast. Travelling there from Stockholm she could think of nothing, she later confessed, but the appalling probability that Fermi's transuranic theory, and along with it her own four years of intense research with Hahn and Strassmann, had been in vain.

In his memoir Frisch remembered how he came out of his hotel to find his aunt Lise studying the letter from Hahn and clearly fretting over it. Frisch was eager to tell her about his own scientific preoccupations, but she would not listen. She insisted that Frisch read the letter there and then, which he did. 'It's content was so startling,' he remembered, 'that I was at first inclined to be sceptical.' Frisch's recollection represents one of the great insightful moments in science in the twentieth century:

Was it just a mistake? No, said Lise, Hahn was too good a chemist for that. But how could barium be formed from uranium? No larger fragment than protons or helium nuclei (alpha particles) had ever been chipped away from nuclei, and to chip off a large number not nearly enough energy was available. Nor was it possible that the uranium nucleus could have been cleaved right across. A nucleus was not like a brittle solid that can be cleaved or broken; George Gamow had suggested early on, and Bohr had given good arguments that a nucleus was much more like a liquid drop. Perhaps a drop could divide itself into two small drops in a more gradual manner, by first becoming elongated, then constricted, and finally being torn rather than broken in two? We knew that there were strong forces that would resist such a process, just as the surface tension of an ordinary liquid drop tends to resist its division into two smaller ones. But the nuclei differed from ordinary drops in one important way: they were electrically charged, and that was known to counteract the surface tension.[16]

Meitner insisted that they talk through the implications of Hahn's letter as they walked in the snowbound woods. Frisch wanted to go cross-country skiing, so they set off together, Frisch on wooden skis and Meitner walking briskly beside him. At one point they sat down on a snow-covered log and Meitner pulled out some scraps of paper and started to do calculations. Frisch remembered:

The charge of uranium nucleus, we found, was indeed large enough to overcome the effect of the surface tension almost completely; so the uranium nucleus might indeed be a very wobbly, unstable drop, ready to divide itself at the slightest provocation, such as at the impact of a neutron. But there was another problem. After separation, the two drops would be driven apart by their mutual electric repulsion and would acquire high speed and hence a very large energy – about 200 million volts in all. Where could that energy come from?[17]

Meitner had the necessary insight for a solution. Recalling the formula for computing the masses of nuclei, she worked out that the two nuclei formed by the division of a uranium nucleus together would be lighter than the original uranium nucleus by about one-fifth the mass of a proton. Whenever mass disappears energy is created, and one-fifth of a proton mass was exactly equivalent to the right amount of energy. 'So here,' wrote Frisch in his memoir, 'was the source for that energy; it all fitted.' In order to achieve their calculation of the energy equivalent to the loss of one-fifth of a proton from one atom of uranium, Meitner employed Einstein's famous equation $E=mc^2$, and was amazed by the result – 200 million electron volts! It showed that the splitting of 1 gram (1/28 of an ounce) of uranium would release energy equivalent to 2½ tonnes of coal. Why had this gone unnoticed in the experiments of Fermi, Hahn and Strassmann and the Paris team – Curie and Savitch? Since only a small number of atoms of minuscule amounts of uranium had split in their experiments, the energy released had clearly gone unnoticed.

On his return to Copenhagen Frisch conducted an experiment in order to measure the force of the fragments resulting from the irradiation of uranium with neutrons and could demonstrate that it was equivalent to Meitner's calculations. Commenting on the result, the late Max Perutz suggested that Frisch's experiment confirmed Karl Popper's ideas on scientific method. 'The violence of the reaction,' wrote Perutz in an essay on the splitting of the atom, 'had remained unnoticed without a hypothesis predicting it; and Frisch detected it by an experiment designed to falsify the hypothesis.'[18]

Meitner and Frisch now wrote two letters to *Nature*. The first, signed by them both, offered a theoretical interpretation of Hahn's and Strassmann's results.[19] The second, by Frisch, published a week later, described his own experiment.[20] The letters argued that the 'transuranes' described by Fermi, and then by Hahn, Meitner and Strassmann, had been the products of the splitting of uranium, although they used the term 'fission' rather than 'splitting'. There was no suggestion about the prospects for harnessing the huge energy involved in the discovery, nor of the potential for weapon making.

Bohr Goes to America

On 3 January 1939 Otto Frisch met with Bohr at his house in Stockholm and reported the news of the splitting of uranium. The conversation lasted only five minutes and Bohr was astonished that he had not considered the possibility earlier. He readily agreed that the disintegration of a heavy nucleus into two larger pieces was an almost classical process that does not occur at all below a certain energy but occurs easily a little above it. On 7 January Bohr and Leon Rosenfeld, a 33-year-old professor of physics from the University of Liège, set sail for New York to begin a five-month stay in the United States at the Princeton Institute for Advanced Study. Bohr had a blackboard installed in his cabin so that he could spend the voyage refining the implications of Frisch and Meitner's proposed joint letter to *Nature*. Rosenfeld would remember:

We had bad weather through the whole crossing, and Bohr was rather miserable, on the verge of seasickness all the time. Nevertheless, we persevered for nine days, and before the American coast was in sight, Bohr had a full grasp of the new process and its main implications.[21]

Bohr and Rosenfeld were now instrumental in spreading word about the fission discovery. The story of their activities and its consequences forms an interesting an instructive tale about how scientists behave: their competitiveness, the pressures that conflict at many points with higher ideals and the need to curb their personal vanities in times of international crisis.

Rosenfeld, much to Bohr's annoyance, jumped the gun by talking about Hahn's result at a club meeting of the Princeton Physics Department. But since the discovery was now being openly discussed, on 26 January Bohr and Fermi (who now had a post at Columbia, New York) addressed the Fifth Washington Conference on Theoretical Physics. They informed the group that Hahn and Strassmann in Berlin had indeed achieved radiochemical confirmation that barium was produced when uranium was bombarded by neutrons, and that Frisch and Meitner had now concluded that these results indicated 'nuclear splitting', releasing a tremendous amount of energy. Among the fifty or so members at the meeting were Gamow, Teller and Bethe. Even before Rosenfeld had finished his presentation physicists were rushing from the room: some to telephone the news to colleagues, others to attempt to replicate the experiments in their own laboratories.

Shortly afterwards, Fermi gave a talk on the radio about fission without mentioning Frisch. Bohr was furious. Rosenfeld reported: 'That was the only time in which I saw Bohr really angry and quite burning with passion.' Bohr, accompanied by Rosenfeld, went to see Fermi to have it out with him. 'I did not witness the interview,' Rosenfeld recollected. 'I saw only their faces when they came out – it was a long time – and both of them were quite pale, harassed, quite exhausted.'[22] Bohr, the man of legendry integrity, was doing no more than enforcing the ethic of proper attribution, a matter of crucial significance in scientific protocol.

At least one scientist saw the military and therefore political implications of the discovery and was shocked to the core. Leo Szilard heard of the fission experiment a few days later when visiting the physicist Eugene Wigner at Princeton. Uranium was clearly the element that he had been speculating about ever since he had paused at that traffic light in London six years earlier. 'When I heard this,' he wrote later,

I saw immediately that these fragments being heavier than corresponds to their charge, must emit neutrons, and if enough neutrons are emitted in this fission process, then it should be, of course, possible to sustain a chain reaction. All the things which H. G. Wells predicted appeared suddenly real to me.[23]

It was essential, he believed, that any further results along these lines should be kept strictly secret, since the world was on the brink of war. He wrote to the British Admiralty, reversing his request before Christmas that his patent on a chain reaction bomb should be released. Then on 25 January he wrote to Lewis L. Strauss, a Wall Street businessman, predicting that fission heralded nuclear energy and more serious 'potential possibilities' leading to 'atomic bombs'.

The following week Bohr went on to Princeton, where he was given temporary lodging in Einstein's office. During a conversation over breakfast on 5 February Bohr realized that the observed fission in uranium was mainly due to the rare isotope U-235, which is present in uranium in a proportion of one part to 139 parts U-238. Two days later, on 7 February, Bohr posted a letter to the *Physical Review* remarking on this fact. Bohr proposed in this and a later paper, published in August with John Wheeler, that U-238 would be an unfeasible source of power, but not U-235, which would be easily fissioned by slow neutrons (the insight led physicists to focus at first on slow neutrons in U-235, rather than on fast neutrons – as occurred later). Bohr nevertheless took part in a discussion at Princeton that month in which he spoke of the difficulty of separating U-235 from natural uranium. 'It would take,' he said, 'the entire efforts of a country to make a bomb.'[24]

The thrilling news of the nuclear fission had nevertheless exerted an impact on many individuals who had contributed to the era of quantum physics and early research into nuclear physics. Within a week of the official discovery and publication, J. Robert Oppenheimer, the polymath New Yorker who had studied under Max Born and shared digs with Paul Dirac at Göttingen, had chalked up on his office blackboard, according to American physicist Philip

Morrison, 'a drawing – a very bad, an execrable drawing – of a bomb'. Meanwhile George Uhlenbeck heard Enrico Fermi say, as he looked down the length of Manhattan from Columbia's physics tower: 'A little bomb like that,' and he cupped his hands, 'and it would all disappear.'

Secrecy, Publishing and Funding

Szilard had understood that the critical next step in research was to confirm how many neutrons are emitted in fission, leading to the potential for a chain reaction and hence a weapon of mass destruction. He knew that two groups of experimental physicists were working on that problem: Enrico Fermi and himself at Columbia University, New York City, and a French group under the physicist Frédéric Joliot-Curie in Paris. Given the amorphous, often unpredictable developments in atomic physics, and given the fact that the world stood on the brink of a war about to be unleashed by Germany's ruthless dictator, Szilard believed that every new discovery in atomic physics was attended by future peril. If discovery of the release of secondary neutrons was essential to building an atomic bomb, he reasoned, it was crucial to keep the discovery out of the public domain. German scientists, after all, were as capable of reading British and American scientific journals. With no way of knowing whether Germany's scientists were already working on secondary neutrons, or chain reactions, or even thinking about the possibility of an atomic weapon, he nevertheless judged that scientists in the democratic world should not publish anything that would assist Hitler's nuclear efforts.

The incidents that followed exemplify the conflict between the pressure for scientists to publish and the imperative for a scientist to deprive a tyrant of the intellectual means to blackmail his enemies. To what extent is a scientist responsible for the misuse of his knowledge by others? Those who doubt that basic science is value-free, morally and politically neutral, may well consider whether Szilard's fears about the effect of publication were justified and whether the scientists involved acted with integrity or otherwise.

Leo Szilard implored both Fermi, in person, and Joliot-Curie, by correspondence, not to publish their results if they found the secondary neutrons they were looking for. 'Obviously, if more than one neutron were liberated, a sort of chain reaction would be possible,' he commented in his letter to Joliot-Curie. 'In certain circumstances this might then lead to the construction of bombs which would be extremely dangerous in general and particularly in the hands of certain governments.' He ended his letter: 'In the hope that there will not be sufficient neutrons emitted by uranium, I am . . .' But he deleted this line.[25] Fermi, who was working with Szilard, agreed not to publish. But Joliot-Curie and his group were non-commital.

In the event, in mid–March 1939 Fermi, Szilard and a post-doctoral physicist, Walter Zinn, and, simultaneously, Joliot-Curie's group in Paris observed the additional neutrons that confirmed Szilard's theory of chain reaction. Their results showed that in addition to the heavy isotopes created as fission products, more than two neutrons on average (2.42 for U-235) were emitted for each neutron absorbed to make fission. A chain reaction was therefore possible, and, although there were still many problems to be solved, an atom bomb was no longer purely in the realms of the imagination. Despite Szilard's pleas, however, Joliot-Curie went ahead and published his observation in the 18 March issue of *Nature* (Joliot-Curie claimed years later that he had awaited a further message from Szilard following a subsequent garbled cable: but Szilard's first letter, quoted above, seems clear enough in its pleading).

The consequences of Joliot-Curie's *Nature* article in Germany were not discovered until after the war, but the train of events justified the principle to which Szilard was appealing: that there are times when scientific rivalry should be tempered by overriding political considerations beyond science.[26]

It so happened that at Hamburg University in Germany there was the 37-year-old physical chemist Paul Harteck. Harteck's research – he was something of an expert on neutrons – had been languishing, starved of funding. He had spent a year in Cambridge in 1932 working with Ernest Rutherford and Marcus Oliphant on nuclear fusion, which involves the combining of light nuclei. After its publication on 18 March 1939 Harteck read Joliot-Curie's *Nature* the article and immediately saw the prospect of an unusual opportunity for himself and his research. He decided to approach the weapons research office of the German Army Ordnance (the Heereswaffenamt) in Berlin to inform them of the possibilities of making a weapon of mass destruction from uranium fission, and hence to solicit funding for his personal research towards that end.

On 24 April Harteck, and his assistant Wilhelm Groth, wrote to Erich Schumann, head of the Ordnance research office:

We take the liberty of calling to your attention the newest developments in nuclear physics, which, in our opinion, will probably make it possible to produce an explosive of many orders of magnitude more powerful than conventional ones . . . That country which first makes use of it has an unsurpassable advantage over the others.[27]

What would induce a physical chemist, an Austrian who claimed to be apolitical, not a member of the Nazi Party, and not especially interested in weaponry (although he seems to have acted as a consultant at some point on chemical explosives to Army Ordnance), to put into the minds of Hitler's weapon makers the idea of making an atomic bomb? After the war Harteck claimed that his only motive was the purely opportunistic quest for scarce funding. He applied to the army not out of patriotism, nor because he supported the Nazi cause, but because the army had plenty of money. 'In those days in Germany we got no support for pure science . . . So we had to go to an agency where money was to be obtained. I was always realistic about such things. The War Office had the money and so we went to them.'[28]

As it happened, Army Ordnance was to take its time to get back to Harteck, but the fact remains that matters involving the fate of entire nations and continents can at times attend the behaviour of scientists. There was a direct connection between Joliot-Curie's article in *Nature* and Paul Harteck's determination to seek funding from whatever source, and with whatever consequences. That the connection did not culminate in Hitler getting the bomb ahead of America has no bearing on the validity of Szilard's arguments. As it happened, other factors were at work that would prove decisive in the race for the atom bomb.

Szilard, Einstein and Roosevelt

While research on the most appropriate means of creating a chain reaction with uranium went forward in the United States, Leo Szilard remained anxious. Of all the scientists in what was to become the Allied camp, after Pearl Harbor, Szilard saw the future significance of allowing Hitler alone to possess an atomic bomb. Since he could not control the tongues and pens of scientists, he could hardly be surprised when the American mass media gave full vent to the news about fission, neutrons and chain reactions. On 30 April 1939, abandoning all caution, the *New York Times* carried an assessment of the significance of atom research developments and the disagreements that reigned between the scientists:

Tempers and temperatures increased visibly today among members of the American Physical Society as they closed their Spring meeting with arguments over the probability of some scientist blowing up a sizable portion of the earth with a tiny bit of uranium, the element which produces radium.

Dr Niels Bohr of Copenhagen, a colleague of Dr Albert Einstein at the Institute for Advanced Study, Princeton, N.J., declared that bombardment of a small amount of the pure isotope U235 of uranium with slow neutron particles of atoms would start a 'chain reaction' or atomic explosion sufficiently great to blow up a laboratory and the surrounding country for many miles.

Many physicists declared, however, that it would be difficult, if not impossible, to separate isotope 235 from the more abundant isotope 238. The isotope 235 is only 1 per cent of the uranium element.

Dr L. Onsager of Yale University described, however, a new apparatus
in which, according to his calculations, the isotopes of elements can be
separated in gaseous form in tubes which are cooled on one side and
heated to high temperatures on the other.

Other physicists argued that such a process would be almost prohibi-
tively expensive and that the yield of isotope 235 would be infinitesimally
small. Nevertheless, they pointed out that, if Dr Onsager's process of
separation would work, the creation of a nuclear explosion which would
wreck as large an area as New York City would be comparatively easy.
A single neutron particle, striking the nucleus of a uranium atom, they
declared, would be sufficient to set off a chain reaction of millions of
other atoms.

Since some of the top physicists, including Heisenberg, Weizsäcker and Otto Hahn, were in Germany, and since information
about the latest research discoveries was leaking continually into
the public domain, what should be done to protect the democratic
world from future calamity?

Even while Szilard was taking a lead in raising funds to conduct
experiments leading to a chain reaction, he was determined to take
positive action in the political and international sphere. Fearing that
Hitler would confiscate Belgian reserves of uranium (mined from
the Belgian Congo), he formed a plan to enlist Einstein to forewarn
Queen Elisabeth of Belgium. But events moved with agonizing
slowness.

It was not until mid-July that Szilard and Wigner received an
invitation to drive to Einstein's summer house on Long Island,
where they informed him of the latest scientific developments. On
their arrival Szilard described to Einstein his recent experiments at
Columbia and his calculations towards a chain reaction in uranium
and graphite. At this point, according to Szilard, Einstein expressed
his surprise as he had neither heard nor thought of the possibility
of a chain reaction. He said: 'Daran habe ich gar nicht gedacht!' (I
never thought of that!)[29] But he was quick to see the implications,
and ready to participate in issuing warnings even if the potential for
a weapon of mass destruction was a false alarm. The three scientists

drafted a letter, which they believed they should first address to the Belgian Ambassador in Washington, with a covering letter to the State Department.

Within three days, and before sending the letter, Szilard sought the advice of Alexander Sachs, a biologist, national economist and businessman. Sachs was insistent that here 'were matters which first and foremost concerned the White House and that the best thing to do, also from the practical point of view, was to inform Roosevelt'. The famous letter from Einstein to Roosevelt on the topic of the atom bomb was dated 2 August 1939: the opening paragraph reads:

> Sir:
>   Some recent work by E. Fermi and L. Szilard, which has been communicated to me in manuscript, leads me to expect that the element uranium may be turned into a new and important source of energy in the immediate future. Certain aspects of the situation seem to call for watchfulness and, if necessary, quick action on the part of the Administration. I believe, therefore, that it is my duty to bring to your attention the following facts and recommendations.

In the event, Einstein's plea, via Alexander Sachs, was not presented to Roosevelt until 11 October 1939, almost six weeks after Hitler's invasion of Poland and Britain's and France's declaration of war on 3 September. Present at the meeting was General Edwin M. Watson, nicknamed 'Pa', Roosevelt's aide, along with various intimates, his own executive secretary and a military aide. Roosevelt greeted Sachs with 'Alex, what are you up to?'

Sachs started with a story of the young American inventor, Robert Fulton, who wrote a letter to Napoleon proposing a fleet of ships without sails that could attack England whatever the weather. Napoleon dismissed the idea: 'Away with your visionists!' Lord Acton, the nineteenth-century English historian, cited this episode as an illustration of French short-sightedness which had worked to Britain's advantage. Had Napoleon showed a little more vision, the history of Europe might have been very different.

With this Roosevelt sent for a bottle of Napoleon brandy and poured a glass for his visitor and one for himself. Sachs had prepared a file containing Einstein's letter and a memorandum by Szilard. He then decided to make his own 800-word presentation, emphasizing the prospects for nuclear energy, radioactive materials for medical use and finally 'bombs of hitherto unenvisaged potency and scope'. The President, he concluded, should take measures to secure the Belgian uranium reserves and encourage research to be financed by industry and private foundations. A committee should be formed to liaise between the scientists and the administration. He ended by quoting from Francis Aston's 1936 lecture, 'Forty Years of Atomic Theory':

Personally I think there is no doubt that sub-atomic energy is available all around us, and that one day man will release and control its almost infinite power. We cannot prevent him from doing so and can only hope that he will not use it exclusively in blowing up his next door neighbor.[30]

'Alex,' Roosevelt said, 'what you are after is to see that the Nazis don't blow us up.'

'Precisely,' said Sachs.

Whereupon the President called on Watson. 'This requires action,' he said tersely.[31]

It would take another three years for that action to be realized.

18. World War II

The path to World War II had begun as early as 1935, when Hitler began to rearm in contravention of the Versailles Treaty, before marching unopposed into the 'demilitarized Rhineland'. By 1938 Hitler had forced through the unification of Germany with Austria, also forbidden by the terms of Versailles, and in the autumn seized part of Czechoslovakia on the pretext of bringing lost territory and German populations back into the Fatherland. Britain agreed to the carve-up of Czechoslovakia in the hope that Hitler would be appeased. But his demands only increased. He seized what was left of Czechoslovakia in March 1939 and in the summer made a pact with the Soviet Union to divide Poland. Hitler invaded Poland brutally and with great loss of civilian life on 1 September 1939. France and Britain declared war on 3 September; but there were no hostilities until the following spring, when Germany attacked Norway and Denmark in response to the blocking of Scandinavian waters by the Royal Navy. British troops failed to thwart the German invasion. The next month Hitler's mechanized forces outflanked the French defences, launching a powerful, high-speed, well-coordinated attack through the Low Countries and thence through France. Britain's fleeing forces narrowly escaped destruction in the Dunkirk evacuation. The armistice between Germany and France was signed on 22 June 1940. Hitler was now master of Europe from the Pyrenees to the North Cape. Britain's plight was perilous; and it would soon become abundantly clear that her fate would depend on technology rather than superior force of arms and men.

In Germany, however, the outbreak of war did not see the recruitment of scientists into appropriate war work. Students of science, postgraduates and senior scientists of every kind, along with engineers and technical experts, were called into the German armed services to fight, with no thought for the repercussions this would have on a war that was to be technologically demanding. There were, however, some exceptions, in particular Heisenberg, who had spent the summer months in the United States. He was required to join the German uranium research effort, which had been in preparation from the spring of 1939.

Heisenberg Mobilized

In the weeks that preceded the outbreak of war, Werner Heisenberg's determination to stay in Germany despite Hitler's regime had become clear as he travelled around America on his regular summer lecture and seminar tour. That year he took in New York, Chicago, Ann Arbor and Indianapolis, and was in frequent debate with peer group physicists on scientific issues. By a stroke of irony the two future leaders of the world's first atomic bomb programmes found themselves locked in dispute when Heisenberg and J. Robert Oppenheimer quarrelled over a technical detail about electrons at the University of Chicago.

Conflicting motives were given by Heisenberg for declining invitations to become an émigré. According to the physicist I. I. Rabi, Heisenberg proffered a less than noble reason. Heisenberg told Rabi that he was afraid that if he emigrated to the United States he would lose his place in the pecking order of German academia.[1] On other occasions he argued that Germany would need him when the Hitler regime came to an end. At Ann Arbor, according to Heisenberg's own recollections, he told Enrico Fermi that in Germany he had gathered together a group of young physicists:

If I abandoned them now, I would feel like a traitor . . . I don't think I have much choice in the matter. I firmly believe that one must be consistent . . . People must learn to prevent catastrophes, not to run away from them. Perhaps we ought even to insist that everyone braves what storms there are in his own country.[2]

It appears that in the summer of 1939 Heisenberg also discussed the possibility of an atomic bomb and the role of the scientist in war. In his post-war memoir Heisenberg remembers observing that in wartime scientists are expected by their respective governments to devote all their energies to building new weapons. But he was convinced in 1939 that the war would be over long before the first atomic bomb could be built. This was clearly a reflection on what he knew to be the technical difficulties at the time. Some old friends noted that Heisenberg assumed that in the event of war a German victory was a foregone conclusion.[3] In the last week of July Heisenberg left a hot and humid New York to set sail for Germany. The following month, the Third Reich was at war.

Throughout the summer of 1939, as the storm clouds of war gathered, Germany's military scientists had been keeping track of the fission discovery and the consequent excitement in the media in the United States. On 30 April 1939, the *New York Times* carried its story proclaiming to the world that the bombardment of a 'small amount of the pure isotope U235' with slow neutron particles of atoms would prompt an explosion sufficient to blow up 'the surrounding country for many miles'.

Meanwhile, in Berlin on the previous day, Abraham Esau, a science bureaucrat and physicist in Bernhard Rust's Reich Education Ministry, convened a meeting to set up a 'uranium club' (*Uranverein*). Esau had been impelled to action by physicists at Göttingen University, who noted the potential for nuclear power in uranium; but the Ministry was not the only Berlin authority interested in the question of the potential of nuclear fission and uranium.[4] As we have seen, on 24 April Paul Harteck had written his letter seeking funding from the research office of Army Ordnance. The army had not been inactive; indeed it had received another approach broaching the possibility of uranium as an explosive from a Nikolaus Riehl, head of scientific research at the Auer Company. (Auer collaborated with the Degussa company of Frankfurt in supplying uranium: Degussa exists to this day and until recently was performing that service for Iraq.)[5] In the event, the army was to take charge of nuclear fission research in its first phase, and oversee Esau's proposed uranium club, demonstrating, as Mark Walker has commented, 'the pecking order of science policy in National Socialist Germany'.[6]

There were two experts in atomic physics at Army Ordnance – Kurt Diebner and Erich Bagge – who convened an initial meeting of the newly formed uranium club in Berlin on 16 September 1939. Ten days later a second, more crucial meeting took place at the office of Army Ordnance on Hardenberg Strasse, Berlin, which included Werner Heisenberg, Otto Hahn, Hans Geiger, Carl Friedrich von Weizsäcker (protégé of Heisenberg), Paul Harteck and an assorted group of nuclear physicists, military men and science bureaucrats. Heisenberg, still young enough to be called into the army, had been ordered to Berlin under the mobilization decrees.

The scientists discussed uranium exploitation, from the point of view of both energy and weapons, and concluded that more research, theoretical and experimental, was needed before aims and means could be defined. Meanwhile, Army Ordnance had informed the general secretary of the Kaiser Wilhelm Society that the Institute for Physics in Berlin was to be requisitioned for war work (its director, Peter Debye, a Dutch national, was forced to relinquish his direction of the institute). The work and the personnel were to focus on nuclear research under the title Nuclear Physics Research Group. Hence the independence of the institute was from the beginning of the war sacrificed to the Nazi state's war aims (a process that had been seen in the First War at Fritz Haber's Institute for Physical Chemistry). As Mark Walker observes, the Kaiser Wilhelm Society, under its new spirit and auspices, also formed links with industry, establishing in microcosm a classic example of the industrial–military complex. Carl Bosch (of Haber-Bosch fame), the IG Farben executive, took over from Planck as president of the Kaiser Wilhelm Society, soon to be followed by Albert Vögler, boss of one of the biggest steel companies in Germany.[7] The uranium club was nevertheless fragmented and not strongly led, its aims not well defined. Diebner did not have a high reputation among the civilian scientists, and in the whole course of the war the strength of the uranium team did not exceed a hundred scientists. Paul Harteck, alone among the scientists it seemed, appreciated the vast industrial requirements to bring such a project to a successful conclusion.[8]

Diebner was appointed administrative head of the KWI for Physics with loose but overall responsibility for the project. Some of the scientists, including Weizsäcker, were to be based at the institute in Berlin and at an army laboratory at Gottow. Others would commute from their bases in eight different locations in Germany – Paul Harteck, for example, from Hamburg, Walther Bothe from Heidelberg and Heisenberg from Leipzig, where he retained his chair in physics. Heisenberg's journey between Leipzig and Berlin took two hours each way. In addition to the dispersal of the personnel, with many hours spent in travelling, there is evidence that, with some exceptions, members of the different groups did not hold each other in high regard.[9]

An insulated shed was built in the grounds of the Kaiser Wilhelm Institute for Biology, close to the Physics Institute, to house a radioactive cylindrical reactor model: it was called 'Virus House' in the hope that the threat to health implied in its title would deter the curious. Meanwhile a second reactor model was prepared at Heisenberg's institute in Leipzig.

The German scientists were undoubtedly in possession of all the nuclear research data and theory in the public domain on both sides of the Atlantic since the Hahn and Strassmann discovery of nuclear fission. The Third Reich, moreover, now enjoyed possession of the world's largest uranium reserves in the Joachimsthal mines in occupied Czechoslovakia. The Nazi state had taken a world lead in establishing a nuclear research programme directly under military auspices, with Heisenberg a crucial figure in that project. Their principal aims, however, exposed a significant deficit. The nuclear research programme focused on two areas: isotope separation and the construction of a reactor. At the same time Walther Bothe, a top experimental physicist in Germany, was intent on measuring properties of nuclear reactions, which required a cyclotron. When he did a survey of cyclotrons throughout the world, he found that there were nine in the United States, with a further twenty-seven under construction, but none in Germany. He would not have a German cyclotron at his disposal until 1944.

The Position Paper

Heisenberg threw himself enthusiastically into writing up a position paper on the theory and applications of nuclear fission. Three months after the September meeting in Berlin, he filed the first of two top secret reports entitled 'The Possibility of the Technical Acquisition of Energy from Uranium Fission', dated 6 December 1939. He could confirm the feasibility of a controlled fission reactor, or uranium machine, and further claimed that if one of the uranium isotopes were obtained in a sufficiently enriched form, it would constitute a bomb of unprecedented explosive force.[10] He recapitulated the state of knowledge. Uranium found in its natural state, he pointed out, is composed of two isotopes, or nuclei of two different masses: uranium-238 (U-238), which is present in abundance in natural uranium, and the much rarer U-235 (being less than 1 per cent of natural uranium). U-235, however, is more fissionable – that is, its atoms are more susceptible to being split by neutron bombardment – than U-238. Hence a prelude to nuclear fission and a successful chain reaction for spectacular release of energy involved the 'enrichment' of the U-235 content of natural uranium by separation from the more plentiful U-238 content.

A crucial aspect of the process, moreover, would involve the use of a 'moderator', which slows down the speeds of the neutrons to reduce their chances of being absorbed by U-238 and thus increases their probability of fissioning the U-235 nuclei. Heisenberg proposed, in theory, two kinds of moderator: carbon and heavy water. In his initial design he provided for the use of both these moderators in a cylindrical pile. At the same time he speculated about the possibility of using enriched U-235 in small mobile reactors as engines for tanks and submarines as well as for a bomb 'the explosive power of which exceeds that of the strongest available explosives by several powers of ten'.[11]

Heisenberg's second report, delivered to Army Ordnance in

February 1940, was more cautious. He made no mention of a bomb, and stressed the engineering difficulties that faced the realization of a reactor for energy purposes.

Reminiscent of the technical problems posed by the notion of fixing nitrogen from the air forty years earlier, identification and separation of uranium isotopes in quantity from natural uranium appeared insuperable, and not only to the Germans, at the beginning of 1940. Not quite, however, since the best prospect for separation – by gaseous diffusion – had been in process of development by Gustav Hertz, who, because of his Jewish descent, had been forced out of his post as head of physics at the Berlin Technical College just across the road from the Army Ordnance offices. Work on his isotope separation solution had ceased with his dismissal. But that very solution became in time a favoured method of separation in the Allies' Manhattan Project.

As the first phase of German research got under way Heisenberg began to nag Diebner for adequate quantities of uranium oxide, insisting that he urgently required a metric tonne for experimental research. He also recommended the construction of a plant to produce heavy water, while urging the need for a cyclotron. In the meantime Heisenberg's research group experimented with paraffin and regular water as potential moderators.

Plutonium

As Hitler's 'Lightning War' got under way, however, German conquests offered the prospect of solving some of Heisenberg's urgent needs. Following the invasion of Norway, the Wehrmacht captured Vermork, site of the world's only heavy-water plant. After the fall of Paris in July, Bothe and Diebner made their way to the Collège de France in Paris, where Joliot-Curie was on the point of completing a usable cyclotron.

The following year, 1940–41, passed for Heisenberg in failed reactor experiments in Leipzig and Berlin, while attempts to produce significant amounts of U-235 proved disappointing. In July of 1940, however, Weizsäcker and his colleagues, working theoretically on the development of a 'uranium machine', or reactor, had published a paper for Army Ordnance entitled 'A Possibility of Energy Production from U-238'. They speculated that the next element beyond uranium, the product of radioactive decay in a reactor using U-238, would have the potential for atomic explosive (he uses, in fact, the term Eka-Rhenium, or neptunium). This would open the way to making atom bombs, precluding the difficult business of separating out U-235. But there were still technical problems that stood in the way of success (the importance of employing fast neutrons was still not understood).[12] Heisenberg was given a copy of the paper.

The following summer, in August 1941, Fritz Houtermans, a brilliant physicist, confirmed theoretically in Berlin the possibility of creating chain reactions from plutonium, the by-product of a reactor working on natural uranium. The appearance of Fritz Houtermans on the German atomic research scene requires some explanation. He was a Göttingen-trained physicist with Soviet socialist leanings who had turned up in Berlin in 1940 from the Soviet Union with the strangest of stories. Against all wiser counsels he had taken a post at the Ukrainian Physics Institute in Kharkov in 1935 amidst Stalin's purges. By December 1937 he had been arrested in Moscow after making an idle remark about Stalin at a party. Under torture he falsely confessed to being a Nazi spy. After two and a half years in Soviet prisons he was released back to the Germans by the Soviet authorities in the summer of 1940, a result of the brief honeymoon of the Hitler–Stalin pact. He was jailed by the Nazis, who continued to suspect that he was a Soviet sympathizer, but was subsequently released through the good offices of Laue, who secured him a research job in the laboratory of another unusual figure, Manfred von Ardenne.

Von Ardenne was engaged in the most unlikely of projects, given the secrecy and monopolistic tendency of the army towards research in atomic physics. Von Ardenne was an entrepreneurial scientist and inventor with wide-ranging interest in new technology. He had built his own electron microscope after the instrument had been invented in 1938, and he had earlier experimented in television technology, which had brought him to the notice of the Post Office (a body naturally interested in new forms of broadcasting and communication) and indeed of the Führer. When von Ardenne expressed an interest in atomic energy, the Post Office funded him to do research in fission studies and electromagnetic mass separators of isotopes with a view to solving the problem of separating U-235. Von Ardenne put Houtermans to work on chain reactions and, by August, basing his thinking on the famous Bohr-Wheeler paper of September 1939, and the report of Joliot-Curie in *Nature* in April 1939, he had come up with the conclusion, theoretical at this stage, that the element 94, plutonium, could be created by a reactor. Was Houtermans aware of the implications of plutonium for nuclear weapons construction? It seems that he was, although it is unlikely that he realized any more than did the Americans in the early stages the technical difficulties involved in making a plutonium bomb.

Discovering the problems associated with employing plutonium in a bomb never arose for the Germans, because they never got a reactor going. In the meantime Fritz Houtermans was apparently appalled by his plutonium discovery and the possibilities that flowed from it. Houtermans, an individual persecuted by both the Nazis and the Soviets, had a thoroughly awakened political consciousness as a scientist and he acted accordingly. Through a refugee who was leaving for the United States he sent a message, without knowing the status of Anglo-American atomic research, to physicists who would understand: it said that they should hurry. The message also claimed that Heisenberg, fearful of creating a bomb, was trying to delay the research.

That item of information, which makes Heisenberg appear heroic, would add to the enigma of his personality and the long-running debates about the part he played in the German nuclear programme.

British Breakthrough

While the scientists on both sides of the democratic–Nazi divide continued with their experiments and speculations on atomic energy and weapons potential, an outstanding obstacle in the path of atom bomb production was the quantity of uranium required for a chain reaction leading to an explosion.

On 9 June 1939 the physicist Siegfried Flügge, a colleague of Heisenberg's at the Kaiser Wilhelm Institute for Physics in Berlin, had published an article in the leading German science periodical *Naturwissenschaften*,[13] calculating that 1 cubic metre of uranium oxide contained sufficient energy to lift a cubic kilometre of water 27 kilometres high. In 1939–40 the amounts of U-235 deemed necessary to make a 'uranium bomb' by scientists in the leading developed countries of the world, including the United States, Britain and Germany, were so enormous as to make the creation of an atomic bomb look highly improbable. Edward Teller, the Hungarian émigré scientist living in the United States, was talking 30 tons in 1940, while Britain's James Chadwick and the French theoretician Francis Perrin were calculating 40 tons.[14] The Italian Fermi, tongue in cheek no doubt, estimated the amount to be equivalent to a small star! How could such a bomb be delivered? Certainly not by an aircraft. Possibly by ship, but would not such a venture carry a catalogue of high risks?

The breakthrough on the question of 'critical mass' involved, once again, that remarkable physicist Otto Frisch, Lise Meitner's nephew. In the summer of 1939 Frisch had left Denmark with two small cases bound for Birmingham University. He had gained a toe-hold in the Physics Department as an 'auxiliary lecturer' under the Australian Marcus Oliphant, who was working in great secrecy on radar technology. Frisch continued to occupy himself, however, with theoretical work on fission. In the grim privations of that first English winter of the war he settled down to write a review of advances in experimental atomic physics for the British Chemical Society. 'I managed to write that article in my bed-sitter where in daytime, with the gas fire going all day, the temperature rose to 42 degrees Fahrenheit (6 c) . . . while at night the water froze in the tumbler at my bedside.' Wearing his overcoat, typewriter on his lap, he finished the article.[15] The article touched on the possibility of an explosion resulting from a chain reaction. Frisch, like others, thought a bomb unfeasible since the amount of uranium required was so enormous.

It was a conclusion, he wrote, that he found consoling. Later he recollected: 'People have often asked me whether that was a piece of deliberate camouflage. I can assure you it wasn't. I really believed what I wrote; that an atomic bomb was impossible.'[16]

At Birmingham Frisch made friends with an another German-born refugee, the mathematical physicist Rudolph Peierls, who had been working in England, mainly in Cambridge, since 1933 on a Rockefeller grant. Their research relationship now became critical. Otto Frisch wrote in his memoir *What Little I Remember* that, using a formula derived by Francis Perrin and refined by Peierls, he discovered that the amount of U-235 required for a chain reaction was 'very much smaller than I had expected; it was not a matter of tons, but something like a pound or two'.[17] Frisch wrote that he and Peierls stared at each other, 'and realized that an atomic bomb might after all be possible'. The mass of U-235, however, had to be a 'critical size', or 'critical mass': great enough for neutrons from one fission to have a chance of striking a second nucleus before escaping outside the mass. They estimated that 5 kgs (11 lbs) of U-235 could produce an explosive power equivalent to 'several thousand tons of dynamite'. This came extraordinarily close to the equivalence of the atomic bombs that would be dropped on Japan.

The groundwork for this historic result had involved a fresh look at the problem of separating U-235 from U-238 in natural uranium. Such were the assumed difficulties of making that separation, or enrichment as it was also called, that nobody had calculated the 'critical mass' required to make an explosion with U-235. To recapitulate: this refers to the fact that there is a minimum amount of the element necessary to sustain a chain reaction; quantities of the material less than the 'critical value' are stable (in the sense that the neutrons escape) and no chain reaction can take place.

Frisch tackled the problem of separation or enrichment by envisaging a technique developed by the German physical chemist Klaus Clusius. The experimental equipment was rudimentary. A tube with a heated rod inside running down its centre had to be filled with a gaseous form of the material to be separated; the tube wall would then be cooled with water, and the enriched lighter isotope would rise while the heavier would sink to the bottom. Using Clusius's separation formula, Frisch concluded that 'with something like a hundred thousand similar separation tubes one might produce a pound of reasonably pure uranium-235 in a modest time, measured in weeks'.

It is often observed that émigré scientists like Frisch and Peierls were at the forefront of atomic research because indigenous physicists were preoccupied with radar and electronic warfare. Frisch commented that in subsequent years, in view of the horrendous nature of the bomb, he was often asked why he did not abandon the project there and then, without telling anybody. 'The answer,' he wrote, 'is very simple. We were at war, and the idea was reasonably obvious; very probably some German scientists had had the same idea and were working on it.' As it was, he and Peierls wrote up their results and sent them in a report to Henry Tizard, the British government adviser on scientific problems and warfare. This 'Frisch Peierls Memorandum' was the first documented claim that atomic weapons were feasible and this memo triggered the formation of the MAUD Committee, the British body assembled to explore the development of an atomic bomb. (The name MAUD was not an acronym; it arose as a result of a cable from Bohr to a friend in England referring to the fact that the governess who had taught Bohr's sons English was living in Kent – 'Maud Ray Kent'. A member of the committee thought it a suitably mysterious and misleading title.) Britain's effort to separate uranium isotopes was now established in a division known as 'Tube Alloys' within the Department of Scientific and Industrial Research.

The MAUD Committee was destined to be part of the top

secret scientific collaboration that had already started between the British and the Americans to develop radar. By the following summer the committee filed a detailed report on a U-235 bomb, stating, 'We have now reached the conclusion that it will be possible to make an effective uranium bomb . . . likely to lead to decisive results in the war.'[18]

It was this document that prompted Vannevar Bush, the creator of the National Research Committee in the United States, to get the American bomb project underway. Armed with the final version of the MAUD report, Bush briefed President Roosevelt about its contents on October 1941.

Anticipating the development of the Allied atom bomb, Roosevelt straight away saw its importance, and accepted that 'a great deal of money would be required'. He told Vannevar Bush 'to expedite the work in every possible way'.[19] Yet it it was to take until the following August, 1942, for the Americans to make the major practical decisions for development. From the outset, control of atomic research was transferred to a Military Policy Committee, which included Vannevar Bush himself and James Conant. A US army office for procurement and plant construction was established, its HQ initially in New York City, hence its code name Manhattan Engineering District or Manhattan Project. The British atom bomb initiatives were transferred to the Manhattan Project. By September Colonel Leslie Groves (later promoted to General), nicknamed 'Greasy', once described by a colleague as 'the biggest sonavabitch I've ever met in my life', was appointed commanding officer of the bomb project. Aged forty-six at the time of his appointment, he had distinguished himself as a builder of barracks and was responsible for the construction of the Pentagon military HQ outside Washington, DC. His construction experience came in useful, since he was obliged to build the equivalent of three small cities at great speed.

19. Machines of War

While the uranium club and von Ardenne's laboratory continued to work on the potential for nuclear energy as a reactor and as a 'reactor bomb', armaments development and production had begun to suffer from the peculiarly chaotic state that reigned in the Third Reich. The social scientist Franz Neumann, writing early in the war, evoked a striking image depicting the relationship between the power structures of the Third Reich and the prospects for science and technology.[1] Neumann saw Nazi Germany not as a regime under the tight executive rule of its dictator, the Führer, but as a cartel of power blocs, a 'Behemoth' (a monstrous beast of prodigious size) or unstate – a 'polycratic' coalition of the army, big business, the civil service and the Party.

The members of the cartel sometimes collaborated, but were mostly in conflict. Ultimately, the blocs exercised power by virtue of Hitler's favour, 'like spokes around the hub of a wheel'.[2] It was incumbent on each of the power brokers to interpret the mind of Hitler and petition him and his close circle for signs of approval. Contrary to the myth of the fascist leadership principle, it meant that Hitler interfered intermittently, but had little real control over the direction of research, development and production in the Third Reich. In an expanded list of Neumann's cartel components, historians have included the navy, the air force, the SS, the Hitler Youth, the Four-Year Plan, the Todt workers' organizations and the Ministry for Armaments and War Production. It has also been noted that the image of discrete spokes of a wheel around the hub of the Führer does not allow for the reality of wasteful duplication and internecine rivalry.

The overlap of interests, as well as occasional cooperation, can be seen in the role of individuals who had assembled a number of executive roles within different blocs of the cartel. An example of this, with significance for science and technology, was the case of SS Brigadier-General Ministerial Director Professor Doctor Rudolf Mentzel, who was a founder member of the Party and honorary member of the SS, while holding down influential positions in the Ministry of Education, the Reich Research Council and the German Research Council, principal bodies for dispensing scientific funding. At the same time individuals were capable of creating conflicting ideological influences within various blocs – for example, the competing thrusts of anti-Semitism, anti-socialism and anti-communism, nationalism and an enthusiasm for technocracy. As the regime's power barons flexed their muscles and responded to what they believed to be the vision of the Führer, or allies close to the power centres, some aspects of science and technology were encouraged, some were oppressed and some flourished and exerted influence without encouragement. Hence the future of Germany's war technology and production depended on a regime that lacked a centralized executive capable of prioritizing the competing demands of labour and matériel. Goering was nominally the country's economic czar and had jealously guarded his directorship of the so-called Four-Year Plan from 1936 onwards, but the proliferation of his other tasks, which included Commander-in-Chief of the Luftwaffe, precluded his ability to run a hard-pressed economy in wartime. Meanwhile Walter Funk, the Economics Minister, in charge of the domestic economy, was a party apparatchik, under Goering's thumb, while military production was controlled by General Georg Thomas, who characterized the chaotic mobilization of resources as 'a war of all against all'.[3]

Conventional Technology

The course of World War II saw remarkable advances in new technology that became war-winners – radar, computerized code-breaking and the atomic bombs that were to end the war in Japan. Germany's rocket project, far ahead of its rivals, was still in research and development until the latter stages of the war, guidance systems and payloads lagging behind the power systems. Despite these innovations, however, much of the fighting in all theatres of war was performed with conventional technology developed in the 1930s and brought rapidly to optimum levels of efficiency.

Germany's tacticians had developed in the pre-war years the combined use of aircraft, tanks, artillery and motorized troop carriers, the *Blitzkrieg*, concentrating their impact in order to punch holes in the enemy's defences and to make rapid outflanking movements. Use of radio communications between land and air forces was crucial to this *Blitzkrieg* effect.[4] Automatic weapons had increased in their efficiency, especially with the design of clips and magazines for rapid fire in rifles; but as historian Gerhard Weinberg comments, 'Any World War I veteran would have had no difficulty recognizing the heavy infantry weapons of 1945.' In one area of German ordnance engineers were to opt for giganticism: this reached its ultimate tumescence in a 31-inch gun named Dora, mounted on a railway track and requiring 4,400 personnel for operation. It would fire only forty-eight shells at the siege of Sevastapol in June 1942 before sinking into rusty oblivion. Another German gun, more than a hundred yards long, known as the V3, working on the principle of a series of explosions along the barrel, would be erected on the Channel coast to fire at London. In the event, it was only used against Luxembourg.

A significant new development in shell technology, employed by both sides, but more effectively by the Allies, was the proximity fuse, designed to explode not on impact but within lethal range of the target. The principle involved the emission of a radio signal which reflected back from the target to the shell so as to detonate the explosive when the signal reached the right strength.

Both sides in the European conflict had developed tank technology during the 1930s beyond the ponderous models of World War I. The Germans had concentrated their production efforts on the Mark II and the Mark IV, although it was the efficiency of their deployment, as massed Panzer divisions, or 'armies', that made for their decisiveness in Poland and France, and early in the invasion of Russia, working best on the open, relatively flat spaces of Northern

Europe and the grasslands of Rùssia. Superior German tactics enabled the Wehrmacht to bring the Red Army to the brink of defeat in the early stages of the invasion of Russia. Three and a half thousand German tanks rolled back some 15,000 Soviet vehicles. The early German onslaught masked the fact that Hitler had a relatively small mechanized army supported by a much larger one, dependent on nineteenth-century means of transport – horses, 700,000 of them, and the railways.

The virtual destruction of Soviet armoured divisions, however, obliged the Russians to design a new generation of tanks, and to employ tactics that mimicked and even surpassed the German formations. The T34 was built for speed, while the KV-I (later surpassed by the 'Stalin') was designed for fire-power. By 1943 they would carry two-way radios. Robust, reliable, relatively basic in design and highly standardized, the Russian T34s were speedily mass-produced. The Soviet tank armies, moreover, were accompanied by skilled and well-equipped motorized maintenance teams. In response, the Germans built the Panther, which equalled the T34, then the notorious Tiger, which supposedly surpassed it. Hastily rushed into service in the second half of 1942, the new tanks were cumbersome, technically complex and difficult to mass-produce and to repair. At the same time, in assembling his army speedily, Hitler had commandeered a prodigious array of vehicles from all over occupied Europe. According to historian Richard Overy, there were 2,000 vehicle models in the assault on Russia: 'Army Group Centre alone had to carry over a million spare parts. One armoured division went into battle with 96 types of personnel carrier, 111 types of truck, and 37 different motor-cycles.'[5] Given the vast distances, the reliance on inferior synthetic fuels and rubber, the poor roads, the autumn rains, turning the terrain into a sea of mud, and, finally, the cruel Russian winter, these motley assemblies of machines and vehicles were rendered vulnerable to breakdown and collapse.

Due to training and tactics the T34 soon proved a match for German armour in the battles of Kursk and Stalingrad. In the early stages of Operation Barbarossa Soviet armoured forces lost six or seven vehicles to every one German; by the autumn of 1944 the ratio was down to one to one.[6] Given the production advantage of the Russians, creating a numerical advantage over the Germans of three to one by 1943, this ratio spelt disaster for Hitler's armies. Both the Americans and the British were dependent on the Sherman tank by 1944, against which the Panthers and Tigers were marginally superior; but the Allies had a vast numerical advantage.

As with other mechanized forms of warfare, aircraft research and development was quickly taken to the zenith of conventional existing technology by all belligerents in the war to produce large numbers of aircraft. Jet propulsion, discussed later, was developed by the Germans and the British; but as with rocket research, the determination to force through a new technology that required steady evolution, special materials and collaborative research sapped urgent resources required in the production of large numbers of standardized conventional fighters and bombers.

The technological challenge in conventional aircraft was, in the first place, to produce single-engined, single-wing fighters that could fly faster and further, carry heavier armour and exert greater fire-power than the enemy. The second challenge was to produce bombers with longer range and the ability to carry ever greater loads. At the same time, the developments required synchronization with radar for air-to-air detection, navigation and accurate targeting, as well as radio communication between bases and other aircraft.

The Germans led initially in speed, but the British soon caught up with the Rolls-Royce Merlin engine, first delivered to the RAF in 1938. The early Merlin engines were considered unreliable, but the manufacturer initiated a quality-control programme which involved taking engines randomly off the assembly line and running them at full throttle until they broke. In this way the engineers discovered which parts had failed. By 1940 the Merlin had developed into an extremely reliable aero engine, at first cooled by an evaporative system working off condensers in the wings, then by Ethelyne Glycol, which was more efficient than water. Eventually the engine could run on 100 octane fuel developed in the

United States, giving aircraft speeds that were further boosted with ever more sophisticated superchargers.

The Luftwaffe could not compete with the high-octane ratings and could only attempt to catch up by producing larger engines. In consequence German planes had a worse power-to-weight ratio. If the Lufwaffe's planes had a partial advantage it was because of their sophisticated fuel injection systems. But the Germans never succeeded in producing an effective bomber, heavy or light, long-range or medium-range.

Apart from the reliance on synthetic fuel, the failure of Germany's aircraft development was not a lack of technological expertise so much as the chaos of the polycratic system and the general climate of corruption. In the critical early stages of the war the technical direction of the Luftwaffe was placed in the hands of an incompetent, who happened to be one of Hermann Goering's favourites. German aircraft research, development and production ran into trouble on a broad front after Goering replaced Erhard Milch, a former director of Lufthansa, with Colonel Ernst Udet in 1939. Once a stuntman in the movies, he was weak in character and in management skills. He had no technical qualifications for the job and had never wanted more than to be a test-pilot. He had a passion for high living, womanizing and hunting and presided over a chaos of competing pressures from power brokers with an interest in aircraft. Insanely committed to the *Blitzkrieg* philosophy of dive-bombing, in order to achieve accuracy by swooping low, he ordered a twin-engined bomber to replace the ageing single-engine dive-bomber (JU-87) which was plagued with technical problems, then a four-engined bomber. In order to make a four-engined aircraft that could dive-bomb its targets, the designers had assigned two engines to each propeller, creating insuperable performance problems. Many of these aircraft caught fire before take off, or in the air, and seldom managed the imperative to dive-bomb their targets.

Battleships

One of the early surprises of World War II was the dramatic demise of the most expensive and impressive symbols of great-power military hardware: the battleship and the battle cruiser. The first sign of the battleship's unusual vulnerability was the sinking of Britain's *Royal Oak* by a lone U-boat on the night of 13 October 1939 while she lay at anchor in Scapa Flow, the 'safe' harbour of the British home fleet. Some 800 hands went down with the ship. Britain was stunned; but the Royal Navy had its revenge two months later when the German pocket battleship *Graf Spee*, trapped in the harbour of Montevideo by the British cruisers *Ajax* and *Achilles*, was scuttled by its commander on 17 December 1939, watched by 250,000 Uruguayans on the shoreline.

Between 1940 and 1941 Germany's active battleships, the *Scheer*, *Scharnhorst* and *Gneisenau*, conducted successful raids on British shipping in the Atlantic and Indian oceans. But the proudest vessels of the Reich at the outbreak of war were the sister battleships *Bismarck* and *Tirpitz* – the largest and technologically the most advanced vessels of their kind in the world. To this day no record exists of the ships' highly accurate, radar-assisted gun-fire controls, although surviving photographs proclaim their gleaming state-of-the-art high technology. They were driven by three massive diesel turbines operating three propellers, producing a top speed of 29 knots, and possessed eight 15-inch guns, with 14-inch armour plating around the turrets. The deck housed a hangar which took four Arado 196 sea-planes, with two more stowed on the open deck, capable of being launched by compressed air-operated catapults with special cranes and derricks for retrieval from the sea; in this sense the vessels doubled as virtual aircraft carriers.[7] At 41,700 tons, built at the Blohm and Voss shipyard in Hamburg, the *Bismarck*'s hull was launched by Hitler in person on 14 February 1939. If Hitler had any doubts about the efficacy of the battleship he gave no hint of it as he lectured the assembled crowds on the spiritual world view of National Socialism and its organizational and techno-logical means of 'defeating the enemies of the Reich now and for ever'.[8] Already planned, in fact, was a generation of monster battleships of more than 100,000 tons each, twice as heavy as the weightiest fighting vessels in existence.

On 19 May 1941, escorted by the cruiser *Prinz Eugen*, the *Bismarck* headed for the Atlantic via Iceland bent on destroying Britain's merchant convoys. At six o'clock on 24 May, HMS *Hood*, the pride of the Royal Navy, and the largest battle cruiser afloat, if somewhat elderly (she was launched in 1920), came upon the two German vessels. After just eight minutes of exchanges, the *Bismarck*'s fifth salvo hit the *Hood*, exploding its magazines. It took just three minutes for the great British ship to disappear below the surface. The cruiser *Prince of Wales*, which accompanied the *Hood*, was also hit, and retired from the battle. The people of Britain were stunned. Churchill issued the order: 'Get the *Bismarck* at any cost.'

Ships of the Royal Navy, including two aircraft carriers, searched the Atlantic for two days in search of the German battleship. Anticipating discussion of radar in the next chapter, it is significant that the *Bismarck* was tracked by the British cruiser HMS *Suffolk* using gunnery-control radar, and that the *coup de grâce* was made with the use of radar in the plane that found the *Bismarck* heading for Brest.[9] On the evening of 26 May a Swordfish aircraft (an open cockpit biplane, affectionately known as a 'Stringbag', with a top air speed of 139 mph) delivered a torpedo which jammed both Bismarck's rudders, crippling the ship fatally so that it could not steer other than to proceed in a wide circle. The planes on the *Bismarck* were never launched, as the catapult mechanism was out of action. The following day four ships of the Royal Navy closed in and finished her off. Only 115 of the *Bismarck*'s sailors survived of a crew of more than 2,200. Thereafter Hitler recalled his capital ships to base and kept them there.

Later that year, following the attack on Pearl Harbor and America's entry into the war, the Royal Navy lost two more great battle cruisers, the *Prince of Wales* and the *Repulse*. Despatched to the northern coast of Malaya, where Japanese troop movements had been reported, the British vessels were attacked by the bombs and torpedoes of a large formation of Japanese bomber aircraft. The two battleships went down with the loss of 840 lives. The day of airpower on the high seas had arrived, and the days of battleships were numbered.

Jet Aircraft

In the early hours of 27 August 1939, four days before Hitler's invasion of Poland, Erich Warsitz, test-pilot for Professor Ernst Heinkel, founder of the German company of that name, walked on to the Heinkel airfield at Marienehe near the German port of Rostock and climbed into a small propellerless aircraft designed by the turbojet pioneer Hans-Joachim Pabst von Ohain. A group of engineers, including Ernst Heinkel, watched anxiously as Warsitz took off. He has recorded an account of this historic maiden flight:

All control surfaces functioned virtually flawlessly, and the turbine sang its loud, monotonous song. It was wonderful flying like that; no little wind stirred and the sun still rested quite low on the horizon. The only thing left was a good landing and the first turbojet flight in the world would be a success . . . Side-slipping in a new airplane is not without its dangers, especially when close to the ground. But the well-built airplane took even this in its stride . . . We justifiably believed we'd conquered a world and ushered in a new age.[10]

Hitler was himself to witness an early jet flight a few weeks later, but he was unimpressed. He asked what possible point there was in flying faster than sound. In the following six years a prodigious number of designs of turbojet planes were sketched out on the drawing boards of Hitler's aircraft designers. But Germany's jet pioneers, imaginative and skilled as they were, laboured under a host of disadvantages, not least the pressure to deliver new technology within unrealistic time constraints, a lack of high-quality metals and raw materials and the absence of skilled labour.

The word 'turbine' comes from the Latin *turbo*, a whirling object, and refers in its earliest manifestation to a means of employing a stream of water to turn a wheel. The first steam turbines appeared in the United States and continued in development in North America and Europe through the century. In Britain, a 'gas turbine' had made its appearance in the 1790s: air and fuel from a gas producer were compressed and directed through a nozzle on to a turbine wheel. The early models were hardly efficient, but the principle anticipated the modern gas turbines of the twentieth century. By 1872 F. Stolze of Germany had patented a more viable machine, known as a 'fire turbine'; in the 1880s the English engineer Charles Parsons designed a turbine for propelling ships, and a French model had been constructed by the turn of the century. The turbine that links the early pioneering attempts with the turbojets of the mid-twentieth century was the work of another German, Hans Holzwarth, who began in 1905 a gas turbine research programme that continued for thirty years. By the early 1930s the turbine principle awaited a designer to make an engine light and powerful enough to propel an aircraft. The principle would involve air being sucked into the engine, where it would be compressed and mixed with appropriate fuel in a combustion chamber, then expanded through a turbine so as to provide sufficient power to drive the compressor before being expelled through power-jet nozzles.

Ideas for turbojet power plants were explored in the 1920s and 1930s in France by Charles Guillaume, who patented an early prototype in 1921, and in Britain by A. A. Griffiths. In the event, it took the genius and determination of Frank Whittle, a young former cadet at the RAF College, Cranwell, to create a feasible engine for Britain. Whittle persevered through the 1930s with little financial backing and no military encouragement. Forming a private company, Power Jets, with several friends and an initial investment of no more than £20,000, his engine started life in a workshop based in a disused shed in Rugby. By June of 1939 he had managed to develop an engine which was tested for twenty minutes. As war now loomed with Germany, the test was observed with interest by the Director of Scientific Research at the British Air Ministry. As a result the Air Ministry placed a contract with Whittle to make an engine for an aircraft frame to be designed by the Gloster Aircraft

Company. The realization of Whittle's initiative, however, was to lag behind the efforts of the Germans.

The origin of Germany's early lead, in actual flight, was largely owed to a young physics student, Hans-Joachim Pabst von Ohain, who financed himself by running a business in Berlin distributing light bulbs while striving to build a prototype turbojet through the mid-1930s with the help of a garage mechanic, Max Hahn. Ohain's fortunes improved when, aged twenty-five, he was introduced to the aircraft manufacturer, Ernst Heinkel, an engineer obsessed with high-speed flight. On 17 March 1936, Ohain demonstrated his prototype to Heinkel and a group of the manufacturer's engineers. By 15 April 1936 Heinkel had signed a contract with Ohain and his mechanic, providing them with funding and a team of engineers. Heinkel built a facility for them fenced off from the rest of his factory at Warnemünde. By the autumn of 1937 Ohain was testing a hydrogen-fuelled engine and planning an aircraft specially designed for jet flight, the He-178, in which Warsitz made his maiden flight in August 1939, two years ahead of Whittle's prototype (the Gloster Whittle E28/39). By this stage, however, Heinkel's model had a keen rival.

The Luftwaffe had not been slow to see the advantages of jet propulsion in the late 1930s. Helmut Schlep, an aeronautical engineer with the German Research Institute for Aeronautics, had been commissioned by the German Air Ministry to develop turbojet technology for the Luftwaffe which led to the placing of a contract for a twin-engine jet aircraft with Messerschmitt, makers of both engines and air frames. The Messerschmitt programme would result in the famous Me-262, which made its prototype maiden flight in July 1942. The world's first operational jet fighter, almost 2,000 were completed before the end of the war; but they were too few and too late to make an impact on the outcome. American and British fighters took to attacking the Me-262s at their slowest speeds, as they took off and landed. They also discovered that the German jet could not make tight turns like propeller aircraft. Hampered by lack of clear direction, and Hitler's interference – he failed to back the early jet aircraft as fighters,

1. The fountains of the 'Château d'Eau' with the 'Palais d'Electricité' behind, at the Paris International Exhibition of 1900

2. Max Planck, one of the founding fathers of quantum physics

3. Fritz Haber (*pointing*) inspecting poison-gas canisters in World War I

4. Fritz Haber

5. Clara Immerwahr, Fritz Haber's wife, who committed suicide in protest against her husband's gas-warfare activities

6. Kaiser Wilhelm II opening a Kaiser Wilhelm Institute in Berlin on 23 October 1912

7. Philipp Lenard, Nazi physicist and anti-Semite

8. Lise Meitner and Otto Hahn, Kaiser Wilhelm Institute for Chemistry, 1913

9. Johannes Stark, Lenard's younger colleague and antagonist of Einstein

10. *(left to right)* Enrico Fermi, Werner Heisenberg, Wolfgang Pauli, 1927, Como

11. Copenhagen Physics Conference, June 1936. First row (*left to right*): Pauli, Jordan, Heisenberg, Born, Meitner, Stern, Franck; second row: Weizsäcker and Hund behind Heisenberg, Otto Frisch (*second from right*); third row (*left to right*): Kopfermann, Euler, Fano; fourth and fifth rows, second from left: Peierls and Weisskopf; standing along wall: Bohr and Rosenfeld

12. Professor Oberth (*centre, in profile*), inventor of the flying bomb, with a group of his staff and a rocket he made in 1931. The group includes Wernher von Braun (*second from the right*)

13. Adolf Hitler inspects the first German U-boats in Kiel in August 1935

14. Hitler and his Deputy Rudolf Hess at the Party Rally, Nuremberg, 1933

15. Sterilization advocate Bernhard Rust, Nazi Science Minister, posing in uniform

16. Hitler at the 1935 Berlin Automobile Exhibition

17. Albert Einstein and Leo Szilard warned President Roosevelt of the dangers of atomic weapons in a letter dated 2 August 1939

18. Frédéric Joliot-Curie (pictured here with his wife Irène) published his research on secondary neutrons in March 1939, alerting Nazi scientists to atomic-bomb potential

19. Otto Frisch and Rudolph Peierls, who at Birmingham University calculated that the amount of uranium-235 required for a chain reaction was 'very much smaller' than had been previously thought

20. After the first A4 success, Armaments Minister Albert Speer (*right, with armband*) moved to take over missile production. Here he watches a V2 launch with Propaganda Minister Josef Goebbels (*centre*), 17/18 August 1943, Peenemünde

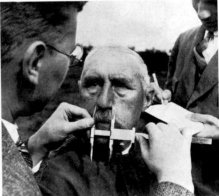

21. (*left*) German architect Professor Albert Speer showing Hitler some of his plans for Berlin's new outline and buildings, 7 February 1938

22. (*above*) Measuring the features of a German

23. (*above left*) German Mk 6 Tiger Tank in front of the British lines near Medjez-el-Bab, 27 April 1943

24. (*above right*) Winston Churchill and his scientific adviser Frederick Lindemann (Lord Cherwell, *left*)

25. (*right*) German signal troops sending coded messages on an Enigma machine

26. V1 rocket over London. Photo taken from a Fleet Street rooftop. The bomb fell in a side road off Drury Lane, blasting, among other buildings, the offices of the *Daily Herald*

27. V2 rocket, 15 August 1947. Britain's security-veiled Rocket Research Department at Westcott, Buckinghamshire, was the central experimental establishment for all applications of rocket propulsion. The staff included twelve German scientists, led by Dr J. Schmidt, formerly in charge of the rocket works at the Walterwerke in Kiel

28. The Mittelwerke facility, where the vengeance rockets were assembled underground with the use of slave labour

29. Messerschmitt Me-262, Germany's jet fighter-bomber

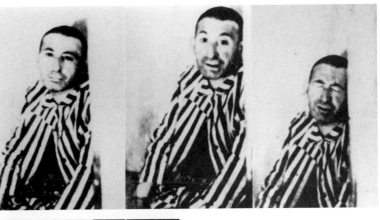

30. (*above*) Experiments on a prisoner at Dachau

31. (*left*) Primo Levi, writer and chemist, was recruited into the Buna chemistry labs at Auschwitz in 1944

32. (*below*) American troops discover Heisenberg's reactor at Haigerloch

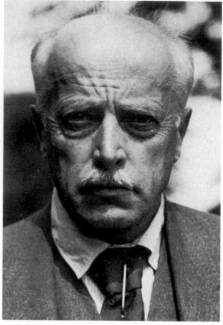

33. Farm Hall, where ten German physicists were detained in 1945

34. Max von Laue resisted the Nazis while remaining in Germany during the war

35. Carl Friedrich von Weizsäcker, Heisenberg's younger colleague in the Nazi bomb project

36. Paul Harteck solicited funding for atomic-bomb research from Army Ordnance in 1939

37. 'Fat Man' nuclear weapon, of the type detonated over Nagasaki

38. General Groves (*right*) and J. Robert Oppenheimer inspect the Trinity atom–bomb site

39. Wernher von Braun surrenders to US counterintelligence personnel, May 1945

40. President John F. Kennedy with Wernher von Braun, November 1963

insisting that they should be deployed rather as bombers – the Reich's jets joined the swelling ranks of useless 'wonder' weapons that promised much, delivered little and sapped the resources of an increasingly beleaguered regime. After the war, both the Americans and the Russians prospered as a result of war spoils in the form of Germany's jet technology.

Submarines

While Britain lacked the ability to detect Hitler's U-boat flotillas with radar and intelligence-gathering at sea, the German navy enjoyed enormous advantages against British merchant shipping in the Atlantic. Even before the Allied technological advances in radar and code-breaking, which began to bite towards the end of 1943, making the Atlantic sea lanes virtual suicide for U-boat crews, Germany had been attempting to push submarine technology to its limit.

For some years before the war a promising forward leap had been expected as a result of the research of a brilliant German marine engineer called Helmut Walter, who designed in the 1930s a system that released oxygen from hydrogen peroxide, thus enabling the diesel engines to work while submerged. Walter's idea was to decompose a supply of pure hydrogen peroxide, using a permanganate catalyst, to produce oxygen. Diesel fuel was injected into the reaction chamber to combust with the oxygen, thus yielding steam and other gases at high temperature that powered a turbine engine. The exhaust gases and steam were expelled outside the boat. In his early experiments Walter was initially interested in speed rather than duration of submersion. An early prototype, in 1940, achieved speeds of over 28 knots submerged when top speeds of submerged conventional submarines were less than 10 knots. Hitler authorized the construction of two Walter prototypes early in the war, but the project was plagued by long-running difficulties, not least a lack of urgency consequent on the huge success of conventional U-boats (the VII and IX series). But there were genuine problems of resources, including a lack of hydrogen peroxide, which was also employed in the A4 rocket programme. A fleet of the new Walter U-boat designs would require the construction of several production plants, which continued to be a low priority. Meanwhile research and development on Walter's first ocean-going test bed, the U-791, would be destroyed in a bombing raid on Kiel in March of 1942.[11] There was, besides, an intractable technical problem. Hydrogen peroxide is a non-renewable fuel (unlike batteries that can be recharged). Writes submarine historian Clay Blair:

Since there was not enough room in the Walter boat for both a big rechargeable storage battery *and* hydrogen-peroxide tankage, and that tankage was not, of course, unlimited in size, the submerged endurance of such a boat was restricted . . . until some means could be devised to replenish this exotic fuel supply at sea.[12]

At the same time, with a combination of wind-tunnel experiments and applied mathematics Walter developed a new streamlined hull model that increased the speed of the vessel. But by this point time was running out for Germany, and Doenitz's order, following Hitler's approval, of 700 ocean-going U-boats to be known as 'electro boats' would not be fulfilled in time to make a difference to the outcome of the war.

In the meantime Walter's engineering expertise was applied to upgrading the existing classes of submarine with an improved *Schnorchel*, or snorkel, widely known among mariners as the 'snort'. The snorkel was composed of twin tubes that could be raised like a periscope when the craft was submerged. One tube sucked in air to assist the diesel engines, the other expelled the exhaust. The idea for such a device was hardly new: designers had experimented with the system since the beginning of the century. The German version (based on a Dutch design developed in the 1930s) suffered major problems. When submerged by waves the snorkel would suck air from the boat's interior and the exhaust would be released inside the boat. The early snorkels, moreover, consumed double the amount of fuel and made a loud noise, making the craft vulnerable to sonar detection.

Walter invented systems of valves and floats to cure the vacuum and exhaust problems, but the results were far from satisfactory. Clay Blair comments: 'Hailed by some historians and engineers as another great technical achievement, the snort was not that by a long shot. Rather it was a miserable, temporary device that the German U-boat crews hated absolutely.' In his memoir of life in the U-boat service, *Iron Coffins* (*Die Eisernen Särge*), Herbert Werner recollects the effect of the air vacuum created within a U-boat when the Walter snorkel float had been jammed:

The men gasped for air, their eyes bulging. The Chief lowered the boat, bringing the snorkel head below surface in an effort to loosen the float. To no avail. Breathing became ever more difficult; suffocation seemed imminent. The Chief gesticulated wildly, trying to tell his men to lie down, which might result in unlocking the float . . . The float cleared with a snap and air was sucked into the boat with a long sigh. The sudden change in pressure burst many an eardrum. Some of the men covered their faces in pain and sagged to the deck plates.[13]

Missiles

In terms of advanced technology and sheer effort of organization and commitment of resources, Hitler's bid to build an arsenal of offensive missiles constituted Germany's boldest attempt to apply high-tech Big Science to the weaponry of World War II. The rocket development facility at Peenemünde – including workshops, labs, launch pads, admin offices and factories – had been started early in 1936, designed and erected by Luftwaffe construction teams in a favoured Nazi neo-classical design. In a hubristic competition between the two services, the Luftwaffe and the army vied with each other to pour funds into the enterprise and its projected research programmes. Hermann Goering laid plans for a rocket-assisted plane; and there would be other schemes, including a pilotless plane, later known as the V1, or the doodlebug, as well as a surface-to-air missile, to be known as the Wasserfall. The army plans, on the other hand, were for the large A-4 supersonic missile to be known as V2.

Dornberger's plan before the outbreak of war was to develop a missile that would carry a 1-ton payload of explosives 160 miles, twice as far as the shells of the massive Paris Gun of World War I. The goal was well within conventional concepts of ordnance, but the technology required was astoundingly innovative. Such a missile would need to travel at almost five times the speed of sound, and this at a time when no army rocket had yet broken the sound barrier.

Wernher von Braun, only twenty-seven at the beginning of the war, had emerged as a dynamic leader of the engineers. If there were occasional complaints, it was a token of his unquenchable enthusiasm – his tendency to digress from the main business in hand. Von Braun's principal and immediate task was to launch the A3, a rocket that weighed 1,650 pounds and stood 21.3 feet high, its motors exerting a 3,300-pound thrust. The development and testing of the A3 was well advanced when the team moved to Peenemünde, but von Braun was also involved with the joint Luftwaffe–army rocket-plane project. There were no plans to pilot the proposed liquid fuel-powered aircraft, but von Braun was all for flying it himself at the research stage, having attended a Luftwaffe flying school. Fears for his safety, however, obliged his superiors to veto his plans to go up in a prototype version.

The A3 had been tested on 4 December 1937 on the island of Greifswalder a few miles offshore from Peenemünde with about 120 personnel in attendance and various high-ranking spectators. The weather was atrocious and the party's tented accommodation and telephone wires had been plagued by mice and rats. The first rocket, known as Deutschland, prematurely deployed its parachute after about three seconds and fell back on to the island, exploding on impact. The second did the same. The third reached about 3,000 feet before coming down out of control, and so did the fourth.

Von Braun decided that it was the guidance system which was at fault.[14]

As far as von Braun was concerned, the failure of the A3 was part of a crucial learning curve. He was not daunted by the prospect of moving on to the much larger challenge of the A4, but could he be confident of securing the sophisticated technology required to realize such a missile? Von Braun had already been playing a key role in bringing a complex array of academic science and technology to the project.

Academic Science and the German Missile

The collaboration between the army rocket programme and academia had begun in the mid-1930s with the contributions of Dr Rudolf Hermann, an assistant at the *Technische Hochschule* at Aachen, where Hermann was working on sub-scale rocket models at the university wind tunnel. Hermann, a Nazi, was eventually enticed to Peenemünde with the sort of prospect scientists and engineers tend to find irresistible: he was invited to construct and manage the largest and most powerful wind tunnel in the world. It was completed in the autumn of 1939 as Germany went to war. By that time other *Technische Hochschulen*, especially those of Dresden and Darmstadt, had begun work on sophisticated guidance technology.

The wind tunnel project formed a highly significant feature of Big Science pioneered at Peenemünde, as its directors attempted to bring every aspect of research, development and production of missiles under one roof. Dornberger later admitted that the budget of the wind tunnel project at 300,000 Marks frightened him. In time this budget would be exceeded again and again. Eventually two large supersonic wind tunnels were built at Peenemünde, and a third smaller one, employing a staff of sixty by the beginning of the war in 1939, rising to 200 by 1943.

The point of these machines was to test the air drag, lift, stability and controllability of airplanes and missiles by subjecting scaled-down models to high wind speeds replicating real conditions in supersonic flight. The tunnels required compressors of great power to move air continuously at supersonic speeds through a circular test duct.

The result of Hermann's work was the refining of the fin and fuselage shape of the proposed A4 so as to create the first fin-stabilized supersonic projectile. Dr Walter Thiel succeeded in increasing the power of the alcohol/liquid oxygen engine used in the A3 to the 25-ton thrust required for the A4.

Before the outbreak of war Peenemünde, with its advanced technologies and no deployment in the foreseeable future, had been awarded the same top-priority status as U-boat and aircraft production, and braced itself to expand the workforce from 5,000 to 9,000. In the meantime the rocket-aircraft project was shelved, and the much vaunted collaboration between the army and the Luftwaffe dwindled to collaboration on guidance systems and a proposed rocket-assisted take-off system for bomber aircraft.

The priority was to complete the technical developments of the A4 and push through the construction of a factory in order to commence production two years ahead of the original schedule. The latter required large quantities of steel at a time of severe ammunition shortages due to the invasion of Poland. In consequence, Hitler intervened in the autumn of 1939 to overcome the 'munitions crisis', relegating the rocket project to its original timetable and scaling down its steel supply to 2,000 tons a month from an earlier increase to 4,000 tons a month. Dornberger nevertheless struggled to keep the start date for production of the A4 to 1941, rather than 1943, with a completion figure of just eighteen rockets a month, rising to ninety a month by July 1942.

As the fateful year of 1940 dawned, with the certainty of the expansion of the war, Dornberger was fighting to maintain the status of his project, using every argument at his disposal, including warnings that the French, the British and the Americans would be attempting to catch up in missile research. Meanwhile the construction task barely staggered on, hampered by shortages of matériel, a harsh winter, power failures and lack of suitable worker accommodation.

Further cutbacks were only prevented by the timely involvement of Albert Speer, Hitler's architect, who had been consulted on the development of the site since early in the previous year. Speer wrote in his autobiography that on visiting Peenemünde in January 1940 the work of the engineers 'had exerted a strange fascination upon me. It was like the planning of a miracle. I was impressed . . . by these technicians with their fantastic visions, these mathematical romantics. Whenever I visited Peenemünde I also felt, quite spontaneously, quite akin to them.'[15] By the autumn of 1940 Speer would have control of the construction of the production plant at Peenemünde, addressing his considerable organizational skills to a project that was dogged by cutbacks and administrative confusion, in an era, as Speer put it, of economic 'incompetence, arrogance, and egotism'.

Minister Todt

Early in 1940, Fritz Todt, the new Minister for Armaments and Munitions, appeared on the scene, with the authority to exercise influence over the missile programme as well as a whole array of weapons production. Todt, like Speer, came from a prosperous middle-class family. He was a qualified engineer and had been an officer in World War I. Tall and self-confident, he was credited with having an imperial profile. He was famous for building Germany's autobahns in the 1930s and for his 'Organization Todt', responsible for building the West Wall (otherwise nicknamed the Siegfried Line), a defensive line 300 miles long opposite France's Maginot Line that ran from Basle to Kleve. Todt's organization employed some half a million men to complete the work and consumed a third of Germany's annual output of cement. His army of workers eventually rose to millions once war had broken out.

His brief, following his appointment on 17 March, was to end the bottlenecks in army weapons production; but such was the competing polycratic nature of the Third Reich, even as it plunged into a war of national survival, he had no authority over the navy and the Luftwaffe. Todt immediately issued orders for cutbacks on projects that could not promise short-term results.

A token of the confusions and chaos in military production in 1940 was the surprise suicide of the Army Ordnance chief, Karl Becker, for so long a high-ranking ally of the rocket project. Discountenanced by Todt's appointment, Becker had convinced Hitler on 8 April to create yet another ordnance bureau to oversee munitions production, which would be run by army military rather than Todt's ministry. But a director of the Krupp armaments corporation persuaded Hitler to change his mind on the very same day, dropping hints as he did so of scandals in Becker's family. Depressed by the endless imputations against Army Ordnance and vitriolic gossip against his own reputation, Becker shot himself.

Todt was confirmed as the man in charge, and the day after Becker's death Dornberger and von Braun paid him a visit and gained his enthusiastic support of the rocket programme. The Peenemünde project received further tokens of support, moreover, from Becker's replacement, General Emil Leeb, who had come up through the artillery.

It was Leeb who that summer, after the fall of France, saw the potential of the A4 as a means of terrorizing Britain into suing for peace. He made a bid to raise the status of Peenemünde to supreme priority above every other kind of weapon, but the scheme was put on hold because his superiors doubted that they could get it past Hitler. As 1940 wore on and Hitler scented victory in the west, including the capitulation of Britain, priorities – for the navy, the Luftwaffe and the army – continued to lurch and overlap, resulting in confusion and lack of momentum. Matters were not helped by the promotion of Goering to Reichsmarschall – putting him above his peer field marshals – who now devised an ineffective priority plan to ration steel according to just two levels of priority.

But the fortunes of war were about to affect Goering's standing and bring the rocket programme into new focus. As the losses of Goering's Luftwaffe mounted in the Battle of Britain in 1940–41, Dornberger recorded in a memorandum for the attention of Field Marshal von Brauchitsch that the A4, the rocket destined to be the V2, should be regarded as a 'significant relief for the employment of aircraft against England, and especially London and the port cities'.[16] He pointed out that the advantage of the missile was its immunity to defence, its accuracy, and its power to land at any time of day or night irrespective of weather: in short, he declared, it would make a crucial contribution to the 'defeat of England'. Implicit in the memorandum was the notion of the A4 as a weapon of terror that would undermine the will of Britain to continue the war.

A major factor in this strategy had been the extraordinary success of Britain's radar, a technology British scientists had developed through the 1930s as part of crucial defence planning against bombing raids. Radar had been developed in Germany through the same period, but its efforts had been fragmented and laggard as a result of offensive rather than defensive thinking.

20. Radar

Radar, which is an acronym for 'radio detection and ranging', coined by a US naval officer (Britain's radar was at first known as RDF – radio detection finding), is widely regarded by historians of technology as one of the most remarkable technological accomplishments of the twentieth century.[1] Its development and defensive deployment early in the war enabled Britain to cheat the Luftwaffe of air domination and so to survive until the Americans joined the conflict after Pearl Harbor in December 1941. Thus radar made the difference between defeat and victory in Europe.

From the outset Germany appeared to be abreast, if not actually in the lead. On 26 September 1935 a party of German naval officials, including Admiral Raeder, Commander of the German Fleet, and various government bureaucrats gathered at a scaffold tower some 40 feet high situated at Pelzerhaken near Neustadt on the Bay of Lübeck.[2] They had come to observe radar equipment, known in German as *Funkmessgerät* (radio measuring device), in action. The equipment, which included turntables, transmitters, receiver arrays, screens and electrical generators, was devised to locate the presence by radio 'echo' of two ships 5 miles out in the bay. A transmitter would send out a radio pulse which would scatter (or 'bounce') off the ships and return to a receiver, which would send a signal to a visual display, creating an image revealing their presence. The attempt, with some tweaking of antennae, was wholly successful and impressed the onlookers.

This early German demonstration was hardly the first use of a transmitter and receiver of radio waves to identify the location of distant static or passing objects. Robert Morris Page had put together a makeshift system in December 1934 near Washington, DC, demonstrating the possibility of radar ahead of Robert Watson-Watt's British experiment conducted in February 1935. Page could also claim to be ahead of the very first German radar effort, conducted in March 1935, when primitive equipment situated on a balcony detected a German warship half a mile distant out in Kiel harbour. Other crude 'radar' experiments were conducted during 1935 in Italy, France, Russia and Japan.

The relatively sophisticated demonstration that took place at Pelzerhaken on the afternoon of 26 September 1935 was principally due to the research of the physicist Rudolf Kühnold, a scientist at the Navy Research Institution at Kiel, and two engineers who were partners in a small young company known as GEMA[3] which specialized in radio transmitters and recording devices. GEMA had close ties with naval research, but the company remained in theory independent of military auspices as it grew to 6,000 personnel during the course of the war. The officials present at the demonstration immediately saw the potential for warfare at sea, but there was no intention of sharing their findings with the Luftwaffe. Other private German companies, such as Siemens, Lorenz and Telefunken, would pursue their own radio location researches in time.

The early story of radar reveals interesting contrasts between the British and German approaches, before America entered the war, to a new technology of critical importance. In the case of Britain, the character and genius of a single engineer proved to be important.

The British physicist Watson-Watt, who claimed ancestry to James Watt, the steam engine pioneer, was fifty-two and head of Britain's National Physical Laboratory's Radio Research Station when in 1934 he was asked by Britain's Air Ministry to look into the possibility of transmitting 'damaging radiation' to be used in defence against air attack. In response, Watson-Watt urged the strategy of using radio signals for early detection of hostile aircraft and received funding to devise a prototype that could perform the task. On 26 February 1935, seven months ahead of the Germans, he and his colleague A. P. Rowe improvised the first test of a radar device at the BBC's short-wave radio station at Daventry in Northamptonshire. Operating at about 50 metres wavelength, it employed a receiver that was already in use as a detector for long-distance radio signals.

Radio signals are generated by a high-frequency alternating

current, and the number of cycles per second is known as the 'frequency'. Radio waves are electromagnetic radiation just like light waves, except that they have a much longer wavelength – the length referring to the distance between the crests or troughs of these waves as they travel. In the use of radio signals for detection of distant objects, a beam is emitted and some of the waves scatter off the target and return as in an echo which can be picked up by a receiver at the point of the original transmission. Although radio wavelengths are relatively large, and those used in radio transmission normally measured in many metres, a relatively short wavelength is required in order to detect detail of the object being targeted. The problem for the early radar experimenters was that waves less than ten centimetres, referred to as microwaves, required immense energy to transmit long distances (since the apparatus is on the scale of the wavelength, a microwave instrument would be intrinsically small, so sufficiently compact to install in an aircraft).

In the case of Watson-Watt's first significant experiment, a bomber was flown over the Daventry BBC radio masts and on the second pass the observers saw 'beats of very good amplitude' on the display screen for a little over two minutes. During a third run signals were registered for four minutes. The bomber had been tracked for almost 8 miles.[4]

Mistakes were made by both Germany and Britain in the early development of radar; perhaps the biggest on the part of Germany, when its technicians decided not to develop a piece of electronic equipment known as a magnetron, which their researchers tested and discarded in the mid-1930s. The view maintained by British historians of radar to this day – that the British had a crucial lead – raises interesting perspectives on the predicament of science and technology under National Socialism and under a democracy.

At a first public conference dealing with radar, held in post-war Germany at Frankfurt in 1953, Watson-Watt lectured his German counterparts:

I believe that [British and American] success in radar depended fundamentally on the informed academic freedom which was accorded, in peacetime radio research, to my colleagues and myself, and to the intimacy and complete confidence between the operational user and the scientific and technical researcher and developer. If I appear immodest in my summary, it is because I believe the most valuable lesson from radar history is that of the intellectual and organizational environment from, and in, which it grew.[5]

A German view of the contrast between American and British advances in radar and Germany's effort accords up to a point with Watson-Watt's.

'An aspect of the German effort that seems to have differed from the Allied,' writes the German historian of radar Harry von Kroge, 'was the degree to which corporate rivalry affected the course of events. The numerous agreements that had to be made concerning licensing and post-war rights in order to smooth production will certainly seem remarkable to American and British readers.' Kroge continues: 'A puzzling aspect of German radar research was the delay imposed by severe secrecy in drawing on the many excellent universities and polytechnic institutes until very late in the war, whereas America and Britain made full use of these faculties from the very beginning.'[6]

The claim is that American and British technology was developed largely but not exclusively under the auspices of the military or the government, with access, where appropriate, to academic and civilian engineering expertise. The first German invention was developed by a private company with the encouragement of a naval research institution. There were serious constraints on their capacity to seek academic consultation because of the requirements of secrecy.

Contrasts between German and British organization and prioritization of research and development will come later in this chapter, but the point made by Watson-Watt and Kroge can hardly be generalized to include all technologies in Nazi Germany. The missile engineers, for example, drew freely and at an early stage on expertise for guidance systems and wind tunnels in the universities. Scientists like Norbert Wiener, moreover, who worked on guidance systems in the United States during the war, complained in 1947 that:

the measures taken during the war by our military agencies, in restricting the free intercourse among scientists on related projects or even on the same project, have gone so far that it is clear that if continued in time of peace this policy will lead to the total irresponsibility of the scientist, and ultimately to the death of science.[7]

Yet it is true to say that British and American attempts to solve specific problems, especially microwave-based radar equipment, were the result (as was the solution to uranium enrichment) of a generally pluralist and diverse research environment, despite wartime restrictions. That pluralist effort, allowing for different avenues to the desired result, and ample scope for trial and error, drawing on academe, industry and military research establishments, eventually found its way – despite many admitted difficulties – to a central research effort and thence to production. In Germany there was not so much a genuine pluralism, or free and yet collaborative diversity, as a series of rival monopolies, hampered rather than protected by secrecy between the armed services, and, as Kroge states, between commercial companies whose concern was intellectual property rights. Nor was there a clear route, as in Britain and, later, the United States, to a supreme command structure which took priority decisions after the scrutiny and recommendations of qualified committees.

Another crucial perspective on Nazi radar technology, outlined by the radar historian Robert Buderi, focuses on the Germans' 'military mindset'. Hitler and his war planners, comments Buderi, 'thought in terms of lightning offensives, and so did not push the development of mainly defensive radars'. This 'one-dimensional thinking, accentuated by the lack of civilian scientists with the clout of a Vannevar Bush, and Frederick Lindemann, had a devastating effect on Germany's war effort'.[8] Finally, there was a reluctance on the part of the German armed forces themselves to be enthusiastic about radar. General Ernst Udet, the incompetent technology chief of the Luftwaffe, objected to research on radar on the grounds that if it worked, 'flying won't be fun any more'.

The historian of science Professor David Zimmerman, however, adds another, more personalized component of technological success and failure. 'Much of the rapid early progress in the earlier years [in Britain],' he writes, 'was a direct result of the drive, energy and leadership of Watson-Watt.' But he adds a cautionary note: 'Paradoxically, it would be Watson-Watt's erratic, almost manic behaviour and lack of administrative skills which would be a significant factor in the failure to mount effective night defences ready in time for the Blitz.'[9] As for secrecy, Britain kept its early developments under wraps with the greatest of difficulty, since Winston Churchill, while still out of office in 1936, threatened to reveal Britain's radar secrets in parliament as part of his attempt to raise public alarm about German rearmament.

Death Rays and the Oslo Report

Radar and code-breaking were aspects of defensive intelligence in which Britain would excel in World War II, but these successes were achieved in the face of prejudice and political obtuseness. The British, in other words, were also lucky. Ironically, for a people who staved off defeat by mastering means of communication – radar and code-breaking – British intelligence officials had a very limited notion of what German weapons technology had achieved in the pre-war era. The failure to discover what Germany was planning, especially in early warning detection in the air and at sea, resulted largely from arrogance – the complacency of the victors of the previous war, as well as lack of judgment. Two instances illustrate the state of mind of British intelligence early in the war.

On 19 September 1939, less than three weeks after the Wehrmacht's invasion of Poland and the subsequent declaration of war on Germany by France and Great Britain, Hitler made a speech in 'liberated' Danzig, which, according to a British Foreign Office translation, warned of a 'secret weapon' against which there was no possible defence. The existence of a 'death ray' had been mooted years earlier by British government scientific advisers. As early as 1917 an operative in the Admiralty's top secret cryptographic department, Room 40, had approached Winston Churchill, then Minister for Munitions, to impart his idea for a 'heat ray', a notion, it seems, that he had picked up from H. G. Wells's *The War of the Worlds*. It came to nothing, as did many other bizarre 'death ray' suggestions down the years, including radio waves to detonate the bombs on board raiding aircraft, 'gas emissions' that were supposed to impede aircraft engines and 'radio waves' that would paralyse air crews.

The notion that the enemy might possess a weapon exploiting an unknown type of scientific wizardry began to haunt Britain's war cabinet before a shot had been fired. Prime Minister Neville Chamberlain urgently demanded information from the British intelligence services, and in consequence the physicist Reginald Victor (R. V.) Jones, appointed that year at the age of twenty-eight as Britain's first scientific intelligence officer, was ordered to scour through available espionage reports to shed some light on the reference. Drawing a blank, Jones eventually went back to the original of Hitler's speech to discover that the Foreign Office had merely misinterpreted a phrase in the text. The Führer had been referring not to a wonder weapon but to the Luftwaffe.

Not long after the 'death ray' panic, British intelligence received from its British Legation in Oslo, Norway, a package containing a prototype proximity fuse which, as we have seen, enables a missile to explode automatically in the vicinity of a target, a technology unknown in Britain at that time. In addition there was a document, which came to be known as the 'Oslo Report'. Seven typed pages in German outlined a list of top secret German weapons innovations. They included information about rocket research at Peenemünde, two new kinds of torpedo (one controlled by radio, the other by magnetic fuse), radio distance-measuring beams for guiding bombers to their targets, and two systems for advanced aircraft warning, or radar. To this day, it is not certain who posted this material, although it is likely that Paul Rosbaud (code name 'Griffin'), the anti-Nazi science publisher in Berlin, had been involved.[10] Hence the information came not from skilful intelligence gathering by the British, but the courage of an anti-Nazi German national. British intelligence officials, however, immediately cast doubt on the information and sought to discount it as a Nazi ploy. They were convinced that all such technologies, with the exception of radar, were beyond the capability of the British, hence they were beyond the capability of Germans.

R. V. Jones was alone in believing that the document had a ring of truth. He was particularly intrigued by the report's claim that an RAF attack on Wilhelmshaven in early September had been detected 75 miles out at sea by a chain of stations emitting pulses of radio energy.

In fact, by 1939 Germany's offensive radar system for directing aircraft to their targets, known as Y-Gerät, or Y apparatus, had been six years in development. Originally invented by Dr Hans Plendl, the system involved intersecting beams transmitted from the continent – Kleve, on Germany's Dutch border, and Schleswig-Holstein, close to the Danish border – capable of guiding German bombers with great accuracy to a target in the British Isles. Confident in the assumption that Britain was well ahead of the enemy, no thought had been given to the need for countermeasures that could one day make the difference between winning or losing the war. R. V. Jones, whose name is synonymous with the 'battle of the beams', now set about finding means to combat the directional technology, but not without resistance.

Churchill and Technology

While Britain stood alone, before America with its prodigious industrial and economic resources joined the war in Europe, Winston Churchill quickly learned that his country was engaged in a conflict in which science and technology was of central importance for survival, let alone eventual victory. The organizational structure which enabled Churchill to screen and prioritize research and development was a key factor in Britain's defence and ability to hold out. The success of Britain's management of technology was by no means a foregone conclusion, however, and had much to do with Churchill's ability to adapt to circumstances and learn from his mistakes.

Ignorant of science and, by his own admission, complacent about the power of the Royal Navy, Churchill's thoughts about a future war in Europe were almost despite himself still shaped by the horrors and fears of World War I. He was nevertheless fortunate in having as a scientific guide an unusual friend, Frederick Lindemann (Lord Cherwell). Churchill wrote in his history of World War II that while he was out of political office from 1931 to 1934, rusticated at Chartwell, his family home, he saw a great deal of Lindemann (whom we last met as secretary to the first Solvay Conference). Known as 'the Prof', he was a physicist by background, but something of a polymath and an able statistician.

Even in the twilight war, and before he was appointed Prime Minister, Churchill was still thinking of ground war in terms of the mud and attrition of the trenches. He was obsessed by a scheme he called 'Cultivator No. 6' and described it as 'a method of imparting to our armies a means of advance up to and through the hostile lines without undue or prohibitive casualties'. It was a machine for creating trenches at speeds up to 3 miles an hour, and could cut a channel sufficiently deep and broad to allow infantry and tanks unimpeded access across no man's land into the enemy trenches. By February of 1940 the cabinet and Treasury had approved the construction of 200 narrow machines of this sort and forty wide ones. In their final form the larger ones weighed more than 100 tons and were 77 feet in length and 8 feet high. They could cut in clay a trench 5 feet deep and 7½ feet wide at the rate of half a mile an hour, shifting 8,000 tons of soil. Three hundred and fifty firms were involved in making the parts and assembling them. It would be used, as Churchill put it, 'as part of a great offensive battle plan'. But he was forced to admit, following Germany's *Blitzkrieg* invasion of France, 'a very different form of warfare' had descended 'upon us like an avalanche, sweeping all before it'. In a typically Churchillian flourish, referring to the huge waste of time, money and energy expended on the 'cultivators', he averred after the war: 'I am responsible but impenitent.'[11]

Unlike the organized chaos of the Reich, with its warlords competing and overlapping for the attention and favour of an indolent Führer, Churchill, on becoming Prime Minister, assumed direct and wide-ranging powers over the technology of war, since his premiership absorbed the job of Defence Minister. The three military service ministers, 'friends of mine whom I had picked for these duties, stood on no ceremony', wrote Churchill.

They organised and administered the ever-growing forces, and helped all they could in the easy, practical English fashion. They had the fullest information by virtue of their membership of the Defence Committee, and constant access to me . . . There was an integral direction of the war to which they loyally submitted. There never was an occasion when their powers were abrogated or challenged and anyone in this circle could always speak his mind . . . the machinery worked almost automatically, and one lived in a stream of coherent thought capable of being translated with great rapidity into executive action.[12]

Casting his mind back to the prospect of a German invasion with which Britain was threatened in the summer of 1940, Churchill remembered after the war how he considered the need for scientific and technical innovation to produce countermeasures. 'This was . . . no time to proceed by ordinary channels in devising expedients.' In order to secure quick action, free from departmental processes, 'upon any bright idea or gadget' he kept in close touch with an 'experimental establishment' formed by a Major Jefferis at Whitchurch, a man 'possessed', according Churchill, 'with an ingenious and inventive mind'. Churchill also appointed Frederick Lindemann his 'personal assistant', and kept him close. Lindemann provided Churchill with a flow of statistics on the various war departments and offered him advice on a prodigious number of scientific topics. Lindemann wrote some 2,000 papers during the war.

Amidst the long-term high-tech developments destined for

success there were some panicky amateurish ones. Faced with Hitler's invasion, Churchill wrote, the best the scientists could come up with was:

a bomb which could be thrown at a tank, perhaps from a window, and would stick upon it . . . We had the picture in mind that devoted soldiers or civilians would run close up to the tank and even thrust the bomb upon it, though its explosion cost them their lives.[13]

He also seriously considered the use of poison gas against invading Germans on Britain's beaches.

In June 1940 Winston Churchill asked for a report on the amount of mustard or other types of poison gas in stock to be used in shells and aerial bombardment.[14] The idea was to bombard invading Germans on the beaches. According to Britain's Inspector of Chemical Warfare,

Low spray attacks on an enemy approaching our shores in open boats or after landing are likely to be effective if frequently repeated, and will ultimately result in 100 per cent casualties among the men hit by the spray. If the enemy are not wearing eye shields, a considerable number will be blinded unless they cover their eyes. They cannot do this and use their weapons at the same time. Low spray attacks are therefore likely to reduce the risk to other low-flying aircraft in bombing and machine gunning.[15]

Needless to say, there were qualms among the military and cabinet members about first use of poison weapons, particularly as the Reich's stocks of poison gas were believed to be five times greater than British stocks; but Churchill insisted on having the poison weapon option in a state of readiness.

By the autumn of 1940, however, Churchill had come to appreciate profoundly the scientific and technical nature of the war against Hitler. In a memorandum prepared for the cabinet in September, entitled 'The Munitions Situation', he stated that the war was not:

a war of masses of men hurling masses of shells at each other. It is by devising new weapons, and above all by scientific leadership, that we shall best cope with the enemy's superior strength. If, for instance, the series of inventions now being developed to find and hit enemy aircraft, both from the air and from the ground, irrespective of visibility, realise what is hoped from them, not only the strategic but the munitions situation would be profoundly altered . . . We must therefore regard the whole sphere of R.D.F. [radar], with its many refinements and measureless possibilities, as ranking in priority with the Air Force, of which it is in fact an essential part. The multiplication of the high-class scientific personnel, as well as the training of those who will handle the new weapons and research work connected with them, should be the very spear-point of our thought and effort.[16]

The Battle of the Beams

Following up the information supplied in the 'Oslo Report', R. V. Jones set about gathering clues to understand Germany's directional beams, including scrutiny of crashed Luftwaffe bombers, interceptions by the code-breakers at the decryption HQ at Bletchley Park and reports from German prisoners of war. By mid-June Jones had in his hands documents which revealed the existence of high-frequency radio signals being transmitted across the British Isles from the continent. At this point he was summoned to a dramatic and historic meeting with Churchill at 10 Downing Street. As Jones was shown into the cabinet room he found himself confronting Churchill, flanked by Lindemann and Lord Beaverbrook, Minister for Aircraft Production; also present were Tizard, Watson-Watt and the then Bomber Command chief Sir Charles Portal. Amazingly, the gentlemen were still heatedly discussing the German radio beams and their possible purpose. After half an hour of this R. V. Jones, still under thirty years of age, interrupted the political and military grandees with a polite opener: 'Would it help, sir, if I told you the story right from the start?' After an astounded pause, Churchill said: 'Well, yes it would.'

It took Jones just twenty minutes to convince Churchill of the reality of the Germans' directional radar system and to assure him that countermeasures could be devised. Churchill was gleeful. Striking the table, he cried: 'All I get from the Air Ministry is files, files, files!' But others around the table, Lindemann included, were still reluctant to believe that radio waves, such as they had identified, could travel around the curvature of the earth from Germany to the British Isles.

By the autumn of that year, however, Jones had discovered definitively that three, not two, radio-navigational networks were being employed as directional aids by German aircraft. Two beams drew the pilot towards the target and a third indicated the precise point for releasing the bomb load. Engineers in Britain's Telecommunications Research Establishment (TRE) consequently devised jamming signals which disturbed the frequencies of the directional beams. As the technicians became more knowledgeable and sophisticated in their countermeasures, they managed to alter some beams so as to lead the bombers astray. The BBC television transmitter at Alexandra Palace, quiescent through the war, was used to send out signals to jam the third bomb-release signal.

German Early Warning Radar Systems

The conviction that Germany did not possess radar as an early warning technology had stemmed partly from the absence of anything resembling the high radio towers so prominent on Britain's coastlines. Watson-Watt, in fact, had spent his vacation in the summer of 1937 looking for tell-tale aerials in Germany; finding none, he assumed that Nazi Germany had not plumbed radar's secrets. After studying the Oslo Report, which mentioned the use of short wavelengths which would have precluded large aerials, R. V. Jones thought otherwise. His suspicions were confirmed in the summer of 1940 when he received information that British aircraft had been intercepted as a result of a German system called 'Freya-Meldung-Freya'. Jones reflected that Freya was a Nordic goddess whose necklace was protected day and night by

Heimdall, a watchman who could see vast distances in every direction.

Meanwhile, in the autumn of 1940, a radar expert with TRE detected radio signals on a wavelength of 80 centimetres in the English Channel which came into operation when British ships were being fired on. He had stumbled on another highly secret German radar development, known as Seetakt, an advanced system integrated with gunfire control.

Jones realized that this gunfire control radar was not the Freya early warning system, which threatened to be of a quite different order of technology. The breakthrough finally came in February 1941, when Jones was shown certain 'curiosities' caught in reconnaissance photos of the Hague peninsula – two circular dishes some 20 feet in diameter on the edge of a field. Further scrutiny revealed that they were being rotated. It was indeed a Freya station. Work could now start on discovering the frequencies on which they operated in order to jam them, as well as the task of finding the sites of all the other Freya stations stretched out across the continent. In the meantime Jones had begun to concentrate on a type of equipment associated with the Freya system, referred to in the Oslo Report and occasionally mentioned in coded German messages as a Würzburg.

The messages spoke of these systems being transported to Romania and Bulgaria, where, in time, British receiving stations traced transmissions on wavelengths of 53 centimetres (the Oslo Report had given the lengths as 50 centimetres). Photo reconnaissance failed to turn up a visual clue as to the nature of the Würzburg equipment until November of 1941, when a small object was detected on a photograph of a Freya station at Cap d'Antifer. Further photographs revealed a parabolic reflector some 10 feet in diameter close by. Thereafter Jones began to deduce the full complexity of the Germans' early warning system, which was composed of the Freya long-range detectors and two types of associated Würzburgs – smaller systems that controlled the operation of searchlights and larger systems, Würzburg Riesen, which directed the German night fighters on to incoming RAF raiders.

He did not rest until he had captured an actual model of a Würzburg, when a Combined Operations raid of 119 British paratroopers dropped on Cap d'Antifer early in 1942 and brought most of the equipment back to Britain with the aid of a small naval force. British intelligence now had all the important components of a Würzburg in their hands and were in a position to calculate the range of wavelengths in order to devise appropriate jamming counter-measures.

German intelligence had shown no such foresight, persistence and cunning during the period in which Reichsmarschall Goering planned, and attempted to carry out, the destruction of Britain's air power by daylight raids as a prelude to invasion. When the Battle of Britain commenced in earnest in July 1940, Britain's early warning radar system was composed of twenty-one Chain Home stations (some of these structures towering 350 feet into the sky) and a further thirty Chain Home Low units giving early warning of invaders approaching the British Isles across the English Channel and the North Sea. Royal Observer Corps units acted as back-up inland, although their visual sightings depended on daylight and the absence of cloud cover.

German intelligence had neglected to identify the purpose of Britain's highly visible radar towers and aerials, even though the structures were recorded two years before the outbreak of war. Unarmed German aircraft had criss-crossed Britain under the pre-text of making weather observations for the Lufthansa airline. In 1939 a Zeppelin codenamed LZ 130, under the command of Luftwaffe signals section, flew along Britain's radar chain attempting to pick up radio signals; but nothing was learned.[17] From July to September the Luftwaffe failed to defeat the Royal Air Force, largely due to the ability of British radar defence to anticipate incoming raiders and respond with interceptors. Although Goering eventually guessed the purpose of the radar towers, he never understood the importance of their function and did not attempt systematically to put them out of action. As described earlier, following the bombing of London in an accidental raid on the night of 24

August 1940, Churchill ordered a bombing raid on Berlin and Hitler retaliated by pounding London night after night for a week. The onslaught on London, destructive as it was, gave the RAF and its bases a much-needed respite. When Goering returned to his original strategy, Fighter Command was ready for a decisive battle. The climax of the battle is customarily dated 15 September, when some seventy-nine Luftwaffe aircraft were downed for the loss of thirty-six RAF fighters. Thereafter Goering switched again to night raids on London and the industrial heartlands, while Hitler postponed his invasion of Britain indefinitely.

Meanwhile, as Britain increased its night raids against targets in Germany and occupied Europe, the Reich's engineers worked on improving radar defences. With headquarters at Zeist in Holland, Major General Josef Kammhuber had by 1941 divided the mainland into a grid of boxes 27 miles wide and 21 miles deep, each containing a Freya early warning dish and two Würzburgs – one to detect individual enemy planes, the other to direct a Luftwaffe fighter. By the end of 1941 the system could detect incoming raiders 200 miles out, and estimate the altitude of aircraft from 150 miles away.

The Magnetron

What gave Britain an inestimable technological advantage in radar, however, an advantage that would be shared with the Americans a year before they entered the war, was the development of a small device, about the size of a saucer, known as a 'resonant cavity magnetron'. Its development tells an instructive tale of military science and technology resulting in incalculable advantage to the Allies, and disaster for Germany.

At stake was the development of forms of radar that would reduce the bulk of equipment needed, and provide flexible and highly accurate information for the users. What was needed was a type of airborne radar small enough to be installed in a fighter aircraft, working on about 10 centimetres wave length so as to pinpoint Luftwaffe night raiders, and yet generating sufficient power to achieve long-range detection. The main goal was to achieve a narrow radar beam so as to pinpoint the target. This is determined by the width of the antenna compared with the wavelength. Hence the shorter the wavelength the smaller the antenna for the same beam width.

German radar historian Kroge claims that in the summer of 1935 German engineers at the private company GEMA were working with a rudimentary form of 'magnetron', originally invented by Albert W. Hull of the American firm General Electric in about 1920. The magnetron consists of a vacuum tube placed in a magnetic field so that electrons follow curved paths while travelling across the tube. German researchers found the magnetron unstable, preventing the expected maximum range of 20 kilometres from being attained consistently: 'only by constant tuning of the receiver could respectable distances be observed at all'.[18] In the end it was abandoned for other solutions.

The British research path to a useable magnetron had started at Birmingham University in 1939 with experiments for generating microwaves, which led to the construction of a 'cavity magnetron', a radical modification of Hull's device. Robert Buderi, the historian of radar, states that the invention came about 'accidentally on purpose'. Two physicists, John Randall and Henry Boot, conducting experiments under Marcus Oliphant at Birmingham University, put together the combined virtues of two devices – the traditional magnetron and a klystron, an American invention (from the Greek *kluzo*, meaning the breaking of waves on the seashore) discovered as part of an attempt to find a microwave power source for blind aircraft landing.

As Buderi puts it, 'The challenge rested in adapting the klystron's doughnut-shaped cavities to the cathode and node structure of the magnetron, which depended on cylindrical symmetry.'[19] As the American physicist I. I. Rabi described it a year later, when an early model was taken to the United States, the cavity magnetron was a kind of whistle operating under electric and magnetic fields. Buderi explains the technicalities:

Not unlike the way air flowing in front of a whistle hole causes a tone to be emitted – the frequency of which is largely dependent on the instrument's size and shape – the electrons oscillated at a specific radio frequency determined by the cavity dimension.[20]

The first test of the completed device was conducted on 21 February 1940, and three days later the researchers could confirm that it operated on 9.5 centimetres and generated an output of 400 watts. By May the output had increased to a prodigious 12–15 kilowatts. It was the dawn of microwave radar, but hard-pressed, battered Britain badly needed a major partner in its development and mass production.

Anglo-American Technological Collaboration

In early August of 1940 Churchill asked one of his top technology advisers, Henry Tizard, to travel to the United States with a basket of top secret discoveries; the cavity magnetron was among them. The device in its completed form appeared like a clay pigeon and could fit in the palm of a person's hand. Tizard and his team demonstrated the operation of the cavity magnetron on 19 September in Washington, DC, to a dumbfounded audience composed of members of America's National Defense Research Committee. In the course of the presentation it was revealed that the tiny device was capable of generating a thousand times more output than America's most advanced instrument on a wavelength less than 10 centimetres. Within a month a decision had been taken to establish a Radiation Laboratory at MIT, Cambridge, Massachusetts, later known as the Rad Lab.

By 1943 centimetric radar, employing the resonant cavity magnetron, was being used by the Allies not only for night interception in the air but in modified form for locating bombing targets 'blind' in Germany. It was only a matter of time before the Germans would discover the guidance secrets of the Allies in a crashed aircraft. On 2 February an aircraft belonging to a Pathfinder force (used to guide bombers on to their targets by dropping flares ahead

of the bomber formations) was shot down near Rotterdam and a
damaged cavity magnetron was retrieved from the wreckage. German engineers, referring to the device as the *Rotterdamgerät*, soon
confirmed that it was a centrimetric radar device and that it aided
night-fighter search or warning as well as bombing guidance.
Herman Goering, on learning of the equipment, declared:

We must frankly admit that in this sphere the British and Americans are
far ahead of us. I expected them to be advanced, but frankly I never
thought that they would get so far ahead. I did hope that even if we were
behind we could at least be in the same race.[21]

Professor Leo Brandt of Telefunken was ordered to reconstruct the
device so that it could be employed by Germany's radar, but on 1
March the equipment was destroyed in a bombing raid. That same
night, however, another magnetron was found in a bomber shot
down over Holland and Brandt went to work again.

German scientists hastened to incorporate cavity magnetrons
into their systems for bomb guidance, anti-aircraft detection and
gunnery and airborne radar, but by the time they were ready to
be deployed the war was approaching its end. In the meantime
Germany's researchers found methods of countering the benefits
of cavity magnetrons, in particular a detector called Naxos and
another device known as Korfu, designed to pick up 10-centimetre
radar wavelengths. Naxos was particularly useful for U-boats
detecting search planes operating on microwave to detect the
presence of periscopes. But as fast as the Reich engineers devised
countermeasures the Allies were developing 3-centimetre wavelength radar to catch their quarries.

Radar, however, important as it was to defend Britain's islands
from attack, was not the only form of information processing that
helped to reverse the tide of war. Equal to the effort, ingenuity and
crucial rewards was Britain's extraordinary success in the field of
deciphering Germany's sophisticated top secret codes.

21. Codes

The art of secret communication and espionage in wartime has always stood to make a significant difference to the conduct and outcome of the conflict. A remarkable feature of German cryptography in World War II, which involved sophisticated state-of-the-art mechanical encryption, was the refusal of the Reich's intelligence mechanical services to believe that its codes could be broken, indicating a fatal lack of respect for the technological achievements of their opponents. On the other hand, German intelligence dedicated considerable intellectual resources to successful decoding of the communications of their enemies and potential enemies.

The story of Germany's highly sophisticated encryption efforts in World War II begins in 1923, when Winston Churchill published his history of the Great War, *The World Crisis*, recounting how in September 1914 the British gained possession of Germany's secret naval codes. The publication of such a story for the whole world to read, let alone the Germans, who had been ignorant of the facts even five years after the war had ended, was unwise. Certainly it was to affect the decisions of the German military towards secret codes during the 1920s and 1930s.

In September 1914, a German light cruiser, the *Magdeburg*, was wrecked in the Baltic Sea. The body of a drowned German sailor was picked up by the Russians a few hours later; clasped to his chest by arms stiff with rigor mortis in the cold sea were the cipher and signal books of the German navy. On 6 September, the Russian naval attaché came to see Winston Churchill, then First Lord of the Admiralty. The official had received a message from Petrograd telling him what had happened, and that the Russian Admiralty with the aid of the cipher and signal books had been able to decode portions of some German naval codes. The Russians felt that as the leading naval power, the British Admiralty ought to have these

books. If the British would send a vessel to Alexandrov, the Russian
officers in charge of the books would bring them to Britain. The
books were eventually delivered to British cryptanalysts in White-
hall's famous Room 40, where they were used to decode German
secret communications routinely. When the Germans came to
write their account of World War I they recorded that 'the German
fleet command, whose radio messages were intercepted and deciphered by the English, played so to speak with open cards against
the British command'.[1]

One consequence of the publishing of the Magdeburg affair
(repeated in the official history of the Royal Navy in that same
year, 1923) was that Germany upgraded her secret codes in the
inter-war period, investing in a complex mechanical system that
would, in time, test Britain and America's decoding resources to the
limit, calling for science, mathematics, technology and engineering
skills far beyond the gentle pencil and paper arts of classical scholars
employed in World War I. Breaking Germany's most impenetrable
codes would eventually involve Britain in the building of Colossus,
the world's first computer, and a staff of some 12,000 workers, including the cream of the country's mathematicians. After the war,
Churchill was keen not to repeat his mistake of 1923. The plans of
Colossus were destroyed and Britain's code-breakers were obliged
to maintain strict secrecy about their activities until 1975. The special
algorithm employed in Colossus remains a secret to this day.

After Churchill's disclosure in 1923, Germany's military authorities turned to a 'cipher rotor' invention, a writing machine that
worked on a system of mechanical 'substitution'. The standard
device, which looked like a typewriter about a foot square and 6
inches deep, weighing under 30 pounds, had been developed by a
German electrical engineer, Arthur Scherbius, for use in commerce,
diplomacy and potentially for the military. He called it the Enigma,
the Greek word for riddle. Describing the machine to the Imperial
German Navy in 1918, Scherbius, then aged thirty-nine and living
in Berlin, had boasted that it 'would avoid any repetition of the
sequence of letters when the same letter is struck millions of times
. . . The solution of a telegram is also impossible if a machine

falls into unauthorized hands, since it requires a prearranged key system.'[2]

Creating a cipher with a rotor machine of the type that evolved into the Enigma involves striking the typewriter keys corresponding to the letters of the message, or the 'plaintext', and noting the 'ciphertext', the successive lit-up letters on a glass screen. Code historian David Kahn explains:

As each letter is enciphered the electrical current passes through the input plate contact for that letter, enters the rotor at the rotor contact opposite, winds through the rotor, emerges at a different position on the other face, passes into the output plate, and goes to the bulb underneath the ciphertext letter.[3]

The receiver of the coded message had to set up the machine with the same key system coordinates in order to retrieve the original plaintext.

The standard Enigma equipment used by the German armed forces included five rotors, from which three were chosen in setting up the machine for each period (normally of a day), giving sixty possible wheel-orders. Each rotor was fitted with an adjustable ring with the twenty-six letters of the alphabet inscribed on it. There were therefore twenty-six cubed (that is, 17,576) different settings of these rings, to give just over 1 million different possible set-ups of the 'scrambler unit' at the core of the machine. But the machine also included a plug board, which effected a substitution before the current entered the scrambler and the same substitution after it had left it to light up the enciphered letter. As this substitution, like that produced by the scrambler itself, was symmetrical and was used at both entry and exit, it preserved the two critical features of the scrambler component of Enigma encipherment: no letter could be enciphered to itself, and the overall encipherment was reciprocal: if A enciphered to P, then P enciphered to AQ. The use of 10 plugged pairs introduced a further factor of some 150 million million, bringing the total number of different ways of setting up the machine to approximately 159 billion billion. The set-up chosen for each day was circulated to all users in each particular network, usually as a monthly key-sheet.

In addition, the start position for each individual message had to be chosen by the originator, and communicated to the recipient without revealing it to an enemy. This information was conveyed in various ways by different users and at different phases of the war. For most of the war the army and airforce used a simple method, namely enciphering the start position (say DQX) on the machine at a start position (say RTG) also chosen by the operator. If DQX enciphered to KLB, the trigrams RTG KLB were sent as part of the preamble of the message. The recipient had merely to set his machine to the position RTG and tap out KLB, revealing the start position DQX of the message proper. So if an enemy knew the the machine set-up for the period in question, he would be able to decipher any message sent in it.

Scherbius, ambitious to market his machine, had also bought up in 1927 a rival system invented by the Dutchman Hugo Alexander Koch. By 1929, the year of his death (he was killed by a runaway horse), Scherbius had sold his invention to the German army and navy, which were to use different versions of it. After Hitler's assumption of power in 1933, and his rejection of the Versailles Treaty with its prohibitions on German rearmament, the demand for cipher machines from the armed services multiplied. Eventually the Luftwaffe, the SS, the Abwehr (German military intelligence and counterintelligence) and the state railways, the Reichsbahn, were all using Enigma machines, and the navy and the army were coordinating their Enigma systems with a view to greater security.

The German high command added a fail-safe security system with the aid of a *Stichwort*, or cue word. Even if the enemy managed to obtain an Enigma machine, and even if he managed to secure the daily key-lists, the *Stichwort* broadcast to operational U-boats would direct the operators to open a sealed envelope in which was a slip of paper containing a key word. The operators would then follow a complex procedure, adding letters of the key word to the settings specified for the daily key.

The Polish Code-breakers

The Germans began to use Enigma machines in 1926, instantly baffling British, French and American code-breakers, all of whom eventually gave up. One country which refused to accept that the new encryption was unbreakable was Poland. Trapped between the Russian giant and a Germany which felt cheated of territory now returned to Poland following the First War, the Polish government was convinced that it could not afford to be ignorant of Germany's secret intentions.

Captain Maksymilian Cieski was in charge of the Polish cipher bureau, Biuro Szyfrów. He was familiar with the commercial Enigma, which would prove helpful in the long run, but the machine employed by the German military was different in the wirings of the scramblers. The Biuro had a lucky breakthrough, however, as a result of the espionage of a disaffected German, Hans-Thilo Schmidt, who passed on copies of code books and settings to the French secret service, which in turn passed them on to their allies the Poles, and hence to the Polish cipher bureau.[4] Schmidt's treachery enabled the Poles to design an Enigma machine, but the cryptanalysts now had to break the settings. The virtual impossibility of the task prompted an important recruitment decision.

Cryptanalysis had traditionally been a task ideally suited, it was thought, to classicists and linguists. In the light of the mechanical complexity of Enigma, the Polish cipher bureau decided to turn to mathematicians, and among their early intake, in 1929, was the 24-year-old Marian Rejewski. Rejewski, a German speaker, had matriculated in mathematics at Göttingen, before proceeding to the university at Poznań, where he studied statistics and cryptology. He found his true vocation attempting to crack the Enigma machine by concentrating on the patterns of the 'message-key' which was repeated twice at the start of every message. Following a laborious process of checking each of 105,456 'scrambler settings', which took a year, Rejewski at last began to penetrate the mystery of the Enigma cipher. As the code historian Simon Singh puts it,

'Rejewski's attack on the Enigma is one of the truly great accomplishments of cryptanalysis.' He adds that the Polish success in breaking the cipher can be attributed to three factors: 'fear, mathematics and espionage'.[5] The Poles in consequence were able to decode top secret German messages through the 1930s and improve their search for the correct scrambler setting with the aid of six parallel mechanical devices known as *bombes*.

As the Germans prepared for war, however, their encoders increased the security of the Enigma system by increasing the number of plug board cables from six to ten, as outlined above: there were now twenty swapped letters before the message entered the scramblers, increasing the number of possible keys, as we have seen, to 159 billion billion. The sudden increase of security was a serious blow for the Poles as Hitler's *Blitzkrieg* warfare depended on communications between air, armour and infantry. Weeks before the outbreak of war, however, the French secret service arranged for the British cryptanalysts to meet with Rejewski and his team at the Polish cipher bureau. Before war broke out on 1 September 1939, the British had received an Enigma replica and blueprints of the *bombes* which had successfully cracked the codes through much of the decade.

The story of the breaking of the Enigma code during World War II and the 12,000 cryptanalysts who eventually worked at Bletchley Park in the English countryside, has been told and retold many times in recent years. Less known are the efforts made by the Germans to crack Allied secret codes.

German Code-breakers

Hitler's ambitions for a speedy war of aggression, and the inherent waste and overlap of polycratic rivalries in the Reich, were to give rise to many areas of negligence in defence, including signals intelligence. Unlike the extraordinary centralization of the code-breaking organization centred at Bletchley Park in Britain, on which Churchill kept an avuncular eye, famously resolving a bottleneck in resources with a resounding memorandum ordering

'ACTION THIS DAY', there were no less than seven code-breaking organizations in Hitler's Reich. These included the Foreign Office, the navy, the army, the Luftwaffe and various security offices. In all there were some 6,000 personnel working on cryptanalysis, but spread through these different organizations. They were often unaware of the work of their 'rivals', and hence had no idea as to overlap and duplication.[6] In 1942 it was suggested to Hitler that the cryptanalysts should be brought under one organization controlled by the Hungarian specialist Major Bibo; but Hitler refused to sanction the move.[7] The most efficient of these organizations was the B-Dienst (Beobachtungs-Dienst), or Observation Service, a naval cryptanalysis unit under the ultimate control of Admiral Dönitz. The B-Dienst was in fact divided into three different parts, a listening service, a decoding operation and an evaluation service.[8] Even before the war began the B-Dienst cracked the fairly simple Royal Navy Administrative Code, or Naval Code, used by enlisted sailors. This made it easier for the B-Dienst to crack the Naval Cypher used by British officers for top secret communications. Those naval systems, which were not mechanical, involved a code book with additives which the Germans reconstructed with little difficulty. By the latter part of 1940 the B-Dienst was deciphering half of all the Royal Navy's signals.[9]

The British and the Americans, however, had developed a mechanical encryption system based on the Enigma model. The British machine was known as Typex, or sometimes Type-X; the American device was called Sigaba. Both machines were complex systems for mechanical substitution and adding using a five-rotor system. The existence of these machines was no secret. There was an open reference to them in an essay written by Abraham Sinkov, a US army cryptanalyst, published in Rouse Ball's *Mathematical Recreations and Essays*. Sinkov commented that 'So far as present cryptanalytic methods are concerned, the cipher systems derived from some of these machines are very close to practical unsolvability.'[10] It is stated by Simon Singh that the Germans did not attempt to decipher the British and American mechanical systems, since they regarded them as effectively invulnerable.[11] Unlike the Enigma machines, moreover, the Allied mechanical encoders, and their successor, a joint British–American system known as the Combined Cipher Machine, were used sparingly for high command signals traffic. German cryptanalysts, however, using IBM-patented Hollerith machines, boasted after the war of a high rate of success (up to 50 per cent by some accounts) in breaking Allied manual codes.

The Hollerith machine was invented by Herman Hollerith, son of migrant Germans, who became the founder of the IBM (International Business Machines) corporation in America. After graduating in engineering in 1879, Hollerith became an assistant at the US Census Bureau in Washington, DC. Taking his cue from the mechanization of player pianos, Hollerith designed a machine for tabulating data with cards punched with holes that could be 'read' by means of adjustable spring mechanisms and electrical brush contacts that could sense the holes as they sped through a feeder. What was to be known as the Hollerith machine could sort millions of items of data at high speed. As Edwin Black, author of *IBM and the Holocaust*, comments, 'It was nothing less than a nineteenth-century bar code for human beings.' Hollerith soon realized that his system could be used to analyse data thousands of times faster than human analysts. His machine had saved the US government $5 million on census labour; now it was in demand for a wide range of accounting, engineering, scientific and actuarial problems. The Hollerith machine, licensed by IBM, was to be widely used in Germany not only to assist the war effort, in operations like code-breaking, but in the task of collecting information on Jews and others destined for the death camps.

By late 1941, after America entered the war, a new Allied code, a hand or book code known as Naval Cypher no. 3 (also known as Anglo-US Cypher or Convoy-Cypher), employed for protecting Atlantic convoys, was in place. But the Germans had broken it by March 1942 to the extent of retrieving up to 80 per cent of the signals traffic. Parts of the cryptanalyst services were so advanced with IBM methods of decoding that at least one operator, Lieuten-ant R. Hans-Joachim Frowein of the B-Dienst, discovered a method of breaking into Enigma using 70,000 IBM cards.

Late in the war the Luftwaffe began to employ a cipher system known as 'Enigma Uhr' (*Uhr* means clock in German). Plugged into the Enigma, the device automatically generated a massive number of extra permutations on the encipherment system, but this was also rapidly cracked.[12]

The Enigma, Typex and IBM technology existed in a kind of limbo between old technology and genuine binary computer technology. Binary systems were eventually exploited and rapidly developed during the war, as we shall see, by Bletchley Park cryptanalysts and by engineers based at Britain's Post Office research division at Dollis Hill, London.

German Computers

A start had been made in Germany on binary computers with the work of a German engineer, Konrad Zuse, but his inventions were never employed in cryptanalysis. While still a student, Zuse had taken Babbage's famous nineteenth calculating machine to its logical next stage. Starting in 1932, aged only twenty-two and working alone, Zuse used a binary, on-off, zero-one representation by means of holes punched in paper, and later pins that could be locked.[13] As an engineer in the aircraft industry, a job he left, to the distress of his parents, he became aware of the 'tremendous number of monotonous calculations necessary for the design of static and aerodynamic structures'. So he was designing and constructing calculating machines suited to solving such problems automatically. Nobody at that time, he commented, 'knew the difference between hardware and software'.[14]

In 1933 a manufacturer of calculating machines had told him over the telephone that 'computer' technology was already exhausted and that there was nothing new to be done. The manufacturer nevertheless came to Zuse's workshop, where the young man converted him by demonstrating the principle of digital, or binary, code calculations.

Thereafter Zuse built his Z1 and Z2, which were test models of an electromechanical relay machine. In 1939 Zuse was called up

into the army, which prompted the manufacturer to write a letter
to Zuse's major informing him that the young inventor should be
given leave to complete work that would be useful for the aircraft
industry. Zuse's commanding officer read the letter and said: 'I
don't understand that. The German aircraft is the best in the world.
I don't see what to calculate further on.'

Six months later Zuse was freed from military service, to work
as an engineer in the aircraft industry. But he persevered part-time
with his computer design, creating by 1941 his Z3, which had
an electromechanical memory composed of relays as well as an
electromechanically arithmetical unit. Z3 has also been described
as the first operational programme-controlled calculating machine.
It was employed by the German aircraft manufacturers to solve
simultaneous equations associated with metal stress. Zuse's proposal
for a computer based on valves, however, was rejected because of
the perception of an imminent end to the war. Meanwhile Zuse
began another computer he called Z4, which survived the war after
being moved from site to site to avoid Allied bombing. Zuse can
lay claim to having made the first programme-controlled model
computer, but it was not employed as a significant wartime
technology.[15]

Colossus Versus the Lorenz Machine

A historic token of the arrogance of the Reich's intelligence services
for what Britain in particular could achieve in cryptanalysis can be
seen today at Bletchley Park near Milton Keynes north of London.
In a wartime shed which is now a museum there is to be found a
humming, clicking, buzzing contraption. The machine exudes an
acrid stench of hot Bakelite and vacuum tubes and appears to be
assembled from giant pieces of Meccano and an assortment of
oddments from a telephone exchange of the 1930s. There are two
cabinets composed of panels, switches and festoons of coloured
wire, and the whole thing stands about 8 feet high, 3 feet wide and
16 feet long. A steel scaffold with gleaming spools runs a length of
telegraph tape past a photoelectric 'eye' that reads off information

at a rate of 5,000 characters per second. In the depths of the machine, 1,500 valves and thousands of plugs and circuits flicker and chatter as they process the information from the tapes.

Thought to have been long extinct, this curious electronic monster is nothing less than a rebuilt Colossus, Britain's supreme secret of World War II, and arguably the country's greatest scientific and technological achievement of the century. Its purpose was to break Germany's most complex secret codes in a war of national survival, and it provides a significant contrast with the conduct of science and technology under Hitler.

Colossus was built to do 'Boolean' calculations: a method of working out logical processes using symbols; in other words expressing ideas such as 'true, false' or 'and, or' purely in binary terms of zeros and ones, or on/off signals. The inspiration behind its making owed much to the genius of Alan Turing, the Cambridge mathematician who 'invented' the idea of the computer as a 'thought experiment' to tackle mathematical problems by mechanical means. Technically known as a 'large electronic valve programmable logic calculator', Colossus was built to decode messages from the Lorenz machine, the next generation of the Enigma, which scrambled messages between members of the Nazi high command, including directives from Hitler himself. According to German historians, the intelligence gathered by Colossus shortened the war by at least two years. The German military writer Hans Meckel believes Colossus prevented Germany having to be bombed into submission, as Japan was, with an atomic bomb. The story of Germany's Lorenz, and Britain's Colossus, illustrates the crucial importance of wartime secret codes and Germany's failure to recognize the vulnerability of its own systems despite their complexity.

The world's single surviving Lorenz machine is also to be found at Bletchley Park. It is a massive precision instrument with twelve glittering steel rotors encased in cast iron; it is heavy and fragile and needs several people to lift it. The machine, which was named after the company that made it, sent its messages mainly by teleprinter land-line (although it could also use the airwaves) using the international system of five-bit digital signals, made up of holes and spaces to represent letters of the alphabet. The Lorenz-type code employed two sets of confusing additives. Each original input character was changed by adding two sets of characters to it that were internally generated by the machine. In this way the Germans managed to obscure each original punched message to the tune of 10^{19} possibilities. Setting the cipher wheels involved moving tiny mechanical 'lugs' – 501 in all – to a specific but regularly altered pattern given by the German encoders. The Germans were convinced that the Lorenz code was unbreakable. A token of their blind faith, not only did they allow Hitler to speak directly to his generals using the machine, but on two occasions they actually sent out the cipher-setting sheets for the next month on the Lorenz machine itself.

In order to decipher a Lorenz-coded message the German recipient needed to pass the received enciphered punched tape through an identical Lorenz machine with its wheels pre-set to the same 'start-wheel' positions ordered on any particular day. These start-wheel positions determined the configurations of the wheels, like a twelve-wheel one-armed bandit that has been 'fixed' to spin and stop in what looks like a purely random fashion. The start positions were changed every month at first, but as the war progressed the Germans altered them each week and eventually each day.

Colossus was designed to discover the start-wheel positions used on the Lorenz machines on any particular day. But the British code-breakers also had to re-create a machine identical in its functions to the Lorenz itself in order to decipher the message. The British version of the Lorenz machine, known as a 'Tunny' (all teleprinter ciphers were nicknamed after fish by the British), was designed by engineers studying the principles of the Lorenz cipher system without ever having seen an example of the German original. How the British managed to improvise the world's first computer, and the accompanying 'Tunny' machine, owes much to a peculiar conjunction of mathematical and engineering genius – and a remarkable element of chance.

Land-line teleprinter traffic was virtually impossible to intercept in occupied Nazi territory, but the same signals were occasionally transmitted across the airwaves in the form of a high signal for a space and a low signal for a mark, creating a binary code. Towards the end of the war the British and Americans, and the Resistance, would deliberately target land-lines and teleprinter transmission stations precisely to force more traffic on to the airwaves, where they could be intercepted.

The first breakthrough in understanding the nature of the Lorenz cipher occurred on 30 August 1941, when a German operator was given a message with 3,900 characters to send by radio transmission. He correctly set up his Lorenz machine and punched the message by hand. Then the receiving end replied that the message had not got through and requested that it be sent again. At this point a fatal mistake was made. Both operators put their Lorenz machines back to the same pattern of start-wheel positions and the transmitting operator tapped out the message once more on the same settings, against all regulations and good sense. He made some different key strokes in the process. After a few characters he put an extra space here or an extra line-feed there, or shortened a word. Thus, when the two versions were compared, the two sets of transmitted characters had slipped out of synchronism.

The phenomenon of a repeated message of this sort was nick-named a 'depth' by the code-breakers. It was like a gift from heaven. Both streams of cipher had been intercepted at the receiving station at Knockholt in Kent and were passed on to Bletchley Park, where they were scrutinized by a top code-breaker who immediately realized their significance. By mulling over both versions of the message for more than two months, he and his team managed to extract what code-breakers call 'pure key stream', that is, the scrambling pattern generated by the Lorenz machine. This was achieved by a complicated process of matching what they speculated to be the possible patterns of lug settings on the wheels of the machine. It was the first clue as to what sort of machine had generated the code. It now took the skills of a second code-breaker, a young Cambridge chemist turned mathematician called Bill Tutte, to work out the structure of the Lorenz machine.

By early 1942 the code-breakers understood the Lorenz machine

sufficiently well to re-create it. The remaining problem, and it was
a daunting one, was how to discover the start-wheel positions for
each message set on any particular day by the German operators.
Solving the start-wheel positions employed on a single message of
a couple of thousand words by hand would have taken fifty or
more code-breakers up to two or three months. At this point the
senior code-breaker at Bletchley, Max Newman, had the idea that
it was going to be possible to mechanize that part of the deciphering
process.

The help of the physics group of the Post Office research centre
at Dollis Hill was enlisted, and within a period of a few months they
devised a series of clumsy machines known as 'Heath Robinsons',
which used a double teleprinter tape to process information. These
contraptions were early versions of the Colossus. But there were
constant problems with keeping the tapes in synchrony and they
failed to do the job swiftly and efficiently. Tommy Flowers, a senior
Post Office engineer, believed, contrary to almost universal opinion
at the time, that electronic equipment using many hundreds of
valves could operate continuously without failure. Acting on this
faith he solved the problem by constructing what was essentially a
computer in the modern sense. He employed some 1,500 valves as
well as other electronic devices to represent on-off, or zero-one,
binary states. This dispensed with the double-tape problem. Flowers
and his team built the first Colossus machine in just eleven months
and it worked within hours of being set up at Bletchley Park. The
ease with which the two organizations, the British Post Office and
the huge cryptanalysis unit at Bletchley, collaborated contrasts yet
again with the wasteful rivalries of the Reich.

The authorities at Bletchley had not been prepared to underwrite
the cost but the Director of Post Office Research had sufficient
faith in Flowers and his team to take it on his own budget. It is
interesting to reflect that what was essentially the first working
electronic computer, albeit with a stored programme, undertook
work which must rank as amongst the most important in comparison with any other of the millions of computers that have been
brought into use subsequently.

The Colossus generated the key streams – that is, the sequence of symbols on the wheels of the German Lorenz machine – internally in its electronic circuits. It read the intercepted message tape at 5,000 characters a second, comparing the tape of the intercepted enciphered text with these internally represented key streams. Then, making some very sophisticated cross-correlations, it found the start-wheel positions for the particular enciphered message.

Information processed by Colossus reassured the Allies that Hitler had swallowed the elaborate deception plan put in place in the months before D-Day. This involved the deployment of 'phantom armies' on the south and east coasts of England, composed of cardboard tanks and bogus landing craft, aimed at deceiving the Germans into thinking that the Allies would invade across the Pas de Calais. As a result the Germans were unprepared for the assault on the Normandy beaches.

Colossus also deciphered messages from Field Marshal Karl von Rundstet after D-Day ordering the German generals to keep the Panzer divisions in reserve in Belgium instead of unleashing them on the Allied armies to the south. This gave the Allies greater confidence to consolidate the invasion without diverting forces to counteract a large-scale tank assault.

PART FIVE

The Nazi Atomic Bomb
1941–1945

22. Copenhagen

On Sunday 15 September 1941, Werner Heisenberg travelled from Berlin to occupied Denmark. Heisenberg's official purpose was to participate in a lecture series on astrophysics at a German propaganda institute in Copenhagen; but he was evidently looking forward to meeting his old mentor and colleague, Niels Bohr.

The war was going well for Germany. With Poland, France and the Low Countries fallen, and the Wehrmacht cutting swathes through the armies and the territories of Russia, it looked as if German hegemony of Europe and the East was a *fait accompli*. Hitler's armies looked unstoppable in Russia, and the United States seemed determined to stay out of the war. Britain might even yet do a deal with Germany, bringing the conflict in Europe to a swift end.

In September of 1941 Heisenberg may well have assumed that the war would not have long to run, that Germany was bound to win and that German atomic physics was at least abreast if not ahead of developments in Britain and America. Given the many unsolved difficulties and imponderables, it probably seemed to Heisenberg that the exploitation of atomic physics for either power or weaponry lay well beyond the war's end. In 1941 no method of separating U-235 from natural uranium had been practically perfected in Germany; but even if it *were* to be perfected at some future date, only negligible amounts could be produced despite the commitment of enormous industrial resources. Indeed Niels Bohr himself had talked in 1939 of transforming the whole of the United States into a factory in order to achieve this fissionable 'enriched uranium'.

Since Heisenberg had no clear idea of the critical mass (scientists, as far as he knew, were still thinking in terms of tons), the question of the vast effort and time necessary to obtain those quantities seemed to put the U-235 path to a bomb beyond feasibility for the foreseeable future. He knew, however, that a bomb might be achieved with plutonium, the by-product of a reactor, but there were many outstanding technical problems involved. No such reactor had yet been built, nor was the construction of one even approaching completion. Then again, even if plutonium became available, the German atomic scientists had no precise idea as to how to make a bomb with it.

The state of Heisenberg's atomic bomb physics at this time was important for what was said, or at least understood to have been said, by both sides in the conversations that took place between Heisenberg and Bohr in Copenhagen. In time, their separate versions of these conversations would be of crucial importance to the conscience of at least one of them: Niels Bohr.

There have been disagreements about when and where these conversations about atomic physics took place. But an as yet unpublished letter written by Heisenberg seems to shed more light on these disagreements.[1]

Heisenberg wrote a letter to his wife while he was actually in Copenhagen during that third week of September. It was written in three stages: on the Tuesday, on the Thursday evening, then on the Saturday before his departure. He told his wife that, when he arrived in Copenhagen on the Monday, he had hurried across the city from the station under a clear and starry sky to Bohr's house.[2] He wrote that the discussion soon turned to human questions and the misfortunes of our time; later, he wrote, he sat for a long time alone with Bohr. It was after midnight when Bohr accompanied him to catch a tram.

Two days later, on the Wednesday, he again visited the Bohrs at their home. He reported to his wife that a young Englishwoman was present who tactfully withdrew during a political conversation when he found himself obliged to defend 'our system'. It is likely that another conversation, the famous conversation over which they disagreed in later years, took place at the end of that evening, when they left the house to walk again to the tram or to Heisenberg's hotel. Then, apart from other meetings that took place at Niels Bohr's institute, there was a third social meeting (unknown until this letter came to light) on the Saturday, when everything was pleasant between the two men. On this occasion Bohr read aloud and Heisenberg played Mozart's piano sonata in A Major, the sonata that ends with the *Alla Turca* movement. I shall return later to the significance of this last evidently cordial meeting.

Heisenberg, accompanied by his younger colleague Weizsäcker, had made this visit to Copenhagen because of the success (in the perception of Nazi propagandists) of an earlier visit by Weizsäcker in March of 1941. On that occasion Weizsäcker had lectured before the Danish Physical and Astronomical Society on a theme far removed from the crises and carnage of Hitler's war, namely: 'Is the world infinite in time and space?' The aim evidently was to win over the hearts and minds of Danish intellectuals and scientists, and in particular the researchers within Niels Bohr's famous institute. In March, Weizsäcker had also accepted an invitation to deliver a lecture entitled 'The Relationship between Quantum Mechanics and Kantian Philosophy'. After some wrangling between Weizsäcker and the Reich Education Ministry, the Party chancellery also gave Heisenberg permission to travel to Copenhagen provided that he kept a low profile.[3]

Heisenberg's reception by the Danish scientists was cool. When he came to give his arranged talk on cosmic radiation, on the Friday, only a few members of the German colony in Copenhagen attended, turning up at the last minute. The scientists from Bohr's institute boycotted the session. At one point during the visit, however, Weizsäcker accompanied the director of the German Cultural Institute to Bohr's institute, where he repeated his symposium lecture on stellar fusion. It was on this occasion that Weizsäcker, so it has been reported by the Danes, contrived to force an encounter between the German Cultural Institute's director and a reluctant Bohr.[4] Bohr quite naturally was at pains not to be seen fraternizing with the Germans.

According to Bohr's assistant, Stefan Rozental, Heisenberg came to lunch several times at Bohr's institute during that week, and offended his hosts by suggesting that it was important for Germany

to win the war. According to some of the Danish scientists, Heisenberg made offensive remarks to the effect that war was a 'biological
necessity'. Bohr remembered in later years that Heisenberg had said
that Hitler's victory was inevitable and that it would be unwise to
doubt it.[5] Bohr was apparently angry at Heisenberg's comments,
but, as we have seen above, he saw Heisenberg socially on more
than one occasion and they had an opportunity to walk in the open
air, where they could talk freely. None of their conversations was
committed to paper, but there is an anecdotal account of uncertain
provenance, and therefore best left out of the picture, that Heisenberg scribbled a drawing of a nuclear reactor – or perhaps a nuclear
bomb – on a piece of paper, which he passed to Bohr.

Heisenberg later insisted that their critical conversation took
place outside;[6] Bohr thought that it had occurred in his study.[7]
Whatever the case, the conversation seems to have got off to a bad
start with an exchange about Germany's destruction of Poland.
Bohr thought the invasion unforgivable, whereas Heisenberg saw
saving grace in Germany's less destructive treatment of France.
Heisenberg went on to say that he thought Germany's imminent
victory over Russia a good thing. Heisenberg then changed tack
to recommend that Bohr make contact with German officials in
Denmark in order to secure a degree of protection in the event of
the Nazi occupation of Scandinavia turning nasty, Bohr being
partly Jewish.

At this point Heisenberg, according to his own account, changed
the subject to inquire whether Bohr considered it right for
scientists to do research on uranium in time of war. To which Bohr
responded: 'Do you really think that uranium fission could be
utilized for the construction of weapons?'[8]

Heisenberg recollected at various times after the war different
reactions to this question. He was consistent in remembering that
he told Bohr of his knowledge of a link between uranium fission
and a bomb, hardly an item of news, since the speculation had been
published two years earlier for all the world to read in the *New York
Times*. He later recorded that he informed Bohr that he was engaged
in research on a nuclear weapon, but writing in 1957 to Robert

Jungk, author of *Brighter than a Thousand Suns*, he declared that he had told Bohr: 'I know that this is in principle possible, but it would require a terrific technical effort, which, one can only hope, cannot be realized in this war.'[9] Heisenberg nevertheless recollected in his memoir *Physics and Beyond* that Bohr was 'so horrified that he failed to take in the most important part of my report, namely, that an enormous technical effort was needed'. Heisenberg commented that the point he was endeavouring to make was 'so important precisely because it gave physicists the possibility of deciding whether or not the construction of atom bombs should be attempted'. Scientists had it in their power, Heisenberg remembered telling Bohr, to advise their governments that 'atom bombs would come too late for use in the present war, that work on them therefore detracted from their effort', or else they could argue that they might be brought into the conflict 'with the utmost exertions'. Heisenberg, according to his letter to Robert Jungk, finally put his cards on the table: 'I then asked Bohr once again if, because of the obvious moral concerns, it would be possible for all physicists to agree among themselves that one should not even attempt work on atomic bombs, which in any case could only be manufactured at monstrous cost.'[10]

Bohr, according to Heisenberg, bristled at this suggestion. Heisenberg interpreted Bohr's reaction almost a quarter of a century later in an interview with David Irving, controversial historian and author of *The German Atom Bomb*. Heisenberg told Irving that the idea of physicists collaborating not to make atom bombs seemed to Bohr 'terribly unreasonable' and tantamount to being a 'pro-Hitler . . . desire on my part'. Heisenberg said that he could see Bohr's point: 'Hitler had driven these good people to America and so he can't be surprised if they make atomic bombs.' He went on to tell Irving, however, that 'if we made atomic bombs we would bring about a terrible change in the world. Who knows what would happen from this? I was scared of everything, also this possibility.'[11]

Bohr's Unpublished Letters

The character of Margarethe, Bohr's wife in Michael Frayn's play *Copenhagen*, asks at the outset: 'Why did he come to Copenhagen?' The historian Paul Lawrence Rose, the harshest of Heisenberg's recent critics, believes that Heisenberg went to Copenhagen to persuade his old mentor to collaborate in the Nazis' atomic research. In 1941, as noted above, it looked as if Hitler was the victorious conquerer of Europe and the East. Hence what upset Bohr according to Rose was not fear that a German atomic bomb was imminent, 'but rather disgust that Heisenberg was planning for atomic research in the imminent Pax Nazica'.[12] Thomas Powers, on the other hand, author of *Heisenberg's War*, insists that Heisenberg had got across to the Allies one piece of information, loud and clear: 'the Germans were interested in a bomb'. In Powers's view, Heisenberg had betrayed the single most important fact about the German bomb programme – the fact that it existed. Heisenberg was telling the Allies, through Bohr, the direction of German research: hence Heisenberg emerges as an unusual kind of hero.

New evidence from the Niels Bohr Archive in Copenhagen tends to show, however, that as far as Bohr was concerned something rather different had passed between them and had given cause to torment him in later years.

During and after the war some of the leading atomic scientists on the Allied side were disturbed by their work and felt the need to justify their research. We have seen how Otto Frisch, for example, wrote that he squared his conscience with the reflection that the Germans were probably racing to be first to make an atomic bomb. We have seen how Einstein, a man of peace, encouraged President Roosevelt to take action on the reasonable basis that Germany would dedicate its physicists to atomic research. Later in this narrative we shall see how the physicist Joseph Rotblat resigned his post at the American bomb laboratory at Los Alamos as soon as he realized that the Germans were nowhere near possessing the weapon. In other words, the justification for the bomb in the eyes of a significant constituency of scientists from Leo Szilard to Joseph

Rotblat rested on the conviction that the Germans might just be ahead. They dealt with their moral scruples by considering the horrifying notion that Hitler might get there first and blackmail the world into capitulation.

How was Bohr affected by these considerations? Bohr, the highly civilized, internationalist scientist, acutely sensitive to political and moral nuances, was clearly deeply affected. But it has taken the release of a series of documents from the Niels Bohr Archive in Copenhagen in 2002 to gauge the precise extent and nature of his thinking.

In the long-running debate over the Copenhagen meeting, reference has often been made to the draft of an unsent letter from Bohr to Heisenberg, which could shed light on Bohr's side of the story. The released papers comprise eleven documents in all, none of them actually sent. The most interesting of these deal with Bohr's fierce objections to the letter Heisenberg sent to the author Robert Jungk, published in English in *Brighter than a Thousand Suns* in 1957. Of great interest are Bohr's repetitions and progressive emphases and deletions, revealing a deeply troubled consciousness about the impressions he had gained from his conversation with Heisenberg and the remote consequences of those impressions.

According to Heisenberg, in the Jungk letter, he had told Bohr:

if [atomic bombs] were easily produced the physicists would have been unable to prevent their manufacture. This situation gave the physicists at that time decisive influence on further developments, since they could argue with the government that atomic bombs would probably not be available during the course of the war.

He told Jungk that he had said, further, that the construction of atomic weapons was 'in principle possible, but it would require a terrific technical effort, which one can only hope, cannot be realized in this war'.[13]

But in draft after draft of Bohr's unsent riposte to this claim he expressed, with subtle, and significant, changes of emphasis, his conviction that Heisenberg had by no means given him to understand that Germany's nuclear weapons programme was lagging, but that it was forging ahead at full speed.

In the earliest draft, however, he concedes that Heisenberg had spoken in 'vague terms' on this topic. The precise text reads:

I also remember quite clearly . . . in vague terms you spoke in a manner that could only give me the firm impression that, under your leadership, everything was being done in Germany to develop atomic weapons and that you said that there was no need to talk about the details since you were completely familiar with them and had spent the past two years working more or less exclusively on such preparations.[14]

In the next draft, however, Bohr's reference to Heisenberg's 'vague terms' disappears, and he hardens the certitude of the 'impression', allowing no room for vagueness or indeed the least misunderstanding:

I remember quite definitely the course of these conversations . . . when without preparation, immediately you informed me that it was your conviction that the war, if it lasted sufficiently long, would be decided with atomic weapons, and I did not sense even the slightest hint that you and your friends were making efforts in another direction.[15]

In a further draft, which is entitled 'notes to Heisenberg', he now mentions the implications of this strong impression. Writing of his escape to Sweden and thence to England in the autumn of 1943, he declares:

The question of how far Germany had come was naturally of the greatest importance both for physicists and for government authorities. I had the opportunity to discuss this question thoroughly both with the English intelligence service and with members of the English government and I naturally reported all of our experiences including in particular the impression I got . . . during the visit to Copenhagen by you . . .[16]

He also mentions that it was in 1943 that he learned 'for the first time about the already then well-advanced American-English atomic project'.

In yet another draft, repeating 'the very strong impression on me', Bohr also declares: 'You added, perhaps when I looked doubtful [about the advance of German atomic bomb science] that I had to understand that in recent years you had occupied yourself almost exclusively with this question and did not doubt that it could be done.'[17]

In an even later draft, however, he declares for the first time:

However, the discussions [in other words, his report in 1943 to the Allies about the German advances in atomic physics] had no decisive influence one way or the other, since it was quite clear already then, on the basis of intelligence reports, that there was no possibility of carrying out such a large undertaking in Germany before the end of the war.[18]

So here he begins to stress that his report – of the strong impression he had gained of German atomic advances – could have had no conceivable consequences for Allied plans, as they knew, according to Bohr, that the Germans were well behind.

The importance of the Bohr drafts, it seems to me, is Bohr's anxiety that his reports in 1943 to the British and the Americans of Germany's uranium programme could have added unwarranted scientific and political impetus to the speed and determination of the Allied atomic bomb programme. Reading between the lines, he appears agonized by the suggestion (clearly stated to the world in Heisenberg's letter to Jungk) that he had misunderstood or misread Heisenberg and that his misunderstanding might have spurred on the Allies.

The most questionable element of the final drafts is his insistence that in the autumn of 1943 Allied intelligence knew that Germany had made no progress in the research and development of an atomic bomb, and certainly would not be capable of completing one before the end of the war. The Allies had no certain knowledge of the failure of the German project until December of 1944, when the intelligence team known as Alsos arrived in Strasbourg. Nor could anyone have guessed in the autumn of 1943 how long the war would last.

My analysis in summary of the Bohr drafts is that the Danish scientist gained the impression that Heisenberg had hinted, in 'vague' terms, that Germany was working on an atomic bomb. As we have seen, the news did not disturb him so much at the time, in 1941, that he could not spend a final afternoon on the Saturday enjoying Heisenberg's company, reading out loud and listening to music. But it was only in retrospect after the war that he had many reasons to ponder the ethics and the politics of the atomic bomb, post-Hiroshima and Nagasaki, all the more so as it became clear that Germany's bomb turned out to be a damp squib. The repetitions and the mounting exculpations in the draft suggest a troubled conscience. Was it possible, he seems to be thinking between the lines, that his reports about his impression of Heisenberg's message, conveyed to the Allies in 1943, might have had an influence on the determination with which the Allies pursued their research and development of the bomb?

Heisenberg's trip to Copenhagen and his conversations with Bohr have raised contentious, long-running debates about whether Hitler's scientists were able, and willing, to make atomic bombs. According to historian of science Mark Walker,

such 'what if' questions have no definite answer, and perhaps exactly for this reason extraordinary and implausible significance has been attributed to a single symbol of the Myth of the German Atomic Bomb: Werner Heisenberg's mythical conversation with his Danish colleague, friend, and mentor Niels Bohr in occupied Copenhagen during the fall of 1941.[19]

Walker is on the whole right, it seems to me, for much of the debate is about intentions and the inner state of two men's unknown and unknowable consciences. And yet, at a subterranean level the significance of the debate is as much about how the Allied and German scientists ought to have behaved as it is about how they actually behaved. It seems to me that in the light of the Bohr letters the interest of the Copenhagen affair now enables us to focus not only on the conscience of Heisenberg, but the troubled ratiocinations of Bohr, one of the most decent scientists on earth, who, unlike Heisenberg, has left evidence of his inner state of mind (although this situation may alter if and when the Heisenberg family release Heisenberg's unpublished correspondence).

The constant drafting and redrafting, as if scratching away at an irritation, indicate that Bohr could not accept that he might have misunderstood Heisenberg and in consequence given impetus to the creation of the atom bomb.

This does not give us a final verdict, however, on Heisenberg. Denmark was not the only country Heisenberg visited in occupied Europe after 1941: he visited, as we shall see, Holland and Poland, where his activities raise many questions about his motives and moral integrity.

At the end of the war Heisenberg and the principal German physicists involved in the Nazi nuclear research programme were held in a house outside Cambridge in England. The rooms were bugged and their conversations are now available for scrutiny. In the light of later events we shall revisit the question of Heisenberg's character and deeds: the question as to whether he was a hero, a villain or a fellow traveller.

23. Speer and Heisenberg

In December 1941, three months after Heisenberg's return from Copenhagen, the head of army research in Berlin, Erich Schumann, told the German uranium scientists that their research would receive continued support provided there was a reasonable chance of 'attaining an application in the foreseeable future'.[1] The scientists undertook to write up a report on their progress and future prospects, and a meeting for assessment was scheduled for February 1942.

The report, when it was completed, stated that

in the present situation preparations should be made for the technical development and utilization of atomic energy. The enormous significance that it has for the energy economy in general and for the Wehrmacht in particular justifies such preliminary research, all the more in that this problem is also being worked on intensively in the enemy nations, especially in America.[2]

The creation of a German atomic bomb, according to the document, depended on isotope-separation techniques or the generation of plutonium from a working reactor, which in turn depended on the acquisition of necessary materials.

At this point there was a dramatic parting of the ways between the Allied atom bomb project and the German programme. The Frisch and Peierls breakthrough in Britain had depended technically on their making connections between fast neutrons and the rapidity of reaction in U-235: it was the Frisch-Peierls initiative that jump-started the Americans, although a completed bomb still lay four years and £500 million, or $2 billion, expenditure away. In Germany, however, the connections made by Frisch and Peierls were simply not envisaged.

Unimpressed by the scientists' progress, Army Ordnance responded by cutting their funding and returning the Kaiser Wilhelm Institute for Physics to its former status under the academic sponsorship of the parent society. By April 1942 Abraham Esau, head of the Physics Section of the Education Ministry's research agency (Reich Research Council), had resumed control of the uranium project. In the meantime there had been two related meetings on nuclear research in Berlin between 26 and 28 February 1942, at one of which Heisenberg delivered a paper entitled 'The Theoretical Foundations for the Production of Energy from Uranium Fission'. Present at the meeting was the science and education supremo Bernhard Rust.

Heisenberg delivered a fairly accessible presentation on nuclear fission accompanied by an optimistic account of the weapons potential. He told the meeting:

If one could assemble a lump of uranium-235 large enough for the escape of neutrons from its surface to be small compared with the internal neutron multiplication, then the number of neutrons would multiply enormously in a very short space of time, and the whole uranium fission energy, of 15 million-million calories per ton, would be liberated in a fraction of a second. Pure uranium-235 is thus seen to be an explosive of quite unimaginable force.[3]

He also mentioned the plutonium path to the bomb. Talking of a reactor, he said: 'through the transmutation of uranium inside the pile a new element is created (atomic number 94) which is in all probability as explosive as pure uranium-235, with the same colossal force'.[4]

Given that there were many imponderables, including the correct critical mass, as we have seen above, it was a strange performance, for it seemed designed to provoke interest and high expectations. Was Heisenberg attempting to solicit the appropriate resources to make a bomb? When virtually unlimited resources were offered by Albert Speer in June that year, Heisenberg spurned the offer. It was possible that he was merely attempting to curry favour in order to remain an important figure in atomic physics, but that surely had its dangers since the German armies were not at this point prospering in Russia, and America had entered the war with its prodigious industrial might.

Yet, while Heisenberg sought, and was given, increasing responsibility for the atomic physics programme, he was taking on a variety of extraneous duties and activities. He was hardly focused on a task that should have absorbed all his energies.

He had turned forty years of age in December 1941 and through 1942 seemed to lose himself increasingly in philosophical speculations rather than demanding experimental physics. During that year he found time to collate into a single text a number of lectures and essays which pondered the helplessness of the individual in the face of national and international conflict and struggle. 'The individual can contribute nothing to this,' he wrote, 'other than to prepare himself internally for the changes that will occur without his action.'[5] He advocated the importance of helping others, while counselling that effectively there was little to be done except to acquiesce in one's fate. The individual is absolved of responsibility for external events:

For us there remains nothing but to turn to the simple things: we should conscientiously fulfil the duties and tasks that life presents to us without asking much about the why or the wherefore. We should transfer to the next generation that which still seems beautiful to us, build up that which is destroyed, and have faith in other people above the noise and passions.[6]

Speer Heads Arms Production

On 7 February 1942, three weeks before Heisenberg delivered his upbeat lecture on the potential of an atomic weapon, Albert Speer visited the Wolf's Lair eastern headquarters, where the Führer was now personally directing the conduct of the war on the Russian front as Commander-in-Chief.

Fritz Todt, Minister of Armaments and Munitions, was already closeted with Hitler as Speer arrived. Shouting was heard within the room and when the Minister emerged, according to Speer, he seemed 'strained and fatigued from a long and – it appeared – trying discussion'.[7] For months there had been tensions between Hitler and Todt. In Todt's view, which included an overview of the economy, munitions supply, men and matériel, the war could not now be won. Germany's only option was to come to a political solution. Hitler, however, continued to believe that victory was not a question of armaments but of will.[8]

Speer and Todt talked for a while, and the Minister offered the architect a seat on the plane that was to fly him to Berlin early the next morning – an offer that Speer accepted. Hitler did not summon Speer to his presence until one o'clock in the morning. The Führer, according to Speer, looked tired and preoccupied. He continued depressed until the discussion turned to Hitler's pet building projects, then he brightened up. The two men sat up discussing architectural dreams until three o'clock.

As a result of the lateness of the hour Speer changed his mind about flying with Todt, deciding to sleep in. Todt was the only passenger and was killed when his plane blew up on take-off. An investigation took place and sabotage was ruled out, but the circumstances, as described by Speer in his autobiography, indicate that Todt might have been assassinated.[9]

That very afternoon Hitler appointed Speer to take over Todt's wide-ranging responsibilities. 'I thought that he had expressed himself imprecisely,' wrote Speer, 'and therefore replied that I would try my best to be an adequate replacement for Dr Todt in his construction assignments. "No, in all his capacities, including that of Minister of Armaments," Hitler corrected me.'[10] Thus Speer, aged just thirty-seven, was thrust into a position of supreme responsibility for science and technology in the Third Reich, as far as it affected the war effort. Yet even as he accepted the order, Goering burst into Hitler's presence (Speer comments that the Reichsmarschall had lost no time rushing to the HQ from his hunting lodge 60 miles distant) to demand that he should take over Todt's responsibilities 'within the framework of the Four-Year

Plan'. The tensions that had existed between Goering and Todt illustrate, again, the wasteful rivalry and competition that Hitler countenanced and even encouraged within his dictatorship. Speer commented, 'As Minister of Armaments, Dr Todt could carry out his assignment from Hitler only by issuing direct orders to industry. Goering, on the other hand, as Commissioner of the Four-Year Plan, felt responsible for running the entire war economy.'[11]

Speer brought organization to the chaotic and hard-pressed armaments production at a point when the war was turning against Hitler. Although lacking Todt's experience, he possessed a combination of aptitudes and a strong technocratic background. Speer's father was an architect with his own practice in Mannheim, and his son Albert enjoyed a privileged childhood which included lessons from a French governess. Speer claimed that he enjoyed statistics even at school, and as a child, he had a passion for cars and for Zeppelins, enjoying a 'technical intoxication in a world that was yet scarcely technical'.[12] He was the best mathematician at school and intended studying that subject at university, but his father 'presented sound reasons against this choice', wrote Speer, 'and I would not have been a mathematician familiar with the laws of logic if I had not yielded to his arguments'. He became an architect instead, studying at the Institutes of Technology at Karlsruhe, Munich and Berlin-Charlottenburg. As a student in Berlin, Speer heard Hitler speak for the first time and was captivated by his combination of 'reasonable modesty' and 'hypnotic persuasiveness'. Speer became a member of the National Socialist Party in 1931 and by stages found himself drawn not so much towards Nazi policies as Hitler's personality and power to expedite his ambitious architectural dreams.

As Minister of Armaments and Munitions Speer saw his task in 1942, the year in which America entered the war and things went from bad to worse on the Russian front, as an innovator charged with the task of freeing up inactive and misdirected production, curbing consumer products for the domestic market, bringing women into industry and recalling technicians from active duty to focus on weapons development and production. He declared in his memoir that 'pressure and coercion' in the regime had 'destroyed all spontaneity'.[13] He wanted to encourage the free-enterprise dimensions of the economy by rewarding initiative. Inevitably this meant a reversal of the *Führerprinzip* methods on which the Third Reich supposedly ran. 'There was more than enough criticism from above to below, but the necessary complement of criticism from below was hard to come by.'[14]

With a characteristic flourish, Speer recollected after the war how his production success depended on the enthusiasm of thousands of technicians who were stimulated by the new sense of responsibility he encouraged in the war industries. 'Basically,' he wrote, 'I exploited the phenomenon of the technician's often blind devotion to this task.' From his post-war vantage point, Speer commented on the dubious ethics of the 'blind' enthusiasm of these technicians. 'Because of what seems to be the moral neutrality of technology, these people were without any scruples about their activities. The more technical the world imposed on us by the war, the more dangerous was this indifference of the technicians to the direct consequences of their anonymous activities.'[15] He might well have been speaking of Heisenberg. What he fails to acknowledge, however, is the murderous exploitation of slave labour in the German economy that had started under Todt, with forced labour from East Europe, expanding to hundreds of thousands during his period of office.

Speer's mild exterior concealed a ruthless drive for power which would in time take over General Thomas's army economics office (pushing him into retirement), and bringing the navy within his ambit. He would never run aircraft production, but he would get on excellent terms with Field Marshal Erhard Milch, Goering's number two.

Speer Meets Heisenberg

Speer wrote in his autobiography of regular working lunches with General Friedrich Fromm in a private room in Horcher's restaurant in Berlin. At the end of April 1942 Fromm remarked that the best chance Germany had of winning the war now lay with new weapons. He mentioned a group of scientists who were working on a bomb that had the power to destroy whole cities and could even throw Britain out of the war. Speer had already been approached by Dr Albert Vögler, boss of one of the German steel giants and president of the Kaiser Wilhelm Society, who had complained of the inadequate funding for nuclear research from the Ministry of Education and Science (which had taken over from Army Ordnance). So Speer suggested that Goering assume direction of the Reich Research Council in the hope that funds for nuclear research could now expand rapidly. Goering was appointed to the post in June 1942.

At about this time Speer arranged a conference at Harnack House, the Berlin headquarters of the Kaiser Wilhelm Society, to meet the nuclear scientists: among them, Heisenberg. Heisenberg had now assumed principal responsibility for the atomic physics programme and been promoted to director 'at', as opposed to 'of', the Kaiser Wilhelm Institute for Physics, with effect from October. He had also been appointed to a chair at Berlin University. Also present were high-level representatives of the weapons departments of the army, navy and air force.

Heisenberg spoke on the theme of 'Atom-smashing and the development of the uranium machine and the cyclotron'. According to Speer, Heisenberg then complained bitterly of a lack of funds and materials, and the drafting of scientists and technicians into the armed services, in contrast to the huge effort that he believed was being made in the United States. Heisenberg told Speer that whereas Germany had been at the forefront of nuclear research a few years earlier, the Americans were now probably well ahead. Heisenberg commented that 'in view of the revolutionary possibilities of nuclear fission, dominance in this field was fraught with enormous consequences'.[16]

According to his own version of the meeting, Speer asked how nuclear physics could be applied to the manufacture of atom bombs. His reply, according to Speer, 'was by no means encouraging'. Heisenberg apparently said that the scientific solution had been found and that, theoretically, there was no obstacle to building such a bomb. But the technical prerequisites, assuming always maximum support for the project, were daunting and would take at least two years to put in place.

Field Marshal Milch asked how much nuclear explosive would be needed to destroy a city, and Heisenberg replied, according to two different accounts, based on Heisenberg's memory, 'as large as a pineapple' or 'as big as a football'.[17] This has been taken by both Heisenberg himself and others to mean that he understood the correct critical mass of U-235 — as kilograms rather than tons — adding to the debate over the true status of Heisenberg's understanding of how to make an atomic bomb.

When Speer asked him what he required, Heisenberg apparently mentioned the lack of a cyclotron: there was one in Paris, he said, but for reasons of secrecy it was not possible to use it to full advantage. But when Speer proposed that as Minister of Armaments he should forthwith order cyclotrons as big as or even bigger than the ones available in the United States, Heisenberg objected that due to lack of expertise in Germany it would be necessary to start with a smaller model. By the end of the meeting Speer asked Heisenberg to present him personally with a list of all that he required in order to push forward nuclear research. Speer recollected the result. After a few weeks the scientists applied for several hundred thousand marks, some small amounts of steel, nickel and other scarce metals. They wanted a bunker, some barrack buildings, and the promise that their experiments would be given highest priority. Speer wrote in his autobiography that he was 'rather put out by these modest requests in a matter of such crucial importance'. He suggested that they take 1 or 2 million marks and correspondingly larger quantities of materials, but the offer was rebuffed since it 'could not be utilized for the present'. In a statement that seems extraordinary in hindsight, and was to lend credibility to the notion that Heisenberg was deliberately sabotaging his own nuclear research programme, Speer reflected: 'I had been given the impression that the atom bomb could no longer have any bearing on the course of the war.'[18]

Speer Abandons the Atom Bomb

Speer spoke with Hitler about the nuclear research programme on 23 June 1942. The Minister had decided to keep his report brief, as he was 'familiar with Hitler's tendency to push fantastic projects by making senseless demands'.[19] But Hitler, who had a tendency to pick the brains of amateurs and gossips, had already received a garbled account of the nuclear prospects from his photographer, Heinrich Hoffmann. Hoffmann was friendly with Post Office Minister Ohnesorge, who had been supporting the nuclear research interests of Manfred von Ardenne, for whom Houtermans was working. Hitler, as Speer put it, was under the impression that the nuclear scientists would turn the earth into a glowing star. In a brief reflection on the theme of an uncontrolled chain reaction, Speer remarked: 'Actually, Professor Heisenberg had not given any final answer to my question whether a successful nuclear fission could be kept under control with absolute certainty or might continue as a chain reaction.'[20] The question of an uncontrolled chain reaction was also raised in 1942 by Edward Teller, who wrote a paper on it with two colleagues. On the day before the first test of the American atomic bomb – the Trinity-test – the military head of the project, General Groves, would express his annoyance with the Italian physicist Enrico Fermi 'when he suddenly offered to take wagers from his fellow scientists on whether or not the bomb would ignite the atmosphere, and if so, whether it would merely destroy New Mexico or destroy the world'.[21]

Speer claimed in his autobiography that 'on the suggestion of the nuclear physicists' the atomic bomb project was scuttled by the autumn of 1942 after he had again asked them about deadlines and been told that they could not count on anything for three or four years. So he authorized the development of an 'energy-producing uranium motor for propelling machinery' which interested the navy for its submarines. Later he wrote that Walther Bothe, on inspecting Germany's first cyclotron, explained that the machine would be useful for medical and biological research.

In the summer of 1943, moreover, Speer gave permission for the use of uranium cores for solid-core ammunition. Long after the war he wrote: 'My release of our uranium stocks of about twelve hundred metric tonnes showed that we no longer had any thought of producing atom bombs.'

Pondering the possibility of a German atom bomb, Speer speculated in his autobiography, with the hindsight of knowing the costs of the Manhattan Project, that it might have just proved possible to make an atomic bomb for deployment by 1945, but it would have absorbed the resources of every other project. Speer was inclined to blame the rocket programme, 'our biggest but our most misguided project', and the ideological waywardness of Philipp Lenard, with his repudiation of nuclear physics and relativity theory. In a final reflection, however, Speer concluded that even in the best of circumstances, Germany could never have matched the 'superior productive capacity that allowed the United States to undertake this gigantic project'.[22] At the earliest, according to Speer, Germany might have produced an atom bomb by 1947; but even so, the consumption of the latest reserves of chromium ore would have prevented Germany from continuing the war beyond 1 January 1946, at the very latest.

Two years after the war was over Heisenberg wrote in *Nature* magazine that German physicists

were spared the decision as to whether or not they should aim at producing atomic bombs. The circumstances shaping policy in the critical year of 1942 guided their work automatically toward the problem of the utilization of nuclear energy in prime movers.[23]

But, while Speer now poured prodigious resources into Wernher von Braun's rocket programme, as the best hope for a wonder weapon, the Allies continued to fret about Germany's potential to be first with an atomic bomb.

Heisenberg's Reactor

On the very day that Speer was briefing Hitler on the progress of German nuclear research, 23 June 1942, there was an explosion in Heisenberg's laboratory in Leipzig which covered a technician with radioactive uranium powder. Hearing of the emergency, Heisenberg had looked in briefly before returning to a seminar; but he was called again when the two hemispheres of his reactor model tore, shearing through hundreds of bolts. A week later Heisenberg left Leipzig to concentrate on his research duties in Berlin at the Kaiser Wilhelm Institute of Physics.

Heisenberg's 'uranium machine' experiments aimed at obtaining a self-sustaining reaction mainly languished, dogged by technical difficulties. The difficulties, exacerbated now by air raids, would continue through to 1943, holding back further experiments until late 1943. Interviewed after the war, Paul Harteck, who was a genuine experimentalist, expressed his frustration when he recollected Heisenberg's direction of the project:

But how can you be a leader in such technological matters when you have never run an experiment in your whole life? That's ridiculous! That's no excuse whatsoever! While Heisenberg is one of the best theoreticians of our age and Weizsäcker, in addition to being a very good theoretical physicist and philosopher, could also expound his views very well, nevertheless both had never been involved in a large experimental venture before. How could they think they could lead the development of a new technology? That was poor judgment; it is almost unbelievable.[24]

Heisenberg continued to commute for the first year of his Berlin appointment. After heavy raids in the spring of 1943 his family moved to a small mountain cottage at Urfeld in southern Germany, where they would be effectively separated from Heisenberg until after the war.

As his reactor research – and hence his work on the Nazi bomb – thus proceeded in fits and starts, Heisenberg was spreading his energies, interests and concentration ever wider, busily fulfilling his role as a Berlin professor, running seminars, lecturing, travelling to make guest appearances throughout Nazi-occupied Europe, directing graduate research and keeping up with his personal research interests in cosmic radiation and the foundations of quantum mechanics. He was also taking an active part in Berlin circles of academics and intellectuals, including the famous Wednesday Society, an elitist Prussian group of the great and good which harked back to the salad days of the Wilhelmine period. It is interesting to compare by contrast the dedicated workaholic preoccupation of his opposite number in the Allied atomic bomb project – J. Robert Oppenheimer, who 'only once in three years found time for an overnight excursion'.[25]

Despite his chair at Berlin University, and his acting directorship of the Kaiser Wilhelm Institute of Physics, Heisenberg appeared insatiable for tokens of honour and further recognition, both for himself and for theoretical physics. 'It was,' writes David Cassidy,

as if recognition and respect, which he always required, had been withheld so painfully long and, when finally given, were so tenuous that no amount of honour accorded him could ever make up for this or reassure him that his status had not somehow declined once again.[26]

In 1943 he won the distinguished Copernicus prize of the Reich University of Königsberg. In March of that year he was asked by the Nazi daily paper *Völkischer Beobachter* to write an article to celebrate Max Planck's eighty-fifth birthday. In October he was recommended by Goering for the War Service Cross, First Class. Throughout the year Goebbels put Heisenberg on the front pages of his propaganda paper *Das Reich*. And while it had been a normal part of his life as a physicist to travel in former years, his journeys now assumed an aura of promotion and propaganda as he responded to the subtle and not so subtle pressures of the regime. Whereas he had travelled under his own impetus in the late 1930s and the first two years of the war, he was now subjected to special 'invitations' from the Ministry of Education, or from the dean of Berlin University, Ludwig Bieberbach, the Nazi mathematician. According to

Cassidy, his visits clearly were 'being used to bolster faltering support for the Reich', but Heisenberg appeared to acquiesce readily whatever the circumstances.

Heisenberg in the Netherlands

Heisenberg made a number of trips to occupied Europe between 1942 and 1944, mainly to fly the German cultural flag; although his defence of 'our system', as he had put it in the letter he wrote to his wife from Copenhagen in September of 1941, could not have excluded the advantages of National Socialism. These trips included a visit to the Netherlands in 1943. The occupation of Holland had been met with large-scale resistance, strikes and non-compliance, followed by brutal suppression and the deportation of Jews. These measures were in turn met with further resistance, notably among students and teachers at the Dutch universities. According to historian Werner Warmbrunn, cited by Cassidy, as many as one-third of those executed during the Nazi occupation were students. Nowhere was resistance more vehement than at the University of Leiden, where students had gone on strike and teachers had resigned in protest against the dismissal of Jewish members of faculty in November of 1940. There was some room for ambiguity in the motives of Heisenberg as he travelled around Holland in October of 1943, lecturing, meeting physicists and even talking with the infamous Dr Arthur Seyss-Inquart, the Austrian Nazi. Cassidy comments: 'Dutch interest in scientific collaboration had been strengthened by the visit – precisely the outcome the official had desired. At the same time, the stated possibility of scientific collaboration was a good argument for the preservation of Dutch laboratories and research.'[27]

After the war Dutch scientists remembered disturbing encounters with Heisenberg. Hendrik Casimir told a member of the Allied science intelligence unit that Heisenberg had admitted knowing about the existence of concentration camps and Germany's plunder of occupied territories. Heisenberg had nevertheless expressed his personal opinion that he wanted Germany to prevail. According to the informant in the intelligence unit, Heisenberg had said: 'Democracy cannot develop sufficient energy to rule Europe. There are, therefore, only two possibilities: Germany and Russia. And then a Europe under German leadership would perhaps be the lesser evil.'[28]

Casimir wrote in his autobiography that he and Heisenberg took a walk and talked about Germany as the defender of the West against the onslaught of the eastern hordes. Neither France nor England, Heisenberg said, would have been sufficiently determined or strong to play a leading role in such a struggle. Casimir goes on:

Of course, I objected that the many iniquities of the Nazi regime, and especially their mad and cruel anti-Semitism, made this unacceptable. Heisenberg did not attempt to deny, still less to defend, these things; but he said one should expect a change for the better once the war was over.[29]

The conversation confirms that Heisenberg thought in terms of the Two-War Idea: that Germany was engaged in two distinct struggles, the first against the Western democracies, the second against Russia, the home of Soviet communism and the eastern hordes. The first war was to be regretted; the other was necessary for the salvation of Western Europe.

While in the Netherlands Heisenberg became involved with another scientist who appeared to be capable of separating Heisenberg's physics, and his acts of kindness, from his evidently pro-German politics and nationalism. This was Hendrik Kramers, who had worked with Heisenberg in Copenhagen in the old days. Kramers, like Casimir, was keen to see Heisenberg in the hope that he would alleviate some of the difficulties of doing science under German occupation. Heisenberg, it seems, helped with countermanding an order of the German authorities to send Dutch scientific equipment to Germany and with easing travel restrictions for scientists and reopening the physics laboratory at Leiden. The friendship, despite Heisenberg's work for the regime during the war, is another ingredient of the complex case of Heisenberg, for Kramers was a

man of integrity and courage. An acquaintance of Kramers once
said that Kramers was one of the handful of genuinely decent
persons that God had sent down to earth to make the miserable fate
of the Jews more bearable. Kramers and Heisenberg talked physics,
and even discussed working together on a problem in fundamental
physics. Kramers was able to write later that year to Heisenberg 'to
tell you once more how happy your visit has made me, stimulating
again old ideals'. But he declined, in the end, to enter into any
form of scientific collaboration, for 'this is not,' he told Heisenberg,
'the time for joint publication.'[30]

Poland and Hans Frank

While Heisenberg's trip to the Netherlands is fraught with ambigu-
ities, another of his trips leaves little doubt that Heisenberg had
become morally anaesthetized to the atrocities of the regime to
which he had devoted his scientific genius. This was the 'cultural'
trip he made to Poland in December of 1943, eight months after
the vicious quelling of the Warsaw uprising.[31]

The Governor-General of occupied Poland was Hans Frank,
who had a connection with Heisenberg and his elder brother,
Erwin: all three were contemporaneously students at secondary
school, the Max-Gymnasium in Munich, and known to each other.
Heisenberg and Frank were also in the Pfadfinder, the German
scout movement, and both became members of the Neupfadfinder,
the 'new pathfinders', which celebrated a mix of Teutonic mysti-
cism and outward-bound activities. Meanwhile Hans Frank graduated in law and started a meteoric rise within the Nazi Party, joining
the storm troopers and acting as legal representative in various libel
suits against Party members in the 1920s.

After the invasion of Poland Hitler planned to turn the country
into a German colony, destroy its culture and reduce its people to
the status of slaves. Some 150 academics at Cracow University were
sent to concentration camps, where about a third of them, mostly
Jews, were liquidated. The university stayed closed until the end
of the war.

Under Frank's administration the Jews of Warsaw were driven into a ghetto in the city. Frank declared to his cabinet in December 1941, 'As far as Jews are concerned, I want to tell you frankly that they must be done way with in one way or another.'[32] At the same time he ordered the arrest and transportation of Poles as slave labourers: some 800,000 Poles were sent to Germany as slave labourers by August 1942. Meanwhile the transportation of Jews to Treblinka had begun in July. By October some 300,000 had been deported.

A movement began to form among the remaining Jews in Warsaw aimed at resisting the inevitable liquidation of the ghetto. On 19 April 1943, some 3,000 German troops launched an attack on the ghetto, killing 14,000 Jews. The remaining 7,000 were transported to Treblinka.

In December of that year Heisenberg made his trip to Poland in order to deliver a lecture and to accept the hospitality of Governor-General Hans Frank. How much did Heisenberg know about Frank's atrocities in Poland? What we know from a memoir, *Inner Exile*, written by Heisenberg's wife about her husband after his death, is that they both were aware of the extermination of Jews in Poland before Heisenberg's visit there. The question is raised in her book, when she comes to describe her father's indignation that people should think so badly of Germans to believe that they could do such things:

I can still see my father standing in front of me. He was a man with a venerable and law-abiding outlook, who actually went into a rage when Heisenberg once showed him a report he had received from a colleague at the institute who had been a witness to the first cynical mass executions of Jews in Poland. My father lost all self-control and started to shout at us: 'So this is what it has come to, you believe things like this! This is what you get from listening to foreign broadcasts all the time. Germans cannot do things like this, it is impossible!' He was not a Nazi, he had prematurely retired from his position following the National Socialist takeover.[33]

Heisenberg was originally invited to lecture in Cracow by Wilhelm Coblitz, director of the Institute for German Work in the East, which had been established by Frank in the spring of 1940. Heisenberg had been keen to come, but he did not obtain permission on this occasion from the authorities. Coblitz again invited Heisenberg in May of 1943, writing in Frank's name, and assuring him that Frank would attend the lecture. Late in September confirmation came again from 'Herr Generaldirektor Dr Frank', inviting Heisenberg and his wife to be his guests. Mrs Heisenberg could not make the journey, but Heisenberg eventually travelled to Cracow some time in the first half of December. He stayed in one of Hans Frank's castles, which had been furnished with artwork stolen from the Poles. According to Professor Bernstein, who has talked with a Polish survivor of the period, some Poles tried to gain access to the lecture but were turned away at the door. Only Germans were welcome. An article appeared in the German-language newspaper *Krakauer Zeitung* on 18 December stating that the physicist lectured to a large audience on quantum physics. The title was 'The Smallest Building Blocks of Matter'.

Knowing what had been done in Poland by his erstwhile acquaintance Hans Frank, and given the duties that weighed upon him at the time, Heisenberg's willingness to go Cracow appears to show that he turned a blind eye to what had been happening there. As for Frank himself, he was found guilty of crimes against humanity during the Nuremberg Trials and was hanged on 16 October 1946.

Heisenberg made another foreign trip in January of 1944, this time to Copenhagen to adjudicate on the question of the fate of Bohr's Institute of Physics. Bohr along with most of the Jews of Denmark had already escaped abroad. Heisenberg was instrumental in returning the institute with its precious cyclotron back to the Danish scientists. The German occupation authorities had placed conditions on the transaction, but it appears that Weizsäcker had prevailed upon his father in the Foreign Office to waive them on 3 February. Heisenberg returned to Copenhagen two months later to lecture at the German Culture Institute and to dine publicly with the vicious local occupation commander SS–Obersturmbannführer

Dr Best. An official Danish verdict on Heisenberg after the war was: 'Heisenberg is not a Nazi but is an intense nationalist with the characteristic deference to the authorities in control of the nation.'[34]

24. Haigerloch and Los Alamos

On 1 March 1943, while Heisenberg and his colleagues had been attending a lecture at the Air Ministry in Berlin on the effects of bomb explosions on the human body, the RAF and the American Eighth Air Force began a massive raid over the capital. The audience repaired to the bomb shelter, where they quaked under the reverberations as the Ministry buildings crashed to the ground above.

After the all-clear Heisenberg emerged to face a ninety-minute walk out to the suburb of Fichteburg and the home of his parents-in-law, where he was lodged while working at his institute in Dahlem. Rubble was strewn everywhere; flaming pools of phosphorus lit up the darkness. He was anxious to get back to his twin children, Wolfgang and Maria, who were visiting his parents-in-law.

The experience of that night stands out in Heisenberg's memories for the conversation he had with another scientist, Adolf Butenandt, who accompanied him.[1] As they walked through the devastated city, Butenandt kept up a tale of woe about the collapse of German science, the destruction of the laboratories, the dearth of research students and the scientists killed. Heisenberg tells us that he himself was mainly intent on saving his shoes from phosphorus fire, dousing them in the puddles and scraping off the inflammable material. The scene he depicts, from the perspective of the post-war era, is designed to show that a hapless scientist could do little in those troubled times except save his shoes, and that was difficult enough.

In the weeks that followed the air raid, decisions were taken to move the atom research programme away from the institute in Dahlem to a more remote and safe location in the south. The need to move south became all the more imperative after the RAF carried out a heavy bombing raid on Leipzig, which destroyed much of Heisenberg's Institute of Physics with all his papers as well as his home of fifteen years.

Heisenberg's colleague Karl Wirtz began to prepare a reactor experiment in a new basement laboratory of the Kaiser Wilhelm Institute for Physics in Dahlem, but the conditions were not ideal in view of the frequent bombing. Meanwhile the long process of removing the institute to Hechingen, a small town near the Swabian Alps, got underway. It was directed by Walter Gerlach, a Munich professor of physics, known to be patriotic, although unsympathetic to the regime. Gerlach had recently assumed control of the nuclear physics division of the Research Council of the Reich.

A high-voltage plant was established in the boiler house of a brewery in Hechingen, while offices and workshops were set up in part of a textile mill. But Gerlach needed to find a safe location for a new reactor, or 'uranium burner', and he eventually settled on a site he had known before the war at a remote village called Haigerloch. Haigerloch is perched on two cliffs above the River Eyach, and it was here that the inn-keeper of the Swan inn leased to the institute a storeroom in an underground cave below the cliffs of Haigerloch castle. The contrast between America's bomb effort with its vast factories and processing sites, its cities in the desert and its many tens of thousands of personnel, could not have been more extraordinary and indicative of the feebleness of Germany's research project. Haigerloch had more in common with a set for a Count Dracula movie than the site of Germany's most advanced power station, but here, over many months, and with much difficulty, the scientists supervised the transportation of their cubes of uranium, heavy water and graphite.

The researchers cycled over from Hechingen every morning, a distance of ten miles. But there were many hours, over the long period of preparation, when Heisenberg whiled away his time, playing Bach fugues on the organ in the chapel attached to the castle high above.

One of Heisenberg's assistants remembers: 'It was the most fantastic period of my life. I have never so often been obliged to think of Gounod's *Faust* and Weber's *Freischütz* as I was in that extravagantly romantic setting.'[2] The enterprise of initiating a reactor, however, lagged through shortages of uranium and other materials.

Heisenberg's Assassin

During the dreamlike Haigerloch period, before Christmas of 1944, while the British and Americans fought their way through France, and the Red Army approached inexorably from the east, Heisenberg was targeted for a bizarre assassination attempt by the chief of the Manhattan Project, General Groves, using the services of the OSS (Office of Strategic Services, the principal US intelligence organization). The idea was to assign an agent called Morris ('Moe') Berg to travel to Zurich, where Heisenberg was scheduled to deliver a physics lecture on 18 December in the presence of specialists. Berg, posing as a physicist (he knew little physics in fact), would be armed with a pistol. If in the course of the lecture Heisenberg betrayed the fact that he was working on an atom bomb, Berg, at the risk of this own life, was to shoot the scientist there and then in the lecture room.[3] Paul Scherer, who was a local US intelligence contact, had informed the OSS of Heisenberg's trip and was aware of Berg's arrival. He would also report on Heisenberg's movements and conversations: but there is no indication that he understood that Berg might attempt to kill the physicist. Among the conversations reported by Scherer to the OSS was Heisenberg's insistence that he had never been privy to an attempt to take over Niels Bohr's physics institute in Copenhagen, but that he had saved the centre by timely intervention. He also told Scherer that Walter Gerlach, the director of the nuclear physics division of the Research Council, had suffered a nervous breakdown.

The seminar took place at the University of Zurich on Ramistrasse at 4.15 in the afternoon. There were no more than twenty people present and the room was intensely cold. Berg sat with Leo Martinuzzi, another OSS agent, his pistol in his pocket.

Berg describes Heisenberg as looking 'Irish' and as having 'sinister eyes'. The lecture was on S-matrix theory – a purely theoretical topic in quantum mechanics with no application for atomic bombs – which Heisenberg delivered walking up and down before the blackboard, occasionally consulting his typewritten notes. When Heisenberg finished, a lively technical discussion followed, which ended at about 6.40 p.m.

After the meeting broke up, Berg talked privately with Scherer. Scherer told Berg that Heisenberg was working on cosmic rays, not on an atomic bomb, and that such a weapons project would take the Germans two to ten years. Their discussion also included a vague plan to kidnap or 'transplant Heisenberg and family to the US'.[4]

Later that week, before Heisenberg's departure from Switzerland, Scherer invited a group of people to his house for dinner, including Heisenberg and Berg. The dinner party was the scene of several conversations about the war, later reported by some of those present including the Dutch physicist Piet Gugelot. Gugelot has in recent years corresponded with the author Thomas Powers about the incident,[5] claiming that a 'very severe argument' took place in which Heisenberg was challenged about Nazi atrocities against Jews in Holland and France. Heisenberg denied all knowledge of such atrocities, but offered the defence, alluding to Germany's isolation after World War I, that 'when you lock people up inside four walls and no windows they go crazy'. On being challenged that he was a supporter of Hitler's government, Heisenberg declared that he was not a Nazi but a German. He then went on to offer an explanation of his position which would remain consistent in years to come: the notion of the Two Wars. According to Gugelot, Heisenberg said that Russia was the real problem, and only Germany could be a bulwark between Russia and European civilization. This apparently appealed to those present, who were Swiss. Gugelot left the party in disgust at this point, which meant that he missed another significant exchange in which the physicist Gregor Wentzel said, 'Now you have to admit that the war is lost.' To which Heisenberg replied: 'Yes, but it would have been so good if we had won.'[6]

The remark would mark him down as a Nazi sympathizer in the

eyes of his post-war critics. Ironically, it was also noted by Berg, whose superiors took it as an indication that Germany had no atom bomb, because Heisenberg evidently agreed that Germany had indeed lost the war.

At the end of the evening Berg, the erstwhile potential assassin, accompanied Heisenberg through the deserted streets of Zurich to the physicist's hotel. They did not talk physics; but Heisenberg would remember that the young man asked searching questions about his regard for the Nazi regime.

Heisenberg left Zurich for his family's temporary home at Urfeld in the Bavarian mountains in time to arrive on Christmas Eve. His son Martin, who was a child at the time, remembers that last Christmas in the isolated 'mountain cabin'. His mother organized Christmas music and a little play in which the children took part. There was very little to eat, but she had managed to save some ingredients for biscuits. When his father returned, he took the children for 'little hikes and sometimes the snow was so high that I could not look over it walking where one could walk'. In the evenings, wrote Martin Heisenberg, his father would play the piano while his mother sang. There was a 'Christmas tree with real candles' and Heisenberg, ever cautious of fire, 'would put a bucket full of water into a corner of the living room'.[7]

Meanwhile, the American Bomb

The Americans, as we have seen, did not begin to lay down practical plans for making an atomic bomb until August 1942. But it was not until July of the following year, 1943, that J. Robert Oppenheimer was appointed director of the Los Alamos laboratory – where hundreds of scientists would live and work. Son of non-practising Jews of New York City, Oppenheimer had travelled extensively in Europe. He had been a student of Max Born in Göttingen in 1927, participating in the exciting early period of quantum mechanics, followed by periods at Leiden and Zurich, after which he returned to the United States to teach at Berkeley and at the California Institute of Technology in Pasadena. He was a brilliant and charismatic teacher; his students paid him the compliment of imitating his speech mannerisms and even his way of lighting cigarettes.

In his book *Oppenheimer: The Story of a Friendship* Haakon Chevalier, who was betrayed by Oppenheimer to the American secret services as a potential spy, wrote a description of the scientist in 1937:

Tall, nervous and intent . . . he moved with an odd gait, a kind of jog, with a great deal of swinging with his limbs, his head always a little to one side, one shoulder higher than the other . . . He looked like a young Einstein, and at the same time like an overgrown choirboy. There was something both subtly wise and terribly innocent about his face.[8]

The first practical stage in the development of the bomb was the construction of a small crude reactor in a deserted squash court in Chicago by Enrico Fermi and physicist Arthur Compton, with Leo Szilard assisting. It was known as Chicago Pile 1 – a contraption, as Richard Rhodes has pointed out, that was potentially 'a small Chernobyl in the midst of a crowded city'. The 'pile' was made of bricks of graphite interspersed with sealed cans of uranium oxide powder and cubes of natural uranium metal. The reactor required a critical size to prevent too many neutrons escaping before they were absorbed by uranium nuclei. The pile went 'critical' on 2 December 1942, demonstrating that it could sustain the production of energy and plutonium.

Making a bomb was a quite different matter. The Manhattan Project to make the world's first atom bomb involved more than thirty different plants across the United States, with the principal research centre at Los Alamos (known for secrecy as Site Y) in the grounds of a boys' school in the deserts near Santa Fe in New Mexico. Here most of the scientists came to work under Oppenheimer. General Groves inititially attempted to maintain secrecy and keep the scientists apart by strict compartmentalization. Oppenheimer was convinced, however, that success depended on the researchers freely communicating with each other.

A huge factory, said to be the longest building in the world at that time, was planned at Oak Ridge in Tennessee for separating U-235 from U-238 by means of gaseous diffusion and electromagnetic techniques. A third plant was established at Hanford on the Columbia river to produce plutonium. A separate Anglo-Canadian-French plant was established under the British scientist J. D. Cockroft at Chalk River, Ottawa, where heavy water was produced.

All these mighty constructions and activities, involving 150,000 workers, were kept secret from the world. The German secret services and the German scientists had no idea what the Americans were planning until the first atom bomb exploded. As fate would have it, Germany's principal physicists would all be together in one room in England when the news of the first weapon used on a civilian population, at Hiroshima, burst on the world on 6 August 1945.

Also

The Allies knew that there was a Nazi bomb programme, but their intelligence services were largely ignorant of the details. General Groves records in his memoir that before the D-Day landings 'we were considering the possibility that the Germans would prepare an impenetrable radioactive defence against our landing troops'.[9] The British, occasionally informed of the fact of the German atomic programme by the Austrian science publisher Paul Rosbaud, were reasonably convinced that the German research was considerably behind the Americans; but the Americans themselves were increasingly anxious in case the Germans had made more progress than evinced by available intelligence.

In preparation for the invasion of Europe and the eventual defeat of Germany, a special mission was proposed by the British and Americans charged with gathering information about scientific know-how, personnel and hardware. The head of this mission was the Dutch scientist Samuel Goudsmit, a physicist and multi-linguist, who had emigrated to the United States in 1927 and who knew

Heisenberg personally. The unit Goudsmit took over was named 'Alsos', meaning 'grove' in Greek, possibly a pun on General Groves's name. Goudsmit, who knew few of the details of bomb developments, has commented that he was an ideal head of Alsos, since he was expendable. 'If I fell into the hands of the Germans,' he wrote in his account of the mission, *Alsos*, 'they could not hope to get any major bomb secrets off me.'[10]

Goudsmit's first target in Europe, following the invasion of Normandy, was Joliot-Curie's laboratory at the Collège de France in Paris. By December 1944 the team entered Strasbourg, where Weizsäcker had taken over the university physics department. Weizsäcker had already fled, but Goudsmit's team found a quantity of documents which revealed that the Germans were far behind the Allies in nuclear bomb research. The Germans, he discovered, had not even managed to construct a successful chain-reacting pile. Groves's fear now, however, was the prospect of nuclear weapons information, or research scientists, falling into the hands of the Russians. Having discovered that the Auer company was manufacturing uranium metal in East Germany, in a region that would inevitably fall into the hands of the Russians, he ordered that the factory be destroyed by bombing raids.

Meanwhile Goudsmit had managed to return to his family home in the Hague, to find it deserted. His memoir provides poignant testimony of his grief. He had not heard from them since receiving a letter in March 1943 bearing the address of a Nazi concentration camp.

Climbing into the little room where I had spent so many hours of my life, I found a few scattered papers, among them my high-school report cards that my parents had saved so carefully through all these years . . . As I stood there in that wreck that had once been my home I was gripped by that shattering emotion all of us have felt who have lost family and friends at the hands of the murderous Nazis – a terrible feeling of guilt. Maybe I could have saved them.[11]

Reflecting, later, on the German talent for systematic organization of the Germans, Goudsmit wrote: 'That is why they kept
such precise records of their evil deeds, which we later found in
their proper files in Germany.' That, he went on, was the reason
that 'I know the precise date my father and my blind mother were
put to death in the gas chamber. It was my father's seventieth
birthday.'[12]

By January 1945 the German atom physicists Walter Gerlach,
Kurt Diebner and Karl Wirtz had abandoned the Berlin laboratories,
taking with them quantities of uranium and heavy water. They were
bound for southern Germany and Hechingen, close to Heisenberg's
reactor pile at Haigerloch. (Evidently Gerlach did not succeed in
transporting the entire stock of uranium oxide, for, according to
historian Antony Beevor, the NKVD, arriving with the Red Army
from the south on 24 April 1945, found at the Kaiser Wilhelm
Insititute for Physics '250 kgs of metallic uranium; three tons of
uranium oxide; and twenty litres of heavy water'. Evidently a prize
that Stalin and Beria had long anticipated.)[13]

By March Goudsmit entered Heidelberg in the wake of the
Allied troops. Here he found Bothe, who talked volubly about
civilian applications of nuclear research, but refused to discuss
German weapons development. By the time the Alsos team reached
Hechingen and Haigerloch, Heisenberg had moved on, and so
had Weizsäcker, although not before hiding the German nuclear
group's research documents, which were eventually found in a
sealed drum lowered into a cesspool.

In and around Hechingen the Alsos team took into custody Laue
and Otto Hahn. They were not connected with nuclear research,
but Goudsmit believed they would be crucial to the reconstruction
of German science in the post-war era. The team also rounded up
Wirtz, Bagge and Horst Korsching in the same district. Gerlach
was picked up in Munich at the beginning of May, and Diebner
was captured the day after.

Meanwhile Heisenberg had fled to his home in Urfeld, travelling
by bicycle through a devastated Germany, a distance of 100 miles
from Hechingen. He was picked up by Alsos on May 3 and taken,

like many of the other scientists, to Allied European headquarters in Rheims. When Goudsmit spoke with him for the first time, Heisenberg said: 'If American colleagues wish to learn about the uranium problem I shall be glad to show them the results of our researches if they come to my laboratory.'[14] Which struck Goudsmit as pathetic.

PART SIX

Science in Hell 1942–1945

25. Slave Labour at Dora

The outbreak of war in September 1939 had exacerbated a shortage of labour on the land and in industry in Germany, especially in mining, which suffered a shortfall of 30,000 workers on the Ruhr. A remedy ready to hand was the employment of foreign labour – an inevitable consequence of war and occupation. Polish workers drafted on to the land were at first described as *Retter in der Not* (saviours in time of need). By 1940 they were being deployed in munitions industries, including the rocket plant at Peenemünde. By the spring of 1941 some 600 Polish workers were joined by 1,000 contract Italians and about 100 French workers. In these early years of highly secret development the numbers of foreigners were limited for reasons of security.

In the years 1942 to 1943, as the A4 rocket project was geared up for production, both at Peenemünde and at the project's new assembly annexes – the Zeppelin plant at Friedrichshafen and the Rax-Werke plant in Austria – the need for unskilled labour became acute. The key figure was Arthur Rudolph, Peenemünde's chief engineer, who was originally preparing to use Russian prisoners of war made available in increasing numbers as the Eastern front swallowed up drafted German troops (in the early days Russian prisoners were mostly left to die of starvation and disease). In April of 1943, however, the rocket chiefs were offered another solution in the form of SS concentration camp prisoners.

In the early days, while the camps were principally a means of incarcerating those suspected of real or alleged opposition to the regime, the SS had attempted to profit from camp-based enterprises. By 1942, however, the primary purpose of the camps in Poland was the extermination of Jews. That deadly shift in purpose coincided with a decision by Himmler to begin a new form of slave enterprise whereby the SS provided camp prisoners to private and government industries at an 'economic' rate. The SS Economic and Administrative Main Office typically charged four marks per day for unskilled labourers and six marks per day for skilled ones. The SS provided accommodation, minimum sustenance and security. Such was the overcrowding, the harshness of the conditions, especially in winter, the starvation rations, the lack of hygiene, the brutality and the workload, that death was an inevitable consequence of the slave labour regime.

Arthur Rudolph preceded his deal with the SS by a visit to the Heinkel factory in Oranienburg, north of Berlin, in April 1943. There he found 4,000 slave labourers – mainly Russians, Poles and French – living in cramped conditions and kept imprisoned behind electric fences and barbed wire. Rudolph wrote a memorandum noting the advantage of exploiting prisoners especially in view of the 'greater protection for secrecy'.[1]

The historian Michael Neufeld points out that the memorandum is indicative of the chief engineer's advocacy (a circumstance well known to Dornberger and von Braun), as well as management, of slave labour 'months before the creation of the infamous Mittelwerke underground facility' in the Harz mountains (in other accounts, including that of von Braun, the blame for slave labour in rocket production was put exclusively on Himmler).[2]

On 11 June 1943 Hitler raised the priority of the rocket programme 'above all other armaments production', and Dornberger was made a Brigadier-General the next day. Rudolph now asked for 1,400 concentration camp prisoners for Peenemünde. The first arrivals came two weeks later from Buchenwald near Weimar together with sixty SS guards. By one account (that of Willy Steimel, a criminal convict employed as a prisoner administrator at Buchenwald) the regime involved an eleven-hour day with one day free a week. Within four months three prisoners had died of disease and two of injuries. One was shot escaping and four others died desperate deaths by drinking rocket fuel.

As if to seal the importance of the vengeance weapon, Hitler now invited Dornberger and von Braun to visit him at the Wolf's Lair on 7 July. Once again Hitler was shown film of the A4 in flight, featuring the successful 3 October launch in the previous year. According to Dornberger, Hitler declared that the rocket should carry a 10-ton warhead and that 2,000 missiles should be manufactured every month. When Dornberger explained that such figures were not possible, a 'strange, fanatical light flared up in Hitler's eyes. Dornberger feared that Hitler was going to break out into one of his mad rages.' It was at this point that Hitler uttered his crazed incantation: 'But what I want is annihilation – annihilating effect!'[3] Before the end of the meeting Hitler awarded von Braun a state professorship and signed the necessary documentation on the spot.

At this meeting Hitler had insisted that foreigners should not be employed on the project for reasons of security. But his order was immediately disobeyed. By early August Dornberger ordered that as a 'basic principle' production in all the assembly plants should be carried out by prisoners. Peenemünde was to have 2,500 detainees from concentration camps, the other plants would have 1,500 between them. By the third week in August, however, the carefully laid plans of some eight years were disrupted by an unexpected visitation from the skies.

The Peenemünde Bombing Raid

Through the course of 1943 British intelligence had been gathering information and taking high-altitude photographs of Peenemünde and other secret weapons plants. The photographs gathered on these missions were an important factor in an Allied operation of sabotage and aerial bombing known as 'Crossbow'.

By the summer of 1943 security had been tightened at every level at Peenemünde; flak batteries were strengthened around the district and SS guards were maintaining strict vigilance at the perimeters. The raid on Peenemünde on the night of 17–18 August was not entirely unexpected. At one o'clock in the morning 600 heavy RAF bombers droned across the night sky, dropping 1.5 million kilograms of explosives on the plant's facilities.

The raid destroyed most of the residential areas of the East

Peenemünde site for the A4 development, including the settlements that housed engineers, and barracks where some 3,000 foreign workers were living behind wire. Walter Thiel, the chief engine designer, was killed along with his family in their air raid shelter. The Development Works buildings were extensively damaged as well as administrative offices: von Braun was seen scrambling around in the ruins trying to recover plans and documents.

There had been about 12,000 workers resident at Peenemünde at the time. More than 700 were killed, 500 of them foreign. The factory buildings where the rockets were assembled were largely undamaged. The destruction looked much worse than the reality and most of the damaged buildings were left in ruins as a kind of camouflage, which discouraged the RAF from carrying out further raids. Historians of the raid, both German and British, claim that the rocket effort was set back by about two months, which meant that some 740 rockets were not launched. The death of Thiel, however, was a significant blow and the anti-aircraft missile Wasserfall lost momentum as well the development of a two-stage rocket that might have reached deep into the British Isles. Martin Middlebrook, whose book *The Peenemünde Raid* exhaustively chronicles the incident, adds to these consequences the profound effect on German morale.

There were many individual tales of shock, from the young woman with a fur coat over her nightdress who had run away from Peenemünde screaming that she wanted to go home, to the Luftwaffe general for whom the raid on Peenemünde 'was the one burden too many and who committed suicide'.[4]

Peenemünde had been found: the 'sleeping beauty', as some had called it, had been awoken and the incident prompted a fateful decision. Himmler moved swiftly to involve himself in the future of the rocket programme now that its removal and dispersal had become inevitable. A week after the raid Himmler arrived at the Wolf's Lair and persuaded the Führer to give the SS a share in the management of A4 production and the brief to move the factory underground and draft in concentration camp prisoners. Testing would be moved to a site in Poland. The Reichsführer-SS had apparently persuaded Hitler that the raid had been the result of espionage: secreting the production plant underground would ensure greater security as well as protection from further bombing.

The construction of the new facilities would be in the hands of SS-Brigadeführer (Brigadier-General) Hans Kammler, an individual of remarkable ruthlessness. Himmler was determined to have a role in the complex jostle for power that surrounded the A4 project. And yet, in the light of his rapidly expanding reputation for producing miracles, Albert Speer still remained ultimately in charge.

By 26 August, just a week after Himmler's talks with Hitler, the site for the underground factory had been chosen: a series of tunnels used as storage for oil and chemical weapons in the Kohnstein mountain near the city of Nordhausen in the Harz mountains: it would be known as Mittelwerke (Central Works). At the time of the decision there were two tunnels, about a mile in length, each large enough for two sets of parallel rail tracks, one of which ran through the length of the mountain. On 28 August concentration camp prisoners were rushed to the site to begin work on the penetration of the second tunnel through the length of the mountain and twenty cross-tunnels between the two.

Meanwhile research and development sites were being dispersed around the Peenemünde district itself, and throughout Germany as far afield as Kochel in the Bavarian Alps, where the wind tunnels were to be reassembled. Despite the difficulties of communication between the separated entities, development on the A4 was resumed within two months; but the involvement of slave labour expanded to an unprecedented level and the brutal treatment of concentration camp workers had few parallels. Four thousand male prisoners were drafted into the Mittelwerke tunnels within six weeks, mostly Russian, Polish and French, but no Jews at this stage (they would be drafted in the summer of 1944). The figure would rise to 8,000 in November. Kammler told his staff: 'Pay no attention to the human cost. The work must go ahead, and in the shortest possible time.' The SS accordingly created a living hell.

A French resistance leader, Jean Michel, wrote an account of Mittelwerke as he experienced in mid–October 1943:

The Kapos and SS drive us on at an infernal speed, shouting and raining blows down on us, threatening us with execution; the demons! The noise bores into the brain and shears the nerves. The demented rhythm lasts for fifteen hours. Arriving at the dormitory . . . we do not even try to reach the bunks. Drunk with exhaustion, we collapse onto the rocks, onto the ground. Behind, the Kapos press us on. Those behind trample over their comrades. Soon, over a thousand despairing men, at the limit of their existence and racked with thirst, lie there hoping for sleep which never comes; for the shouts of the guards, the noise of the machines, the explosions and the ringing of the [locomotive] bell reach them even there.[5]

The work went on day and night and the tunnels were frequently racked with explosions as the rock was dynamited to extend the tunnels, filling the atmosphere with choking dust. Hygiene and washing facilities were non–existent and the men cut oil drums in half to create latrines. Michel writes: 'Some deportees are too weak and collapse. They have dysentery. They foul their trousers. They no longer have the strength to sit over the barrels, even to get to them.'[6] In the first seven months of the operation 6,000 prisoners died (including those transported back to the death camps).

In December of 1943 Albert Speer toured the Mittelwerke and subsequently wrote to Kammler praising him for an achievement 'that far exceeds anything ever done in Europe and is unsurpassed even by American standards'.[7] After the war he took credit for improvements to the barracks, known as Dora, that were being erected for the prisoners outside the tunnels. As in other instances, his self–serving post–war remarks are unreliable.[8]

Speer clearly bore responsibility for the horrors of Mittelwerke, which he shared with Himmler and Kammler. There is evidence that Dornberger and von Braun had also advocated the use of slave labour as part of a productivity calculation.[9] Von Braun, however, was soon to have a curious brush with the SS that would distance him from Himmler and the SS in decades to come, providing him with an alibi of sorts against accusations that he was implicated in slave labour exploitation.

In February of 1944, according to a manuscript article written by him after the war, von Braun received a phone call to report to Himmler's headquarters in East Prussia. He recollected that he felt scared when he was shown into the Reichsführer's office. Himmler, who reminded von Braun of a schoolteacher rather than a monster steeped in blood, greeted the young rocketeer politely. According to von Braun, Himmler now said, 'Why don't you come over to us?' It was a plain invitation to leave the service of the army and dedicate his services to the SS. Von Braun replied that in General Dornberger he had the best boss he could wish to have. That seemed to be the end of the affair.

The following month, however, von Braun and his close associates Klaus Riedel and Helmut Gröttrup were arrested and charged with having stated that the main task of their research was to 'create a spaceship'. They had also been overheard, their accusers alleged, commenting that the war was turning out badly and that the A4 was an 'instrument of murder'. In addition, and more seriously, they were guilty, it was said, of being associated with communist cells.

Only by the energetic services of Dornberger and, it appears, the direct intervention of Speer with Hitler was von Braun eventually released after two weeks. The others were set free a short while later. Even though von Braun appears to have been in some danger, for he might well have been executed had the charges stuck, his arrest turned out to be a stroke of fortune for his subsequent career as head of space research in the United States in the post-war era. It bolstered his image as a scientist who had doggedly maintained a non-political stance and who was even credited with having been persecuted by the SS. The truth of the matter was that Himmler had perpetrated the allegations in order to take over the A4 project.

26. The 'Science' of Extermination and Human Experiment

The spread of the ideology and practice of pseudo-scientific 'racial hygiene' in Germany in the 1920s anticipated, as we have seen, the gradual promotion and acceptance of forced sterilizations, culminating in the policy of forced 'euthanasia', which began in earnest after the outbreak of war. It was a short step, thereafter, to the extermination of Jews and others killed on the grounds of race hatred.

Candidates for sterilization, according to the special act of June 1933, included a catalogue of congenital mental and physical conditions, as well as hereditary blindness, deafness and alcoholism. The poor state of psychiatry and mental hospitals in the aftermath of World War I led to attention being given to patients worth treating, minimum provision for the rest and general sterilization of those discharged beyond the walls of their asylums.

Nazi groups were routinely taken to mental institutions to view the pointless predicament of the inmates; schoolchildren were exposed indirectly to 'euthanasia' by being set questions in the classroom about the cost of maintaining useless existences. In the propaganda of the times the prospect of national crisis, such as a war of national survival, when the normal were being called upon to make huge sacrifices, would justify the elimination of lives that were 'not merely absolutely valueless, but negatively valued existences'.[1]

Meanwhile the reformists had by their very initiatives – advocating occupational therapy and research into 'abnormality' in the wider community – encouraged, in the words of Michael Burleigh, not 'questions concerning the socio-economic environment', but 'the control function of registering widespread deviance in primitive data banks'.[2] The widespread practice of occupational therapy had actually contributed to the 'creation of a psychiatrically defined sub-class within a group of people already consigned to the margins of society'.

The scope of sterilization, organized and administered by the medical profession, widened with the passing years, taking in convicts, prostitutes and even children considered uncooperative in orphanages. In time, even social problems like poverty were attributed to heredity. From 1941 the policy was extended to Sinti, Gypsies and Jews. Complications and fatalities as a result of the procedures were widespread. Between 1934 and 1944 it is estimated that between 300,000 and 400,000 people were forcibly sterilized in the Greater Reich.[3]

Next came the legalized killing of children with mental and physical deformities. The process began in late 1938, when Hitler ordered Karl Brandt, one of his personal physicians, to travel to Leipzig to look into a request on the part of parents for the 'mercy killing' of their child. Brandt was thereafter authorized to initiate a programme involving the murder of handicapped children in hospitals by means of lethal injection. The initiative was a prelude to the expansion of 'euthanasia' to include large numbers of adults with mental illnesses, given the force of law by a Führer decree in October 1939. The decree was backdated to 1 September, as if to coincide with the outbreak of war, lending the weight of national crisis to the measure. Hitler's eventual decree allowing the process of 'euthanasia' would charge Philipp Bouhler and Karl Brandt 'with responsibility for increasing the authority of physicians, to be designated by name, to the end that patients considered incurable according to the best available human judgment of their state of health, can be granted a mercy death'. A bureaucracy of murder was established, known as the Euthanasia Office, or Aktion T4, as we shall describe later, whereby patients were assessed by participating doctors and transported to centres for gassing. The process became a precursor for the Final Solution of Jews, which gathered pace from 1942.

Typical of the synergy between the Euthanasia Office and 'academic' science were the activities of the Kaiser Wilhelm Institute for Brain Research at Berlin-Buch, which had enjoyed large grants from the Rockefeller Foundation. After the T4 operation got under way in 1939 Professor Julius Hallervorden at this centre saw an opportunity to increase the institute's neuropathological collection of brain material. His main access to specimens was a euthanasia killing facility in Brandenburg. The victims, mainly from the Görden Hospital, a local mental institution, were gathered into a room with the appearance of a bath-house, where they were killed with poison gas. Hallervorden was present at some of the murders and had the brains removed speedily and specially treated in a laboratory at the hospital before being taken to the Kaiser Wilhelm Institute (these specimens remained in the renamed institute – Max Planck – until as late as 1990).

The T4 project, so named after the address of its headquarters at Tiergarten 4, 'Reich Work Group of Sanatoriums and Nursing Homes' in Berlin, involved special killing facilities spread throughout Germany. The facilities were supervised by qualified physicians who selected the victims for elimination in gas chambers made to look like shower rooms, anticipating the exterminations in the death camps. The prelude to mass killing of Jews, and other 'lives not worth living', was the '14f13' programme (otherwise known as Operation Invalid, or 'prisoner euthanasia'), after the label on the file which recorded the bureaucratic arrangements to kill 'excess' prisoners in concentration camps. According to Robert J. Lifton, the '14f13' initiative constituted two bridges, one ideological and the other institutional, whereby early concepts of 'racial hygiene' led via the medicalization of killing to the Final Solution. Nazi physicians thus supervised the murder of millions of victims of the death camps, selecting on the ramps those who should be murdered and those who should be spared for work and overseeing the smooth operation of the gas chambers. The medical doctor thus, finally, embodied the depraved Nazi vision of mass murder as a form of hellish racial therapy.

On 20 January 1942, a meeting took place at number 58 am Grossen Wannsee, a villa overlooking the Grosse Wannsee, a lake outside Berlin.[4] There were fifteen high-ranking officials present and Reinhard Heydrich took the chair. Heydrich asked all present to cooperate in the implementation of 'the solution'. Reading from a draft prepared by Eichmann, Heydrich ordered that Jews should be brought under appropriate direction in a suitable manner to the East for forced labour. Separated by sex, the Jews that could work would be pressed into service; it was expected that large numbers 'will fall away through natural reduction'. According to the statistics prepared by Adolf Eichmann 11 million Jews would die. The deportations began in March 1942 and would continue until 1944. Death camps were designed and staffed in remote areas of former Poland – Auschwitz-Birkenau, Treblinka, Belzec, Sobibor, Chelmno and Majdanek. Transportation would become a priority, involving a complex bureaucracy of timetables, rented railway cars, shunting arrangements and provision of guards. Eichmann's representatives were despatched for these purposes to France, Belgium, Holland, Luxembourg, Norway, Romania, Greece, Bulgaria, Hungary, Poland and Czechoslovakia.

Technology of Extermination and Disposal

In recent years information has come to light showing how 'normal' technology employed in the cremation of human bodies was transformed from 1942 onwards and raised to unprecedented levels of 'productivity' as a result of the pressures for mass killing in the Final Solution. The scope and efficiency of the engineering of incineration, it is clear, gave impetus to the policy. Technical ingenuity and skill, careful planning and organization and the drive for individual and corporate financial reward combined to make the killing rates possible.

Earlier histories of the crematoria at Auschwitz were based on archives in Germany, Poland and Israel. To this material has been added sources in Moscow, made available in the early 1990s as a result of the fall of the Berlin Wall. The historians Jean-Claude Pressac and Robert-Jan Van Pelt brought their expertise and knowledge together in 1994 with the publication of a remarkable essay, 'The Machinery of Mass Murder at Auschwitz', making the chronicle of crematorium creation in Auschwitz-Birkenau an examplar for the grisly technological 'progress' throughout the Third Reich.[5]

State-of-the-art technology of human cremation by the early 1930s in Germany involved a burner, a complex economizer that recycled heat from the combustion gases and the 'crucible' – the core of the furnace which took in the coffin. The problem with this model was the expense and size, since the circuits of the economizer took up as much as two-thirds of the entire furnace. German engineers designed a new system of cremation technology by employing compressed cold air, dramatically reducing fuel consumption and speeding up the incineration process. The design was patented in 1928. In 1935 the compressed air innovation was further developed by Kurt Prüfer, an engineer at the noted furnace makers Topf and Sons at Erfurt.

In May of 1937 the SS invited tenders for a crematorium at Dachau concentration camp, since the numbers of deaths were outstripping the capacity of local crematoria, where, in any case, the presence of camp corpses was inviting unwelcome publicity. A bid for the SS tender came in from the firm of Walter Müller of Allach. It employed a cold air compressor, but it had only one crucible and was cased in neo-Grecian style marble: the SS decided not to go ahead.

In 1939, with the growing need for camp crematoria, the SS central budgetary office in Berlin invited tenders once more. This time Prüfer at Topf and Sons proposed a robust, oil-fired, mobile, utilitarian model with a double crucible, again using the compressed air technology. The 'yield' was two bodies an hour, instead of just one, and at today's exchange the price was £23,000, instead of Müller's £25,000 for a single crucible model. The SS ordered one of Prüfer's designs for Dachau and another for Buchenwald, which was close to the furnace factory at Erfurt. Early the following year the SS ordered a further two models, one for Auschwitz, at the site of a former army barracks where 10,000 Polish prisoners were to be 'quarantined', and another for the concentration camp at Flossenburg.

The vicinity of Auschwitz, a town of some 12,000 inhabitants, had originally been selected by the Reich as a model colony of eastern expansion, a settlement exhibiting fanciful medieval architecture and ideal recreational facilities for dancing, music and a range of artistic expression with bucolic resonances. Workers would live within an enclosed fantasy reminiscent of the Victorian Arts and Crafts communities. The scheme, emanating from the Blood and Soil fantasies, would within months degenerate into a living hell of barbed and electronic wire, starvation, disease, torture under the guise of medical experiment and genocide. Among the 'design' features of the camp were single latrines to be shared by more than 150 inmates, many of them suffering from dysentery.

Prüfer was delighted with the orders, as he was personally receiving a 2 per cent commission on the profit of each item: but he was also aware that Kori, a rival company with links to the SS, had designed a similar cheap and mobile crematorium fired by coke at a time when oil was rationed. Stimulated by the competition, Prüfer redesigned the new Auschwitz crematorium to fire on coke and enhanced its efficiency by equipping it with an electric forced-draught fan capable of removing 4,000 cubic metres of smoke an hour and an electric blower to force cold air into the crucible. He estimated that the machine could incinerate thirty to thirty-six bodies in ten hours, or seventy bodies for a twenty-hour cycle. He could also report that the machine would require only three hours' maintenance a day. It came to be known as Auschwitz's crematorium I and it worked flawlessly. In consequence Prüfer was now commissioned to construct a similar crematorium for the Mauthausen camp, which up to this point had been a Kori client.

On 1 March 1941, the SS chief Heinrich Himmler visited Auschwitz and made an on-the-spot decision to expand the camp so as to house 30,000 prisoners immediately and to create a further camp at Birkenau for 100,000 prisoners of war. From these human 'resources' he would provide an IG Farben-owned plant with 10,000 forced labourers for constructing facilities for production of synthetic fuel and rubber. Karl Bischoff, a former warrant officer in the Luftwaffe, once charged with preparing airfields in France, was to head the building work at Auschwitz. Bischoff calculated the need for unusual crematorium capacity and called for Prüfer. Together they designed a crematorium which boasted five furnaces, with three crucibles to each furnace; that is, fifteen crucibles in all, capable of despatching sixty bodies an hour, or 1,440 bodies every twenty-four hours. Later in 1941 a confident Prüfer assured the SS that a four-crucible furnace configuration would also be possible – indicating, in a configuration of five furnaces, twenty corpses burning simultaneously and thus incinerating 1,920 bodies every twenty-four hours.

In the meantime the pesticide Zyklon B, prussic or hydrocyanic acid, was used, not on lice or insects, for which it was intended, but to murder 250 'incurable' camp inmates and 600 prisoners of war. The use of the poison was not entirely efficient; some victims, by one account, were still alive after two days.[6] But in these early months there were large numbers of deaths at Auschwitz because of an outbreak of typhoid fever in consequence of contaminated water. A decision was taken made by May 1942, four months after the Wannsee conference, to site a new fifteen-crucible crematorium at Auschwitz-Birkenau.

The paperwork proliferated, as the Final Solution went into top gear, and Auschwitz-Birkenau was chosen in June 1942 as a principal centre for genocide. The number of commissioned crematoria, with adjoining gas chambers disguised as shower rooms, increased in subsequent years from two to three, then four, then five facilities. The orders for a fourth and a fifth crematorium were worth at today's exchange rate £230,000. But the correspondence between Topf and the SS indicate that the client picked over bills and had to be sent constant reminders for bad debts.

The surviving documentation and the torrent of engineers' blueprints, moreover, tell a story of technical struggles against time and strains on 'capacity', as well as bitter squabbles over materials, design features and standards, delays in delivery, costings and profit margins. There is an impression of constant crises and furious arguments over cracked chimney stacks, inefficient ventilation, insulation and blower systems, as the ever-growing demand for incineration constantly outstripped the engineers' ability to deliver. As the capacity increased, so did the expertise: no less than eleven engineering sub-contractors were involved in the supply of components and specialized construction for crematoria III, IV and V, with large numbers of furnace and ventilation engineers working on site. Civilian sub-contractors were also involved in problems of fireproofing, drainage, roofing, waterproofing, elevators and prussic acid detectors. In the case of crematorium III, the SS asked for a feasibility study inquiring whether a furnace could be used to heat 100 showers.

The final gassings and incineration of the estimated million victims murdered at Auschwitz-Birkenau alone involved the massacre of Hungarian Jews in the months of May and June 1944 in crematoria II, III and V. According to the records crematorium V was quickly overwhelmed by sheer weight of numbers and the corpses were burned in pits dug outside the gas chambers. When Zyklon B ran short at the end of the summer, victims were thrown into the pit and burned alive.

While destruction of Hungary's Jews went ahead, Topf was settling its accounts with the SS, leaving a long and querulous correspondence over minor items, such as the use of an oxygen cylinder and a 'loan' of a few litres of engine oil.

In October 1944, with the Red Army approaching, crematorium IV was set on fire by its Sonderkommando operators; the building was torn down after the insurrection had been violently suppressed by the SS. By the end of November Himmler had ordered an end to the gassings and crematoria II and III were dismantled. The camp was evacuated on 18 January and two days later the SS blew up the remaining hard structures. Crematorium V was dynamited at one o'clock on the morning on 22 January and the Red Army turned up the following day to find nothing but snow-covered rubble.

Prüfer was arrested by the Americans after the German surrender on 8 May 1945. His boss, Ludwig Topf, committed suicide at the end of the month. Prüfer was soon released, but before he left captivity he managed to secure a contract for a furnace from the American military. He was never heard of again; but it is likely that he was rearrested by the Soviets and ended his days in the Gulag.

Human Experiments

J. B. S. Haldane, distinguished physiologist and geneticist, one of Britain's most enthusiastic poison gas pioneers, related in an interview with a journalist how, when he was nine, his father (the Oxford physiologist John Scott Haldane) sent him down a coal mine shaft to test the state of the air. On another occasion his father shut him in an air-tight coffin that left only his head free: the purpose was to ascertain the effect on the boy of certain mixtures of gasses. At age twelve, he was put by his father into a leaking diving suit and let down to a depth of 40 feet beneath the surface of a freezing lake, where he was kept for half an hour and almost drowned. 'It was not altogether a pleasant experience,' Haldane told his interviewer, 'by the time I was pulled up I was wet to the neck and most bitterly cold.'[7]

The case of Haldane father and son is a long way from the experiments in the Nazi death camps, but it seems important to observe from the outset that the temptation to experiment on human subjects (in this case, even one's own precious son) has been shown to be extremely strong and widespread in medical science. Nazi science, as we shall see in the latter part of this book, was by no means unique in practising experiments on human subjects. The impetus is ever-present: the importance is where the lines are drawn and whether the prevailing norms of law, regulating such experimentation, are in force.

Medical research scientists do not repudiate the use of human beings as guinea pigs. Without tests on human subjects, whether in self-experiments conducted by researchers or clinical trials of untested drugs on volunteers (often medical colleagues or courageous and penniless medical students), there would have been no progress in medicine. Certainly J. B. S. Haldane, as an adult, became one of the most famous self-experimenters in the history of British medicine. But science and scientists, as this book has repeatedly emphasized, do not operate in a social and moral vacuum. The key conditions for any form of human tests or experiments, in guidelines drawn up as a result of the post-war Nuremberg doctors' trials, are appropriate consent (one's under-age son hardly seems a suitable candidate, even if he did give his consent), absence of undue influence, humane conditions, avoidance of unnecessary pain and authentically scientific aims.

Nazi scientists who exploited concentration camp inmates for their potential as human guinea pigs were guilty not only of blatant disregard for ethical norms of medical experiments on humans, but were invariably involved in inflicting sadistic injury with no possible scientific purpose in view. These activities, beginning in 1939 and continuing to the war's end in 1945, also formed connections with the Nazi racial hygiene movement, the 'euthanasia' operation, the slave labour policy and the Final Solution itself. They are to be understood, to emphasize the point yet again, as an aspect of a world view that saw certain ethnic and 'medical' groups as possessing worthless lives – Jews, Gypsies, *Untermenschen*, the retarded, homosexuals, those suffering with incurable diseases.

Such people were expendable in the interests of the 'purity' of the *Volk*, or to aid the health and safety of German troops on active service. It was to be expected that groups within the Nazi medical world, aided and abetted by Himmler and his SS, would subject camp inmates to forms of 'medical' research, without consent and with no regard for suffering and risk of life. If this was science, it was a dark and depraved inversion of it.

The groups of perpetrators were not confined to warped pseudo-scientists working in hidden and isolated situations within the death camps. Some 350 qualified doctors (including university professors and lecturers) were involved in concentration camp experiments, which means one out of every 300 members of the German medical community.[8] Professor Kurt Gutzeit, gastroenterologist at the University of Breslau, conducted hepatitis experiments on Jewish children from Auschwitz. Professor Heinrich Berning of Hamburg

University carried out famine experiments on Soviet prisoners of war, carefully noting their symptoms as they starved to death. Professor Julius Hallervorden of the Kaiser Wilhelm Institute for Brain Research in Berlin-Buch ordered, as we have seen, hundreds of brains of the victims of euthanasia for his neuropathological researchers.[9] Professor Otmar von Veschuer of the Kaiser Wilhelm Intitute for Anthropology in Berlin engaged in extensive collaboration with Josef Mengele in Auschwitz.[10] Although the camp doctors sought to exonerate themselves after the war by claiming that they were acting under orders, these 'scientists', of their own volition, played an active and leading role in the organization and execution of such 'research' programmes. In this context, the perpetrators were not always acting under outside pressure, from the regime itself, or from Himmler's Ahnenerbe, but were enthusiastic facilitators of programmes they themselves had initiated.[11] With the norms of civil and criminal law widely suspended throughout Germany, doctors who had a mind to venture into these activities knew that they could do so with impunity.

The facts are well known and have been recounted in a number of key works over the past four decades.[12] Some of the programmes were related to improving the survival prospects of German fighting men. Prisoners of war were immersed in freezing water to estimate how long a pilot or seaman might survive freezing temperatures after being ditched or shipwrecked in a winter sea. Inmates were forced to drink seawater to test the human physiological limits to its consumption. Prisoners were placed in special low-pressure chambers to test endurance at high altitudes. Others were subjected to phosgene and mustard gas contamination, or infected with diseases that might be contracted by German soldiers in Africa. In experiments related to plans for the repopulation of Eastern Europe, castration and sterilization procedures were performed on healthy men and women. Hormone injections were forced on homosexuals, deemed to be a danger to the health of the German *Volk*. In a procedure that has become a byword for Nazi experimental atrocity, the eyes of men, women and children were injected with dye. In Dr Josef Mengele's laboratory at the Birkenau camp, twins were selected from the camp for 'genetic' and 'germ' experiments.

In one experiment at Buchenwald, prisoners were shot with poisoned bullets in order to see how swiftly the poison would work. The intention to kill and mutilate runs through the details of much of the Nuremberg trial evidence, even when the type of experiment appeared to be usefulness to survival research. Seeking a word to characterize the science of such killing for its own sake, Telford Taylor, prosecutor at the Nuremberg doctors' trial which opened on 9 December 1946, chose *thanatology*, the science of death.[13]

The following chapter sections give an impression of a range of experiments in just five camps and the links between the camp experimenters and academic science in universities, hospitals and scientific institutions in Germany.

Dachau[14]

High-altitude and low-pressure experiments were conducted at Dachau concentration camp in 1942 under the leadership of a Dr Sigmund Rascher, a captain in the Medical Service of the Luftwaffe and an SS officer. At the time of the Nuremberg trial Rascher was missing and presumed dead. Rascher had prompted the use of prisoners for experiments in a letter to Himmler in May of 1941 by regretting a lack of data based on 'human material'. He asked for human subjects to be put at his disposal, warning that such human guinea pigs might die. Himmler's adjutant, Rudolf Brandt, responded, confirming that prisoners would be made available. The experiments were carried out in the spring and summer of 1942 with the use of a mobile pressure chamber. The tests duplicated the effects of falling from great heights without parachute or oxygen. In one report, quoted at the Nuremberg doctors' trial, Rascher cited three successive experiments on a '37-year-old Jew' who was subjected to 'falls' from an altitude of 12 kilometres. After the third such 'fall', the victim died in evident agony. A further section of

Rascher's report contained an autopsy describing severe trauma in the pericardium, the brain, the heart and the liver.[15]

Other victims in other experiments included Poles, Russians and Jews accused of *Rassenschande* (racial shame), indicating that they had been guilty of marriage or intercourse with an Aryan, or in the case of an Aryan victim, a non-Aryan. Other reports vividly describe the physical and psychological agonies of the victims in the pressure chamber.

Dachau was also the site of freezing experiments, which followed on the high-altitude tests and continued until the spring of 1943. Victims were forced to stand or lie naked in the open air in the depths of winter from nine to fourteen hours. Some were forced to remain in a tank of iced water for up to three hours, sometimes longer. The organs of victims who died were removed and despatched to the Pathological Institute at Munich. A variety of rewarming 'remedies' were employed. One method urged by Himmler, based on anecdotal stories handed down from fishing communities, was to bed the victims with women volunteers from the camps – mostly Gypsies.

In these experiments, authorized by Deputy Reich Physician Führer Kurt Blome, some eighty of about 300 inmates died.[16] A prisoner nurse called Walter Neff recorded the following eye-witness event:

Two Russian officers were brought from the prison barracks. They arrived at about four o'clock in the afternoon. Rascher had them stripped, and they had to go in the vat naked.

Hour after hour went by, and whereas usually unconsciousness from the cold which set in after sixty minutes at the latest, the two men in this case still responded fully after two and a half hours. All appeals to Rascher to put them to sleep by injection were fruitless. About the third hour one of the Russians said to the other: 'Comrade, please tell the officer to shoot us.' The other replied that he expected no mercy from this fascist dog. They shook hands with each other and uttered the words 'Farewell, Comrade.' . . . The experiment lasted at least five hours before death supervened. The two bodies were taken to the Schwabing Hospital in Munich for post-mortem examination.[17]

The data on freezing experiments became known outside Nazi medical circles. At the end of October 1942 a conference on pilot survival methods was held at the Deutscher Hof Hotel in Nuremberg with ninety-five participants including senior Luftwaffe officers. One speaker gave a presentation on 'Prevention and Treatment of Freezing' and another spoke on 'Warming-up after Freezing to the Danger Point'. It was obvious from the details of experiments that victims had died. But there was no report of objections from the audience.[18]

Also at Dachau some 1,200 inmates were subjected to malarial infection by exposing them to mosquitoes or by injections from the glands of mosquitoes. Victims had a little box containing mosquitoes fixed to their hands so that they would be stung. After contracting the disease victims were treated with a variety of strong drugs, including quinine, neosalvarsan, pyramidion, antipyrin and combinations of such drugs. Many died from overdoses of these medications. According to evidence placed before the judges at the Nuremberg trial, malaria was the direct cause of thirty deaths and 300 to 400 died as a result of later complications.

These experiments were under the general direction of a Dr Klaus Schilling, emeritus professor of parasitology in the Faculty of Medicine at Berlin University, director of the Malaria Commission of the League of Nations and recipient of grants from the Rockefeller Foundation in New York and the Kahn Foundation in Paris. Already aged seventy-three, his research had run into the sand when he made contact with Himmler, who sent him to Dachau to research the possibility of immunization against malaria, a problem for soldiers serving in Africa. According to evidence brought out of Dachau, Schilling was directly responsible for the deaths of ten prisoners.[19]

Dachau was also the scene of research into methods of making seawater potable. Representatives of the Lufwaffe, the navy and I G

Farben met in May of 1944, and it was agreed to conduct experiments on human guinea pigs, some of whom would be obliged to drink seawater. The Medical Inspector of the Luftwaffe approached Himmler for forty suitable experimental subjects; the Reichsführer ordered that Gypsies would be made available.[20]

The victims of this experiment, which was devised and run by Wilhelm Beiglboeck, Consulting Physician to the Luftwaffe, were divided into four groups. The first received no water; the second was required to drink ordinary seawater; the third to drink seawater processed by the 'Berka' method, which disguised the taste of salt water, but was nevertheless saline. The fourth drank seawater which had been desalinated by a method employing a substance called 'Wofatit'.

The tests were carried out in the autumn of 1944 and the victims endured terrible agonies. According to an eye-witness prisoner nurse:

It happened frequently that these patients drank from the slop buckets of the orderlies, or in unobserved moments drained water from the air-raid protection buckets in the hall. Some patients actually lapped up the water poured out on the floor for mopping. I had to weigh the men taking part in the test every day, and noted that the daily loss of weight was up to two pounds.[21]

Sachsenhausen and Natzweiler

Professor Karl Brandt, personal physician to Adolf Hitler, Reich Commissioner for Health and Sanitation, instigated in 1943 a series of jaundice experiments at Sachsenhausen and Natzweiler concentration camps. Hepatitis had become a problem for fighting troops, especially in southern Russia. In some companies 60 per cent of the troops had gone down with the disease. Inmates from Auschwitz (eight Jews of the Polish Resistance Movement) were forcibly brought to Sachsenhausen and Natzweiler camps for the experiments which later culminated in torture and death.

At these same camps wounds were inflicted on victims and the lesions infected with mustard gas, the poison gas widely used in World War I. Others were forced to inhale the poison, or to imbibe it in liquid form. Some were injected with the substance. The purpose of the tests was to discover an antidote. Himmler made his Ahnenerbe laboratory at Natzweiler available to the chief experimenter, Professor August Hirt of the Strasbourg Medical Faculty. A former prisoner testified as to the effects after their arms had been treated with the gas in liquid form:

After about ten hours . . . burns began to appear, all over the body. There were burns wherever the vapour from this gas had reached. Some of the men went blind. The pains were so terrific that it was almost impossible to stay near these patients.[22]

Inmates at Natzweiler were infected with typhus, yellow fever, smallpox, paratyphoid A and B, cholera and diphtheria. Hundreds of them died. The research was conducted by a Dr Eugen Haagen, also professor at the University of Strasbourg. In the case of anti-typhus experiments a group of 'healthy' inmates was selected that had some resistance to the disease and injected with an anti-typhus vaccine. Then all the persons in the group would be infected with typhus. Meanwhile, other inmates, known as the 'control group', with no vaccination, were also infected. At the same time, other inmates were deliberately infected with the disease simply in order to keep the typhus virus alive and available.[23]

Ravensbrück

At Ravensbrück camp women were subjected to transplant experiments involving bone, organs and nerves. In one case the scapula of an inmate was removed and transplanted into a patient at Hohenlychen Hopital. Wounds were created and deliberately infected with gangrene and other infections. Bullet wounds were simulated and subsequent infections encouraged. The prime mover in the sulphonamide experiments was Dr Kerl Gebhardt, surgical consultant to the Waffen-SS and Himmler's personal physician. Gebhardt was blamed for failing to use sufficient sulphonamide to save the life of Reinhard Heydrich (Himmler's second in command) after he had been attacked on 27 May approaching Prague. Four Czech patriots had grenaded his car and fragments of leather, horsehair and springs had been driven into his spleen. He developed peritonitis and died of infection on 4 June. Professor Karl Grawitz began an intensive research programme and Gebhardt took charge of the human experiments employing sulphonamide at Ravensbrück.[24] Some seventy-five camp inmates were given deep wounds, deliberately infected with bacteria and wooden splinters. Sulphonamide was then administered to some and not others. An unspecified number of patients died as a result.

Sterilization experiments were conducted using X-rays, which caused extensive damage to the victims' vital organs. In a letter from SS administrator Viktor Brack to Himmler in June 1942 the policy behind mass sterilization was made evident:

Among 10 millions of Jews in Europe there are, I figure, at least 2 to 3 millions of men and women who are fit enough to work. I hold the view that these ... should be specially selected and preserved. This can, however, only be done if at the same time they are rendered incapable to propagate.[25]

Auschwitz and Mengele

In his book *Murderous Science: Elimination by Scientific Selection of Jews, Gypsies, and Others in Germany, 1933–1945*, the German geneticist Professor Benno Müller-Hill tersely describes the extermination and 'scientific' experiments at Auschwitz under Josef Mengele and others, based on extensive archival material.

Planned as a slave-labour camp for IG Farben, Auschwitz, as is well known, became a death camp in 1943. As many as 10,000 prisoners arrived each day by train. The victims were killed in sealed gas chambers by means of Zyklon B, the hydrogen cyanide poison manufactured by Degesch, which was almost half owned by IG Farben. The bodies were burned in crematoria that were eventually capable of destroying some 5,000 corpses a day.

The existence of Auschwitz with its large number of 'lives not worth living' presented an opportunity to Professor Otmar von Verschuer, director from 1942 of the Kaiser Wilhelm Institute for Anthropology, Human Genetics and Eugenics. Verschuer and his predecessor Eugen Fischer were both enthusiasts for the racial hygiene movement. Verschuer's special interest was genetic analysis of pathological and normal traits of humans, especially based on twins. As the war progressed, however, Verschuer was no longer able to travel through Germany to investigate twins with rare hereditary disease destined to be put to death by 'euthanasia'.

Josef Mengele was born in 1911, the second son of a well-to-do devoutly Catholic family. Joining the SA in 1934, he studied medicine, anthropology and genetics at Munich, Bonn, Vienna and, finally, Frankfurt, where he worked under Verschuer in the Institute of Hereditary Biology. He became chief medical officer of Auschwitz after military service on the eastern front. On his arrival Verschuer secured funding for him to carry out work on heredity.

Mengele and other doctors and anthropologists had ample opportunity to 'select' prisoners on the railway ramp as they arrived. Here the Jews who arrived at Auschwitz were divided into two main groups: mothers and children and the old were sent to the left (to Birkenau) to be gassed; the able were sent to the right (to Monowitz) as slave workers for the IG Farben plants. Mengele selected some hundred pairs of twins and about a hundred families of dwarfs and deformed prisoners.

The dwarfs and handicapped individuals were examined and underwent psychological tests; those that did not die of disease were usually despatched by Mengele with a lethal injection, after which their organs were preserved and sent on to appropriate laboratories at the Kaiser Wilhelm Institute for Anthropology.

Mengele pursued various research projects involving twins at Auschwitz, in association with Verschuer and others. Some of this work can be reconstructed from a report made by his Hungarian Jewish slave-assistant, Dr Miklos Nyiszli, as well as accounts by other inmates. Earlier research on Gypsies had discovered two families with hereditary anomalies of the eye (that is, heterochromatism, or partial discoloration of the iris); Mengele developed this work by using twins selected among the inmates. Nyiszli described how he prepared eyes of four pairs of twins Mengele had murdered with intracardiac injections. The eyes were sent to the Kaiser Wilhelm Institute of Anthropology, where they were studied by a Dr Magnussen, who was writing a paper on the subject.

In 1944 Mengele began a project on twins which was of considerable interest to Verschuer. Were there reproducible, racially determined differences in serum following an infectious disease? Mengele infected identical and fraternal Jewish and Gypsy twins with typhoid bacteria, took blood at various times for chemical analysis in Berlin and followed the course of the disease to the point of death.

27. The Devil's Chemists

On an icy dawn in Poland early in 1944 Primo Levi, chemist, writer and Auschwitz inmate, was summoned from the murderous routines of hard manual labour to attend an oral examination. The purpose was to pass or fail him as suitable to work as a technician in one of the laboratories of the synthetic rubber plant known as Buna, owned by the IG Farben chemical industry conglomerate. The examiner was a Dr Pannwitz, a 'tall, thin, blond' man, who sat 'formidably behind a complicated writing table'. Levi reflected after the war,

From that day I have thought about Doctor Pannwitz many times and in many ways. I have asked myself how he really functioned as a man; how he filled his time, outside of the Polymerization and the Indo-Germanic conscience; above all . . . I wanted to meet him again . . . merely from a personal curiosity about the human soul.[1]

The examination went well. Pannwitz asked Levi on what subject he wrote his degree thesis. Levi had to make a violent effort to recall a sequence of memories so deeply buried, as if he were trying to remember the events of a previous incarnation. Luckily Pannwitz was interested in one of Levi's special subjects, 'measurements of dielectrical constants'. He had passed the test. In the preface to his book *If This is a Man*, Primo Levi wrote:

It was my good fortune to be deported to Auschwitz only in 1944, that is after the German Government had decided, owing to the growing scarcity of labour, to lengthen the average lifespan of the prisoners destined for elimination; it conceded noticeable improvements in the camp routine and temporarily suspended killings at the whim of individuals.[2]

For the next nine months Levi was obliged to continue manual labour outside, lifting and carrying sacks of chemicals: 'The [chemicals] seeped under our clothes and stuck to our sweating limbs and chafed us like leprosy; the skin came off our faces in large burnt patches.'[3]

Although Monowitz, Levi's camp, was one of the closest to the Buna factory, the journey to work involved a march of 4 miles to begin an eight- to twelve-hour working day. At midday each slave labourer received a litre of soup, containing a few scraps of cabbage or turnip in hot water. In the evening, the same amount with bits of rotten potato or swede and chick peas. The bread, 350 grams per portion distributed each morning, was supplemented with additives, including sawdust. A prisoner at Monowitz consumed little more than 1,000 calories a day of a diet that lacked proteins and fats.

A token of the mental resilience of these brutalized inmates: his companion Jean Samuel belonged to an amateur mathematical group which had formed under Jacques Feldbau, a brilliant French mathematician. They talked mathematics on the long marches to and from work, and bartered their precious bread for books purchased by their guards on logarithmic equations.

It was winter again before Levi was invited into the chemistry laboratory and ordered to start work in the polished, spotless confines. But there was little work to be done; and the days of the Buna plant were numbered:

The ravaged Buna lies under the first snows, silent and stiff like an enormous corpse; every day the sirens of the *Fliegeralarm* wail; the Russians are fifty miles away. The electric power station has stopped, the methanol rectification columns no longer exist, three of the four acetylene gasometers have been blown up.[4]

But, at last, Levi has gained entrance to those laboratories with their 'weak aromatic smell of organic chemistry.' The smell made him start back 'as if from the blow of a whip.' He felt that he had been transported back into the world of his youth in Turin, to 'the large semidark room at the university'.

As it was, Levi had little to do except erect experimental tubes and instruments, then dismantle them when bombing raids threatened; put them up, then take them down again, waiting for the arrival of the Russians.

IG Farben at Auschwitz

Germany's chemical industries had sunk to abysmal depths of inhumanity amidst the frozen mud and ruins of Buna's factory for synthetic products at Auschwitz. IG Farben, which had once led the world in the creation of synthetic substances, was now using slave labour in its chemical laboratories. How had this great industrial enterprise, the largest company in Europe (fourth largest in the world after General Motors, US Steel and Standard Oil), sunk so low? What was this conglomerate, a world leader in the technology of chemical processes, doing in the heart of the Auschwitz complex using men such as Primo Levi as brute slaves? The story of the rise and fall of IG Farben takes us back through developments at the beginning of the century in Germany, revisiting great German figures in science who had won Nobel awards for their inventiveness.

As Germany expanded industrially and militarily through the nineteenth century it enjoyed unlimited energy supplies in the form of coal, which served as sources of energy for steam power and domestic heating and provided the raw material for synthetic products. By the first decade of the twentieth century, and the arrival of gasoline and diesel engines, Germany became economically and strategically vulnerable for want of oil deposits. Once again, the talent and experience of its chemists came to Germany's aid. Just as the country's lack of nitrogen, and dependence on the good will of the British Empire and its navy, had stimulated research in the quest for synthetic ammonia, so Germany's scientists and engineers embarked on research to discover forms of synthetic petroleum from coal and its derivatives.[5] One method was hydrogenation, by which different varieties of coal are reacted with hydrogen gas at high pressure and high temperature. The process, known as 'Bergius', involved splitting the complex molecules of coal and introducing hydrogen with massive pressure so as to make liquid oil molecules. The leading pioneer of this method, Friedrich Bergius, learned techniques of high-pressure technology from Fritz Haber in Karlsruhe and Walther Nernst in Berlin before establishing a plant in Hanover, where he began his first experiments in synthetic fuel using artificial coal developed from cellulose.

By 1913 Bergius filed a patent for synthetic petroleum developed from hydrogenation of brown and bituminous coals from which he gained conversion yields of 85 per cent. By 1915 he had begun a new plant in Rheinau near Mannheim. Germany's need for petroleum during World War I was critical, but Bergius's new technology lagged and the capture of the Romanian oilfields made up the deficit temporarily.

Meanwhile, in 1914, two chemists, Franz Fischer and Hans Tropsch of the Kaiser Wilhelm Institute for Coal Research at Mülheim on the Ruhr, discovered a different process, later known as 'F-T synthesis', based on an original idea developed by the chemical giant Badische Anilin- und Soda-Fabrik (BASF). The process involved passing water gas (a mixture of carbon monoxide and hydrogen, produced by treating coke with steam) over a hot catalyst to produce a mixture of hydrocarbons. These were called Kogasin and were used as fuels. A year before the outbreak of World War I, BASF filed a patent for the hydrogenation of carbon monoxide to produce hydrocarbons. The following year, in contrast to the BASF process, Fischer and Tropsch developed a 2:1 hydrogen-carbon monoxide volume mixture they called synthesis gas, which became the basis for future synthetic fuel research.

The eventual success of Germany's synthetic fuel industry was owed to the discovery of efficient catalysts at IG Farben in the 1920s. The technology involved the splitting of coal and tar ring structures into smaller molecules in the liquid phase, and the discovery of alternative catalysts that hydrogenated the smaller molecules in the vapour phase (or gaseous state). Fuels could then be distilled into different octane fractions.

Most of Germany's synthetic fuel was made by the Bergius hydrogenation process, which employed huge compressors generating pressures of up to 10,000 pounds per square inch. The Bergius method produced high-octane aviation fuel and high-quality gasoline, whereas the F-T synthesis method was more suited to diesel oil, lubricants and waxes.

Synthetic fuel development in Germany, ten times more expensive than normal oil refining, was artificially stimulated and sustained by the Nazi regime. Government support based on prodigious orders and subsidies drove the expansion of the processes after Hitler's declaration of the Four-Year Plan under Hermann Goering in 1936. As the prospect of war loomed, vast quantities of fuel suitable for tanks, trucks and aeroplanes were stockpiled in hardened bunker sites around Germany.

Meanwhile IG Farben had made a decision to produce synthetic rubber by a procedure that involved the molecule butadiene and the element sodium (Na), hence 'buna'. The first buna rubber came out of a new plant at Schkopau in 1937 and won the gold medal at the International Exhibition at Paris. But the expense and the availability of natural rubber meant that production remained slow. The outbreak of war in September 1939 altered everything. Hitler went to war with no more than two months' supply of rubber in stock. There was to be no deal with Britain, and Germany was in danger of being starved of rubber. The drive to produce synthetic rubber now became a major priority and IG Farben set about selecting sites for new plants. Germany was looking eastwards and IG Farben was already pondering the huge potential of new markets across the landmasses to Asia.

In an extraordinary twist of fate, while Himmler had been musing over the advantages of the Auschwitz site as a model eastern colony, Dr Otto Ambros, an executive in charge of rubber and plastics at IG Farben, had settled independently on precisely the same area of the map. Foremost in his mind was the need to build a factory which required more than 525,000 cubic metres of water each hour in a region that enjoyed good rail connections. His scrutiny of areas in occupied Poland had brought him in late 1940 to focus on the conjunction of three rivers: the Sola, the lesser Vistula and the Przemsza. The closest town to this site was Auschwitz. Ambros now wrote to the mayor of that town asking for details about the region and its outlying villages, the schools provision and the number of inhabitants. The mayor replied with a wealth of facts and figures in January of 1941. The timing coincided with the circumstance of two crucial interests. Himmler (who had been a fellow student of Ambros at school) had it in mind to build up an eastern colony of Germans and to exploit large numbers of slave labourers that were becoming available. IG Farben's managers faltered for a time, concerned that the town of Auschwitz would never fulfil the conditions it proposed for its German workers. 'Auschwitz and its villages give and impression of extreme filth and squalor,' complained a reporter for the company.[6] Within weeks, however, these problems dissolved with confirmation of the symbiotic relationship between SS provision of slave labour and IG Farben's commitment of money and construction materials.

Because of acute labour shortages Germany's major industrial companies had come to accept with enthusiasm the principle of employing both foreign workers and slave labourers from the concentration camps. Meanwhile, Himmler's SS saw industry and the slave force as a means of building its kingdom of resettled Germans.

The mutual interest leading to the collaboration between IG Farben and the SS became a symbiosis of cruelty and violence. To build the Buna plant quickly, the non-German labour force was urged on with violence and threat of death. A worker, Rudolf Vrba, transported to Auschwitz in June 1942 describes the scene:

Men ran and fell, were kicked and shot. Wild-eyed kapos drove their blood-stained path through rucks of prisoners, while SS men shot from the hip . . . quiet men in impeccable civilian clothes [were] picking their way through corpses they did not want to see, measuring timbers with bright yellow folding rules, making neat little notes in black leather books, oblivious to the blood-bath.[7]

Sergeant Charles J. Conrad, a member of a work party composed of 1,200 British prisoners of war, testified to the active brutality of IG Farben staff:

I saw several civilian employees of the Farben firm beat six inmates while they were working in the factory while three or four civilians looked on. They beat them with pieces of iron and wood for not doing their work properly.[8]

The Fall and Rise of Farben

IG Farben's exploitation of forced and slave labour in its synthetic fuel and rubber plants throughout Germany would rise from 9 per cent of the labour force in 1941 to 30 per cent by the end of the war.[9] There would be some fifty sub-camps in the Auschwitz area. Auschwitz III was the largest, serving the vast Buna plant designed mainly for the production of synthetic rubber. Primo Levi described the plant as 'as big as a city'. It covered 12 square miles.[10] Some 300,000 concentration camp inmates were involved in the slave labour programme by 1944 and some 30,000 of them died, although, as with the synthetic rubber production, not a drop of synthetic material, except methanol, ever left the plants.

Germany's vast effort in producing synthetic fuel, fibres, rubber and other products involved clusters of interrelated plants that eventually became vulnerable to aerial bombing as the Luftwaffe lost control of the skies. Nitrogen, methanol (which many inmates drank, with fatal results), ammonia and calcium carbide were all crucial for the war effort along with a variety of other products. Nitric acid, made from ammonia, was essential for explosives and propellants. Calcium carbide (made from lime and coke in electric furnaces) was the basis of acetylene, which was in turn the basis of butadiene, from which synthetic rubber was made. The process required small proportions of natural rubber, which were brought to Germany via submarine from Japan.

The production of these materials depended on availability of types of coal, coke, coal tar, power generation and transportation. There were six principal complexes (excluding the Auschwitz plant), two of which, Leuna and Ludwigshafen, turned out enormous quantities of crucial chemical products in addition to synthetic fuel and rubber. Leuna, in addition to making some 46 per cent of Germany's synthetic rubber, was producing heavy water for the Nazi atomic programme.

After the war an indictment was filed against twenty-four members of the board of IG Farben on 3 May 1947 on behalf of the United States; among them was Otto Ambros. The charges included waging war, plunder and spoliation, but the crucial crime was 'slavery and mass murder'.

The precise language of the slavery and murder charge was as follows:

Farben, in complete defiance of all decency and human considerations, abused its slave workers by subjecting them, among other things, to excessively long, arduous, and exhausting work, utterly disregarding their health or physical condition. The sole criterion of the right to live or die was the production efficiency of the said inmates.[11]

The trial, which opened on 27 August 1947 at the Palace of Justice in Nuremberg, began in the style of an anti-trust suit, with many hours of organizational exposition. It was not until the slavery and murder charges were introduced that the court began to hear eye-witness accounts of former inmates. Typical of the testimonies was this by Rudolf Vitek, a physician and inmate: 'The prisoners were pushed in their work by the kapos, foremen, and overseers of the IG in an inhuman way. No mercy was shown. Thrashings, ill-treatment of the worst kind, even direct killings were the fashion.'[12] One after another of the witnesses spoke of IG Farben's participation in selections that would mean death for those not selected, of company workers witnessing the hanging of prisoners, of Farben people being aware of the gassing and cremation of inmates in other parts of Auschwitz.

At least one of the judges, Paul M. Herbert, wanted to draw an equivalence between IG Farben's production drive of buna and the deaths of workers as a direct result of the lethal pace of labour. Herbert declared:

It was Farben's drive for speed in the construction of Auschwitz which resulted indirectly in thousands of inmates being selected for extermination by the SS when they were rendered unfit for work. The proof establishes that fear of extermination was used to spur the inmates to greater efforts and that they undertook tasks beyond their physical strength as a result of such fear.[13]

The majority view of the Nuremberg tribunal of judges was that the cruelty and inhumanity at Buna was the responsibility not of the corporate people at Farben but of the Third Reich regime which had imposed the regulations that led to those crimes.

The sentences handed down by the court to the twelve IG Farben executives ranged from eight years (for Otto Ambros) to one and a half years. Only five of the twelve were found guilty of slavery and mass murder (and all received sentences ranging from six to eight years). The chief prosecutor, Josiah Du Bois, remarked that the sentences were 'light enough to please a chicken thief'. He declared that he would write a book to tell the world of the evidence that had been produced in the court. Within four years he made good his promise with his grim account of German industry and the Third Reich, called *The Devil's Chemists*.

Judge Herbert commented on the trial: 'It is important not only to pass judgment upon the guilt or innocence of the accused, but also to set forth an accurate record of the more essential facts established by the proof.'[14]

Primo Levi, for whom the telling of the 'essential facts' became a consuming purpose, wrote after the war: 'one must want to survive, to tell the story, to bear witness . . . to survive we must force ourselves to save at least the skeleton, the scaffolding, the form of civilization.'[15]

It had been General Eisenhower's determination after the war that IG Farben should be broken up as 'one means of assuring world peace'. The Allied Control Council had decided that all the Farben assets should be seized and the legal title vested in the Control Council. The ultimate aim was to deprive the company of its war-making capacities and to break it into separate businesses. Nothing happened until after the Western Allied High Commission replaced the Control Council in June 1949. From that point onwards the plan to break up the conglomerate into its forty-seven units was thwarted. Instead the businesses were consolidated into three of the old major companies: Bayer, BASF and Hoechst. In 1955 the successors to the great IG Farben held their first share-holders' meeting, which voted to allow owners of Bayer stock to be anonymous. The other two followed suit thereafter. In September of 1955 Freidrich Jaehne, sentenced to a year and a half in prison at Nuremberg, was elected chairman of Hoechst. The following year Fritz ter Meer, convicted of plunder and slavery, was elected the chairman of the supervisory board of Bayer.[16]

The Buna plant at Auschwitz survived the war and is productive to this day. Primo Levi himself could note with bitter irony in 1984 that its infamous Carbide tower still loomed over the plant, and the former Buna plant had become the largest synthetic rubber factory in Poland.

28. Wonder Weapons

In the course of 1942, with America in the war and increasing reversals in Russia, Hitler had ordered a total mobilization of science for the war effort.[1] In response weapons designers embarked on a range of uncoordinated and competitive forced innovations in a desperate bid to retrieve the initiative.

There is today a genre of popular books and videos that cite the more exotic of these weapons as indicative of superior Nazi technology which might just have won the war. I have in mind a book which shows on its cover an artist's impression in full colour of a delta-wing six-engined jet aircraft, Nazi emblems on the wings, swooping over the city of New York city with the caption:

A view of what might have been. The Arado E555/1 Flying Wing was designed to operate as a long-distance bomber at altitudes in excess of 40,000 feet. Had the war continued nearer 1950, then this scenario of two examples being pursued by a P-80 Shooting Star over Manhattan might just have become a reality.[2]

But even this 'reconstruction' dwindles to the conventional by comparison with attempts to propose that Germany had developed in the early 1940s an advanced technology, kept a closely guarded secret even to this day by the CIA. For example, a species of aircraft described as a 'flying saucer'. Alleged expert witnesses had seen during the war over Germany an unidentified flying object probably obtaining 'its effect by discharging and instantaneously igniting a blue plasma'. It was, as one observer said, 'a luminous disc spinning round its own axis' and 'looked like a burning balloon' by night.[3] There was also an 'anti-gravity' machine, and a device that wrecked the electronics of aeroplanes (of which more later).

Germany had attained high levels of rocket and jet propulsion technologies that would prove enviable to the Allies, but these fantasies of a futuristic Third Reich science and technology, narrowly cheated of victory by Hitler's poor leadership and the constraints of time, form a popular species of pseudo-history. Apart from the rocket developments and the sluggish, mercifully ill-fated attempt to create an atom bomb and possibly a 'dirty' radioactive bomb, there were, indeed, many hasty research programmes which continued until the final weeks of the war; but most were wasteful and impracticable diversionary tactics aimed at staving off the inevitable. Some were designed to be awesome and vengeful rather than strategically effective. Whether they were specifically Nazi, apart from reliance on slave labour, is debatable. Certainly they revealed the desperation of a chaotic regime moving further and further from reality and rationality.

There was, for example, the prototype of a Porsche super combat tank weighing 185 tons, sporting a gun with a prodigious 15 centimetre calibre, as large as the guns on the battleship *Bismarck*. In the final days of the Reich, moreover, a 1,000-ton tank, nicknamed 'the mouse', was in development.

Among aeronautical innovations there were prototypes of vertical take-off and landing aircraft on the drawing boards of the Heinkel company, dubbed the 'wasp' and the 'lark'. There were exotic plans, too, from the factories of Daimler Benz and Mercedes, for high-speed intercontinental bombers which would be taken part of the way by a carrier aircraft, before being launched to continue the rest of the journey to the target area; the concept involved the crew baling out (presumably on the shores of the continental United States) to be picked up by a waiting submarine. Meanwhile BMW came up with a design for a jet aircraft in which the pilot would lie prone so as to be able to endure forces of up to 14 'g' (pilots in a seated position were normally capable of withstanding only 9 'g' before blacking out).

As resources became ever more scarce there was a spate of experiments with new materials, or the application of traditional ones to new purposes. Hermann Goering, for example, was keen to make locomotives out of cement. One of Himmler's favourite projects was a 'new' substance called 'durofol', which was supposedly a kind of 'miraculous' non-inflammable material, a replacement for iron and non-ferrous metals. In fact it was merely a form of compressed wood. Himmler ordered that a prototype motor car should made of durofol, and urged its use in a variety of other vehicles and buildings. It proved to be highly pliable and useless for anything more than gear levers, car fenders and window frames.

After his release from Spandau prison, Albert Speer compiled in his book *The Slave State* a catalogue of 'wonder weapons' and other 'advanced technological' programmes, mainly initiated or encouraged by Himmler. Himmler's wide-ranging schemes illustrate the chaos of overlap and competition, often backed by Hitler, as powerful members of the regime indulged their dilettante fantasies.

Among Himmler's initiatives was a programme to produce so-called 'oxygen turbines' for U-boats, a project to be headed by an individual totally unqualified in submarine engineering. Himmler also set up a Reich Agency for High-Frequency Research at Dachau, with the use of concentration camp 'scientists' to invent ways of bringing down enemy aircraft 'electronically'. He promoted a 'wonder' combat pistol which never shot straight; a high-speed motor boat – unsolicited by the navy – which never took to the water; and a 'Zisch' flying boat which never took to the skies.

Following the attempt on Hitler's life on 20 July 1944, Himmler's influence and the flow of schemes and peremptory orders for new research knew no bounds. Not only was he Reichsführer-SS, head of the Gestapo and of the police, but he began to exert influence over every area of the Wehrmacht as Commander-in-Chief of the army reserves, as well as Commander of army divisions on the Upper Rhine and another group on the Vistula.

As the war drew to its inevitable conclusion, the 'miracle weapons' continued to flow from eccentric inventors. Himmler rarely let an idea go unexplored by his uninformed enthusiasm. One scheme involved a 'remote control' machine that would switch off electrical devices by means of exploiting the 'insulating material of the atmosphere'.[4] This material, it was hypothesized, is the 'insulating foundation of all electro technology', and by removing its insulating effect it should be impossible for any electrical device of a 'familiar construction and implementation to function'. This bizarre notion, proposed by an amateur, was taken seriously by Himmler, and in consequence by his underlings, who knew that to thwart his 'research programmes' could cost them their lives.

As late as January 1945, with the Reich crumbling, Himmler spent time and energy on a scheme to exploit fir-tree roots as a source of gasoline and other fuels. Nor was this merely the latter-day panic of a crazed leader of the vanquished. It was precisely on a par with an earlier notion of Himmler's that alcohol could be somehow 'caught' from the exhaust fumes of bakery chimneys. In May of 1943 he had sent a directive to his underling Oswald Pohl: 'Bakeries like our bakery in Dachau could supply 100 to 120 litres of alcohol of this sort daily. Please look into this matter and see whether we can do the same thing with all our bakeries.'[5]

When one of his deputed scientific officers, an SS Haupsturm-führer Niemann, dismissed the notion, a directive hurtled from Himmler to Pohl:

Employ a different SS commander for these experiments . . . Niemann strikes me as being absolutely negative toward the entire question. I don't like the tone of his report either. I am of the opinion that in wartime, the yield of even small quantities of alcohol is important.'[6]

Six months on, Himmler was pursuing the ludicrous notion that precious fuel for running the Nazi war machine could be obtained from geranium flowers. He ordered that fields of geraniums should be planted to test the potential of the idea, warning that it should be pursued 'systematically'.

The People's Fighter

Towards the end of the war the new weapons initiatives aimed at delaying inevitable defeat were low-tech rather than high-tech, and assumed a lack of skills as well as a premise of virtual suicide. Teutonic pride had entered an era of sloppy technology and a disregard for the lives of the Reich's pilots. As the Allied bombing campaigns increased in intensity, Germany's leading aircraft manufacturers were invited in September 1944 to tender for the design and production of a single-engine jet plane, cheap to build, easy to fly, and to be produced in tens of thousands, with just twelve days to submit. The idea from the outset was to exploit Germany's Flying Hitler Youth (*Flieger-Hitlerjugend*), boys as young as sixteen, with minimum training, to fly mini-jets in a final defence of the Reich. Reichsmarschall Goering was thrilled, it was reported, at the prospect of thousands of people's fighters, their brave young pilots taking off from the autobahns against the Allied bombing formations.[7]

A token of the unrealistic helter-skelter of development: Heinkel won an order for the mass production of their 'successful' people's fighter design one week after submission. The prototype, He-162 Salamander, took to the air on 6 December 1944; a subsequent test-flight, on 10 December, proved fatal for the pilot as the wooden wing came away — a result of faulty glue. The centre fuselage to which the engine was attached was made of metal; the ailerons of plywood. The production schedules called for 1,000 planes to start production on 1 January 1945, while Junkers would commence another thousand in various underground facilities, and the Mittelwerke underground factory, where V1s and V2s were in production, would start assembly of a further 2,000 units. It was envisaged that the plants would eventually produce these numbers every month. The manufacture of components was farmed out to dispersed factories and workshops, many of them underground. The parts were mostly brought to assembly points by truck and some were even brought by messengers in backpacks. Normal testing and controls were to be set aside in the interests of speed;

the contract for the plane warned that 'serial production will be based on the first drawings and that the risk of a potential failure escaping notice has to be accepted'.[8] The design teams were to be rewarded with 10,000 cigarettes and 500 bottles of vermouth for keeping to their absurd targets.

In a technical critique of the fighter, the historian Ulrich Albrecht comments:

Actual experience showed that the people's fighter was far from being a truly operational fighting vehicle. When the machine gun was fired, muzzle pressure tore away parts of the aluminium covering. The landing gear showed weaknesses. The ribs of the leading wing edge were under-dimensioned. Lateral stability was poor. Stalling characteristics needed to be improved.[9]

Many pilots were killed trying to land the plane: typically, the rudder was caught by the jet exhaust, resulting in collapse and the loss of the aircraft. The scheme was a failure even before events overtook its potential deployment.

Suicide Missions

The ultimate weapon of desperation in the air was the design for an aircraft which assumed the death of the combatant in a suicide mission (*Selbstopferung*). The notion was originally devised as a result of volunteers expressing their willingness to crash their planes on to the landing craft of an Allied invasion fleet approaching the shores of Nazi Europe. One of its keenest exponents was Hanna Reitsch, a test-pilot, who appealed to Hitler to sanction the scheme. The Führer agreed to further work on the proposal, but did not give it his blessing. Reitsch became famous for her daring exploits, which included flying and crash-landing a version of a pulse jet flying bomb – the V1 pilotless aircraft intended for a new Blitz on Britain. Wernher von Braun was also associated with the scheme, having designed a rocket plane which involved a suicide pilot in the early stages of the war. Another plan was to develop a rocket plane which could take off as raiding Allied aircraft came into view. The aircraft of choice was known as the Natter (or adder), designed to carry an array of rockets: unfortunately, once these weapons were despatched, the plane would become unstable and could no longer be flown; it was assumed that the pilot would bale out. On the maiden flight of a prototype the test-pilot's neck was broken as the aircraft was launched. Experiments were still continuing at the end of the war.

As Germany ran out of fuel in the final months and weeks, the regime designed a type of armoured glider which was supposed to ram a raiding bomber or fire at it from close quarters. Either way, suicide was implicit in the tactic. The tactical concept was for the pilot to fly the heavy explosive payload close to an enemy bomber formation, to detonate the charge and bale out. 'Because parachuting during the approach to the enemy was theoretically possible, but practically excluded, one could label this a "suicide bomb".'[10] Another prototype glider, made of wood and designed by Messerschmitt, was an aircraft that was to be ferried above the bomber formations then released to glide down and attack. It had poor manoeuvrability and never entered combat. Yet another suicide concept, based on the carrier idea, involved a manned 'flying bazooka', a glider carrying two unguided rockets, to be fired directly at a bomber before the pilot made his getaway.

Dirty Bombs

Boris Rajewsky, director of the Institute for the Physical Foundations of Medicine and involved in health and safety standards in Germany's mines, sought funding in February 1943 to explore the potential for 'biological effects of corpuscular radiation, including neutrons, with regard to the possibility of their use as a weapon, but above all the biological basis of radiation protection'.[11] It appears that the research was not in the event funded, but the application has given credibility to the idea that Nazi weapons researchers were developing a 'dirty bomb' to target civilian populations in Britain, and ultimately the United States.

Some historians have suggested that a radioactive bomb could have formed the warhead of a giant rocket (A9/10), plans for which were at least on the drawing boards at Peenemünde; and that when these schemes appeared too far in the future for feasibility, the engineers began to consider delivery by smaller rockets launched from submarines.

Evidence for a plan to make and deliver radioactive bombs is scarce and largely circumstantial, but, given the wide range of outlandish and murderous projects, hardly beyond the realms of possibility.[12] There are occasional references in the latter part of the war to a *Uraniumbombe*, by, for example, Hitler's stenographer, Dr Henry Picker, 'a prototype of the German uranium bomb', which was surrounded with 'greatest secrecy'.[13] According to Julius Schaub, chief ADC to Hitler, who had been informed about such a bomb by SS officers working at the Mittelwerke, it was the 'size of a small pumpkin' and composed of 'several small uranium bombs set around a conventional explosive'.[14]

If there is any truth in the suggestion that Hitler might have possessed such a weapon it would have had the same status as other poisonous gas weapons and nerve agents available on both sides during the war. Its use by Hitler, moreover, would have been constrained by his caution in relation to first use of other chemical weapons: that retribution would have been swift and massive. The only people likely to have been harmed by these dirty bombs would have been the slave workers conscripted from concentration camps to work on such products.

Hitler's Poison Gas

The story of Nazi Germany's development and stockpiling of poison gas began in December 1936 when Dr Gerhard Schrader, an IG Farben research chemist working on insecticides, discovered a highly lethal substance that attacks the human nervous system. It came to be known as 'tabun', which is in the range of organic phosphorus compounds.[15] Tabun disrupts a neurotransmitter, a natural chemical in the human body, known as cholinesterase, which reacts with the neurotransmitter acetylcholine to allow normal muscular movement. When acetylcholine collects uninhibitedly in the central nervous system, it causes severe contractions and rigor, especially in the respiratory system. The victim literally chokes to death.

Schrader was ordered to army headquarters in Berlin, where he demonstrated the power of the new agent on dogs and monkeys, which died within twenty minutes of exposure. The substance became a top secret potential 'weapon'. The head of army poison gas strategy, Colonel Rudriger, commissioned Schrader to continue refining his discovery at a factory at Elberfeld in the Ruhr, and put in hand plans for a special poison gas plant at Spandau.

The following year, however, Schrader came up with a second discovery in the form of isopropyl methylphosphorofluoridate, which proved even more lethal in animal tests than tabun. It came to be known as 'sarin'. In September 1939, a decision was taken by IG Farben to build a plant for the production of tabun and sarin in Silesia, at a place called Dyherfurth. Funding ultimately came from the army, but the chief facilitator of the programme was Otto Ambros of IG Farben. Dyherfurth would eventually expand into a vast factory more than a mile and a half in length and half a mile wide, with underground facilities, employing 3,000 workers under conditions of the highest secrecy and security. Work at the plant was psychologically oppressive, because of its remoteness, as well as the extreme danger of handling toxic substances. Workers lived in special enclosed barracks. Despite protective clothing, operatives were regularly affected and at least ten died after accidental contamination.

In the meantime, at other plants around Germany, alternative forms of gas warfare, of the World War I variety, were being developed, including phosgene, chlorine and mustard gas. Animals were widely used in experiments, and so were human guinea pigs within the concentration camps, as we have seen. Baron Georg von Schnitzler, a director of IG Farben, testified after the war that Ambros knew of experiments on human beings.[16]

In addition to many tens of thousands of tons of stockpiled

mustard gas, chlorine and phosgene, an estimated 12,000 tons of tabun were discovered at the end of the war. A number of delivery systems had been developed including types of personnel mines, hand grenades, hand sprays and poison bullets delivered by machine gun.

Vengeance Weapons – V1s and V2s

The ultimate deployment of the advanced technology of missile technology at Peenemünde and the slave labour tunnels of the Mittelwerke was in an irrational and wasteful strategy of revenge, or retaliation. In German they were called *Vergeltungswaffen* (revenge weapons). The first to land on the south-east of England, the V1 ('Vergeltungswaffe 1'), were popularly known by the British as the 'doodlebugs', 'buzz-bombs' and 'flying bombs'. It was essentially a pilotless monoplane made of thin pressed steel and plywood. Catapaulted from a ramp by steam piston, or fired from an aircraft, the missile could soon reached a cruising speed of about 300-plus miles per hour at altitudes as high as 4,000 feet or as low as the tree-tops. The pulse jet engine was fuelled by a tank of 150 gallons of gasoline with compressed air as the oxidizer. The guidance system was operated automatically by a gyroscopic device and a pre-set directional compass. An air log driven by a nose airscrew ceased to turn once the allotted distance had been completed, thereby instructing the elevators to depress and so bring down the craft and its explosive payload to the ground.

The first V1s landed on London on 13 June 1944; by the end of the month 2,452 had been launched. A third was destroyed or crashed before reaching the English coast, another third crashed haphazardly in open country, but the remainder, about 800, landed in the London area or in the vicinity of Southampton. The worst incident occurred on Sunday 18 June, when a flying bomb exploded on the chapel of Wellington Barracks in central London, killing 121 people, sixty-three of them soldiers, while they were at worship.

In the autumn of 1944 the V1s were also fired in large numbers at Antwerp to harass supplies destined for the invading Allied armies. In all some 10,000 V1s were launched against Britain. About 2,500 landed in London; more than 6,000 people were killed and 18,000 injured.[17]

As we have seen, the V2 went into production in May 1944 at the underground Mittelwerke plant in the Harz mountains supported by slave labour. The missile, which was supersonic and against which there was no defence, was ready for launching in September 1944 during a lull in the flying bomb attacks. The first two, launched from a site near The Hague in Holland, landed without warning on 8 September almost simultaneously in the evening, at Chiswick in west London, and at Parndon Wood, near Epping, east of London. Each missile carried a payload of a ton of explosive. Three people were killed and seventeen injured at the site of the Chiswick explosion. Between that date and 27 March 1945 some 1,054 rockets fell (about five a day), killing 2,700 Londoners. More than 900 were fired during the last quarter of 1944 at Antwerp.

If Hitler had intending bringing Britain to its knees with these terror devices it was, of course, a vain hope. From the German point of view, the vast effort and ingenuity applied to the V2s in particular was irrational in its motivation. Interrogated after the war, Field Marshal Erhard Milch, Goering's deputy in command of the Luftwaffe, said:

The primary reason why so much manpower was tied up in the production of V weapons was simply that great promises about miraculous weapons had been made to the people, and they now wanted to redeem these promises in some way or other.[18]

This stated motivation was confirmed in the interrogation of Dr Karl Frydag, an aircraft production chief:

[The V2] has been pursued because of the propaganda effect and it was a foolish thing. Speer, the armaments minister, stated that the purpose of the 'V' weapons was to 'counter the British night attacks with something similar, without the expensive bombers and practically without losses. The main reason was therefore a psychological one for the benefit of the German people.'[19]

The campaign, which drew away resources from aircraft production with very little result, probably did more for the morale of Hitler in the final months of the war than for his people.

PART SEVEN

In Hitler's Shadow

29. Farm Hall

On 30 April 1945 Hitler committed suicide with Eva Braun in his Berlin bunker. He was fifty-six years of age and he had been in power twelve years and three months. The Third Reich officially surrendered a week later and the war was officially over in Europe. Germany was in ruins, its communications shattered, millions of its people homeless, hungry and wandering the country as refugees.

Two months later, on 3 July 1945, as thousands of incendiary bombs rained down on the cities of Japan, which continued to fight, ten leading German physicists, most of them involved in the Nazi wartime nuclear research programme, arrived at a mansion close to the river Ouse ten miles west of Cambridge in England. One of their number, the physical chemist Paul Harteck, recognized Ely Cathedral as their Dakota circled to land at a military base near Huntingdon.

Farm Hall, a red-brick Georgian house on the edge of the village of Godmanchester, was surrounded by private parkland and a high wall. It remains to this day much as it was, although it is now close to the busy A14 connecting the Midlands with East Anglia. During the 1940s it was owned by Britain's Foreign Office and used by MI6 as a 'safe house' for members of resistance groups waiting to be parachuted into secret dropping zones.

The rooms of the house had been rigged in preparation for the scientists' arrival with hidden microphones connected to shellacked metal recording discs, which would be monitored by a team of army interpreters. Transcripts of the recordings were to be sent on to London and thence to General Groves, chief of the Manhattan Project. This eavesdropping exercise was known as 'Operation Epsilon'.

Groves's interest in these scientists in the aftermath of the European war was not so much to discover how much the arrested

German scientists knew about the making of an atomic bomb, for that had already been ascertained, but their attitude towards the Russians. Would they go over to the Russians, given the opportunity, and work on a Soviet atomic bomb? The transcriptions of the recordings, which remained secret for fifty years, constitute an unusual historical record. Here were Germany's top physicists, caught in a limbo between the end of one era and the beginning of another, discussing among themselves, unconscious of being recorded, how they saw their goals, achievements and failures.[1]

Two of the leading figures were veterans: Otto Hahn, the eminent radiochemist who had discovered nuclear fission, and the mild-mannered Max von Laue, who had won the Nobel prize for his work on X-rays and who had managed to stay aloof from the regime. Neither played a crucial part in Germany's wartime nuclear research, but Goudsmit, as we have seen, had wanted these older distinguished scientists in American and British hands, since he saw an important role for them in Germany's post-war reconstruction.

Bespectacled Erich Bagge, at thirty-three, was the youngest of the group; he had worked on isotope separation (on a return visit to Farm Hall in the 1980s he boasted to the current owners that he had sometimes climbed over the wall to make assignations with local girls).[2] The rest of the group included Kurt Diebner of Army Ordnance, described by his captors as 'an unpleasant personality'; Walter Gerlach, white-haired and of military bearing, a key figure in the German atomic research programme and with connections inside the Gestapo; Paul Harteck, who had written that first letter to Army Ordnance on reading the Joliot-Curie article on secondary neutrons in *Nature*; Karl Wirtz, an expert on heavy water; Horst Korsching, with dark cinematic good looks, described by his captors as 'a complete enigma'. Finally, there were Carl Friedrich von Weizsäcker and Werner Heisenberg.

For six months the scientists would live hidden from the world; their greatest discomfort, lack of communication with their families back in Germany. Their daily lives were regulated by periods for study and exercise. They had newspapers, and for evening recreation they listened to the radio, or attended a piano recital by

Heisenberg, before playing cards up till midnight. German prisoners of war acted as orderlies and serving staff.

Within a week of their arrival Diebner said to Heisenberg: 'I wonder whether there are microphones installed here?' Heisenberg laughed: 'Microphones? Oh no, they're not as cute as all that. I don't think they know the real Gestapo methods; they're a bit old-fashioned in that respect.'[3]

The Farm Hall tapes, quoted below from Professor Jeremy Bernstein's edition of the transcripts, reveal that the men displayed no pangs of conscience, and readily exculpated themselves of official association with the regime. They did not think of themselves as Nazi scientists in any meaningful sense whatsoever. Erich Bagge and Kurt Diebner had been full members of the party, whereas Otto Hahn, Max von Laue and Werner Heisenberg were the only internees who had not been members of any Nazi association whatsoever. Diebner maintained that he had suffered under the Nazis and that he had only joined the Party because he wanted to be assured of a decent job after the war. He had used his Party membership, he claimed, to prevent the arrest of Norwegian colleagues. Erich Bagge claimed that he had been made a member of the Party by mistake: his mother had sent in his name since she thought it would be good for him. Heisenberg commented to a visiting British scientist, Professor Patrick Blackett, that Bagge was 'in some ways' a 'proletarian type', and that is 'one of the reasons why he went into the party, but he never was what one would call a fanatical Nazi'.[4]

Gerlach nevertheless claimed that no one had to join the Party against their will. Bagge retorted that Gerlach was protected because he was a personal acquaintance of Goering's and that he had a brother in the SS; then he commented that Nazi excesses, including concentration camps, were a consequence of the stresses of war.

Hiroshima

On 6 August 1945 news reached the seclusion of this rural prison house that was to shatter the calm routines of the ten German physicists. From this point onwards all discussion about Nazi affiliation ceased. That evening, shortly before dinner, the British officer in charge told Otto Hahn that it had been reported on the news that the Americans and British had dropped an atomic bomb on Japan. The officer evidently provided various details: that hundreds of thousands had died, that many thousands of workers had been involved in the creation of the bomb, that £500 million had been spent and that the explosion was equal to 20,000 tons of TNT.

Hahn, according to the British officer's report, was 'completely shattered by the news', since he believed that his original discovery had made the bomb possible. He said that he had originally contemplated suicide when he understood the 'terrible potentialities of the discovery' and that he now felt these potentialities had been realized and he was to blame. The British eavesdroppers reported to their superiors that Hahn calmed down 'with the help of considerable alcoholic stimulant'.

When Hahn passed on the news to the scientists who had assembled for supper, their reactions alternated between puzzlement and disbelief. Then Hahn offered the observation: 'If the Americans have a uranium bomb then you're all second-raters. Poor old Heisenberg.'

Laue rubbed it in by scoffing: 'The innocent!'.

Heisenberg said: 'Did they use the word uranium in connection with this atomic bomb?'[5]

His companions replied 'No!' in unison.

'Then it's got nothing to do with atoms,' Heisenberg went on. 'But the equivalent of 20,000 tons of high explosive is terrific.'

When Gerlach suggested that the Allies had achieved this through a plutonium bomb, or what he described as a bomb made of neptunium, the product of a reactor, or what the Germans termed an 'engine' (93, as we have seen, decays into 94, or plutonium), Heisenberg expressed his incredulity: 'All I can suggest is that some dilettante in America who knows very little about it has bluffed them into saying: "If you drop this it has the equivalent of 20,000 tons of high explosive" and in reality it doesn't work at all.'

While the scientists continued to speculate on how the bomb had been made, the question of scientific responsibility arose.

Wirtz said: 'I'm glad we didn't have it.'

Heisenberg snapped back: 'That's another matter.'

A few moments later, Weizsäcker said: 'I think it is dreadful of the Americans to have done it. I think it is madness on their part.'

'One can't say that,' insisted Heisenberg. 'One could equally well say "That's the quickest way of ending the war."' The reflection, of course, coincided precisely with the rationalization that would constitute the Anglo-American defence of the atom bombing of Japan in succeeding decades.

'That's what consoles me,' said Hahn.

The question of the amount in weight of pure '235' necessary for a bomb, and the means of achieving that weight, was discussed. At one point Hahn said to Heisenberg: 'But tell me why you used to tell me that one needed 50 kilograms of 235 in order to do anything. Now you say one needs two tons.' Heisenberg said he did not wish to commit himself at that moment, but Hahn's question was significant, for it demonstrated that the Germans had discussed the making of a bomb – a fact that they later denied in public.

Again, they returned to the morality of the bomb. Hahn said: 'Once I wanted to suggest that all uranium should be sunk to the bottom of the ocean. I always thought that one could only make a bomb of such a size that a whole province would be blown up.'

Back and forth went the arguments. Had the Americans achieved a breakthrough by an invention or a discovery? And why had they – the Germans – failed? Self-doubt and self-castigation were in the air.

At nine o'clock the guests gathered around the radio to listen to the BBC main news broadcast, which contained the following report:

Here is the News: It's dominated by a tremendous achievement of Allied scientists – the production of the atomic bomb. One has already been dropped on a Japanese army base. It alone contained as much explosive power as two thousand of our great ten-tonners.

After more headlines, listing statements from President Truman, Winston Churchill and General Eisenhower, the newscaster added with consummate bathos: 'At home, it's been a Bank Holiday of sunshine and thunderstorms; a record crowd at Lord's has seen Australia make 273 for five wickets.'[6]

The newscaster now enlarged on the details of the dropping of the bomb. Some 125,000 workers had built the bomb factories, and 65,000 had worked in them. All was achieved with maximum secrecy: 'they could see huge quantities of materials going in, and nothing coming out – for the size of the explosive charge is very small'. Uranium had been used in making the bomb, and scientists were convinced that the bomb could be 'developed still further'. The fact that atomic energy could be released on a large scale meant that it will 'ultimately be used in peacetime industry'.

The newscast ended with the reading of a statement written earlier by Winston Churchill, who had by this date lost the first general election at the end of the war in Europe to the Labour Party. It spoke of the long scientific path to the bomb, and of the breakthrough that first occurred in Britain, followed by collaboration with the Americans, and the construction of the bomb in the United States in order to avoid the susceptibility of research and development to German air raids over Britain. Then he allowed himself a spasm of British and American pride, heavily contrasted with German scientific failure.

By God's mercy British and American science out-paced all German efforts. These were on a considerable scale, but far behind. The possession of these powers by the Germans at any time might have altered the result of the war, and profound anxiety was felt by those who were informed.

He ended with the solemn reflection:

We must indeed pray that these awful agencies will be made to conduce to peace among nations, and that instead of wreaking measureless havoc upon the entire globe they may become a perennial fountain of world prosperity.

After the broadcast the guests launched once more into contention. Harteck speculated that the the Americans had managed it 'with mass spectrographs on a large scale or they have been successful with a photochemical process'. Such a process was possible, said Heisenberg, 'seeing as they had 180,000 people working on it'.

'Which is 100 times more than we had'.

'The Americans are capable of real cooperation on a tremendous scale,' said Korsching. 'That would have been impossible in Germany. Each one said that the other was unimportant.'

'How many people,' Weizsäcker asked, 'were working on the V1 and V2?'

'Thousands worked on that,' said Diebner.

Then Heisenberg remarked: 'We wouldn't have had the moral courage to recommend to the government in the spring of 1942 that they should employ 120,000 just for building the thing up.' Clearly, he meant that he would have been afraid to make promises to the regime that could not be kept.

Then came a proposition that was to become the basis of the 'myth' of the German nuclear weapon project – the promotion of a declaration of exoneration, connecting the self-exculpations of the Nazi bomb scientists with Robert Jungk's *Brighter than a Thousand Suns*, Tom Powers's *Heisenberg's War* and even aspects of Michael Frayn's *Copenhagen*. As Jeremy Bernstein summarizes the myth: 'We could have done it, we knew how to do it, but we didn't do it on principle.'

The myth was to make its most dramatic, public and early appearance in Jungk's book: 'It seems paradoxical that the German nuclear physicists,' he wrote,

living under a sabre-rattling dictatorship, obeyed the voice of conscience and attempted to prevent the construction of atomic bombs, while their professional colleagues in the democracies, who had no fear, with very few exceptions concentrated their whole energies on production of the new weapon.[7]

Three years after the publication of Jungk's book, in a correspondence with Paul Rosbaud, Laue would report that Weizsäcker had attempted to create this myth. Laue called it the *Lesart*, or the version. 'The leader in all these discussions was Weizsäcker,' he wrote. 'I did not hear the mention of any ethical point of view. Heisenberg was mostly silent.'[8]

On the evening of the broadcast, at Farm Hall, Weizsäcker said: 'I believe the reason we didn't do it was because all the physicists didn't want to do it, on principle. If we had all wanted Germany to win the war we would have succeeded.'

Hahn, the veteran scientist, already agonized by his sense of responsibility for the creation of the bomb, said flatly: 'I don't believe that but I am thankful we didn't succeed.' It was significant that nobody contradicted him – at least, not at this juncture.

The discussion now returned to the question as to whether the scientists had intended to make a bomb or not. Weizsäcker said: 'It is a fact that we were all convinced that the thing could not be completed during the war.'

'Well that's not quite right,' objected Heisenberg. 'I would say that I was absolutely convinced of the possibility of our making a uranium engine but I never thought that we would make a bomb and at the bottom of my heart I was really glad that it was to be an engine [reactor] and not a bomb, I must admit.'

In the midst of these exchanges Hahn left the room. Then Weizsäcker continued with his one-track self-exculpation: 'I don't think we ought to make excuses now because we did not succeed, but we must admit that we didn't want to succeed. If we had put the same energy into it as the Americans and had wanted it as they did, it is quite certain that we would not have succeeded as they would have smashed up the factories.'

Eventually Heisenberg said: 'I think we ought to avoid squabbling amongst ourselves concerning a lost cause. In addition, we must not make things too difficult for Hahn.'

Then Wirtz plunged in with a strident comparison of German inventiveness and American rashness: 'I think it is characteristic that the Germans made the discovery and didn't use it, whereas the Americans have used it. I must say I didn't think the Americans would dare to use it.'

As Hahn and Laue discussed the event, describing the news as a tremendous achievement without parallel in world history, Gerlach, who as we have seen was in charge of the nuclear programme with Speer's backing, left for his room, where the eavesdroppers, and his companions, too, heard him sobbing aloud. Harteck and Laue eventually went to comfort him. Gerlach said: 'I never for a moment thought of a bomb, but I said to myself: "If Hahn has made this discovery let us at least be the first to make use of it." When we get back to Germany we will have a dreadful time. We will be looked upon as the ones who have sabotaged everything. We won't remain alive long there.'

Later Hahn went along to see Gerlach and asked him: 'Are you upset because we did not make the uranium bomb? I thank God on my bended knees that we did not make the uranium bomb. Or are you depressed because the Americans could do it better than we could?'

Gerlach said: 'Yes.'

'Surely,' said Hahn, 'you are not in favour of such an inhuman weapon as the uranium bomb?'

To which Gerlach responded, 'No. We never worked on a bomb. I didn't believe that it would go so quickly. But I did think that we should do everything to make the sources of energy and exploit the possibilities for the future.'

Could Gerlach have been entirely honest in this? It was not strictly true that the Germans did not have a bomb in mind during various stages of atom research planning during the war. Hahn nevertheless said: 'I am thankful that we were not the first to drop the uranium bomb.'

So the arguments went on, ranging over the internecine conflicts and failures of the years, including Heisenberg's predilection for theory over experiment, and returning again and again to technical details.

Later yet in the evening Hahn and Heisenberg spoke alone together. They talked about Gerlach's distress, which Heisenberg put down to the man's sorrow at Germany's defeat, despite his disapproval of Nazi crimes. Hahn responded that although he loved Germany it was for this very reason that he had hoped for his country's defeat. Both men, according to the notes taken by the eavesdroppers, maintained that 'they had never wanted to work on a bomb and had been pleased when it was decided to concentrate everything on the engine [reactor]'.

In the notes that completed the day's report, the eavesdroppers recorded:

Although the guests retired to bed about 1:30, most of them appear to have spent a somewhat disturbed night judging by the deep sighs and occasional shouts which were heard during the night. There was also a considerable amount of coming and going along the corridors.[9]

Earlier that evening the British commanding officer called Laue in to request that the scientists make sure that Hahn did not do any harm to himself. Laue replied that he was more worried about Gerlach, who seemed to have had 'a real nervous breakdown, with many tears'.

The Moral High Ground

The next day the bid to seize the moral high ground continued apace, with Hahn venturing: 'If Niels Bohr helped, then I must say he has gone down in my estimation.'

Then Weizsäcker came out with his full-blown assertion of moral superiority. He said: 'History will record that the Americans and the English made a bomb, and that at the same time the Germans, under the Hitler regime, produced a workable engine.

In other words, the peaceful development of the uranium engine was made in Germany under the Hitler regime, whereas the Americans and the English developed this ghastly weapon of war.'[10]

Many years later Laue commented: 'during the table conversation, the version [*Lesart*] was developed that the German atomic physicists really had not wanted the atomic bomb, either because it was impossible to achieve it during the expected duration of the war or because they simply did not want to have it at all'.[11]

That day the internees agreed, some of them reluctantly, to draw up a memorandum outlining the nature of their nuclear research project. It was drafted by Heisenberg and Gerlach in an army exercise book, and signed by all present. The memorandum, dated 8 August, starts with the discovery of fission of the atomic nucleus in uranium by Hahn and Strassmann in Berlin in 1938, and ends in 1945 with the anticipated 'building of a power-producing apparatus' at Haigerloch. The scientists noted that nuclear fission was the 'result of pure scientific research, which had nothing to do with practical uses'. The memorandum states that the realization that a chain reaction of atomic nuclei had become feasible, and consequently the exploitation of nuclear energies, dawned 'almost simultaneously in various countries'.

The document goes on to claim that 'at the beginning of the war a group of research workers was formed with instructions to investigate the practical application of these energies'. Their work showed by 1941 that it would be 'possible to use the nuclear energies for the production of heat and thereby to drive machinery'. It did not appear feasible at the time 'to produce a bomb with the technical possibilities available in Germany'. Their research required not only uranium but also heavy water; however, production of heavy water was stopped by Allied attacks on the plant at Norsk Hydro. By the end of the war, as the memorandum puts it, experiments were in hand to try to obviate the use of heavy water by the concentration of the rare isotope U-235.

Quite apart from the neglect of Lise Meitner's part in fission discovery, the document represents the first attempt at self-exculpation by the German physicists intended for public consumption. In the event the document was not published, but it accords with the spirit of the campaign conducted by some of them in subsequent years. Ironically, the very fact of their detention provided the opportunity for them to agree on the story they would tell to their fellow countrymen and the world. Absent from the document, moreover, is Paul Harteck's approach to Bernhard Rust for funding to push forward uranium research with the aim to create weapons of mass destruction. Absent too is Wiezsäcker's report in 1940 on the neptunium path to a bomb, and Houtermans's speculations about plutonium. Absent too was the theoretical research on graphite as a moderator.

Heisenberg's Lecture

Between 8 and 22 August the detainees' discussion turned to whether they would be prepared to cooperate with the Allies, an issue of great interest to their captors; an issue of significance too in how these scientists regarded the relationship between their future research and its auspices.

They agreed that the British and Americans, whom they often referred to as the 'Anglo-Saxons', would do everything to keep them from the Russians. Weizsäcker continued to express his 'antipathy' for the bomb; and while admitting that he had 'a lot in common with the Anglo-Saxons', professed to despise their governments. Heisenberg said that 'each of us must be very careful to see that he gets into a proper position'. Weizsäcker told Heisenberg: 'If you are in Germany, a great deal of the responsibility for the continuation of physics in Germany will be yours.' The comment seemed to reveal a throw-back to the mindset of scientists in a former age, when the great Herr Professor was responsible for the whole discipline: a system that was not to survive.

Following several days of discussions about the technical aspects of the building of an atomic bomb, Heisenberg was eventually asked by his companions to give a lecture on the subject; it took place on 14 August.[12] But despite a week of reading newspaper accounts and broadcasts, as well as thinking about the problem, it is clear that Heisenberg had not understood the key difference between a reactor and a bomb. 'Nearly everything Heisenberg says in this lecture,' writes Jeremy Bernstein, who is both a historian of the period and a former practising physicist in this area of atomic physics, 'misses essential points, and the comments of his colleagues are worse.'[13]

Bernstein's assessment, in the light of the Farm Hall transcripts, is important, for there have been assertions to the contrary – by the Germans themselves, by Jungk in *Brighter than a Thousand Suns*, and notably by Thomas Powers in his book *Heisenberg's War*. Powers writes in his biography of Heisenberg: 'The general discussion prompted by Heisenberg's lecture . . . made it clear that only some of the scientists really understood bomb physics – Heisenberg, Harteck, von Wiezsäcker and Wirtz – while the others were evidently hearing much that was new to them.'[14] Bernstein's rejection of the implications of this statement – namely, that Heisenberg knew how to build a bomb, but deprived Hitler of his knowledge – is vehement and exemplifies one side of the divide in the long-running debate, contradicting the view that Heisenberg knew how to build a bomb but frustrated its realization.[15]

The importance of the publication of Heisenberg's Farm Hall lecture is that it should enable us to determine just how close Heisenberg was to understanding the principles and the practical problems of making an atomic bomb. This is less a question of historical judgment, according to Bernstein, and more a matter of grasping the physics. This is how the physicist Bernstein combatively states his verdict on Heisenberg's knowledge of the physics necessary to build an atom bomb at the time of this lecture:

The notion that this lecture showed that the four scientists mentioned 'really understood bomb physics' is so ludicrous that one wonders if Powers has any understanding at all of the physics contained in this lecture and the comments made during it. Moreover, the fantasy that Heisenberg understood how to make a bomb all along but kept 'the secret' to himself is equally absurd. This lecture was given to show that the 'professors' could figure out how a bomb was made. It represented the high-water mark of their understanding.[16]

Much as I admire the assiduous research and the narrative skill of Thomas Powers, I am inclined to concur with Bernstein's verdict as the more accurate interpretation of the physics.

30. Heroes, Villains and Fellow Travellers

The Farm Hall transcripts make for a strange scientific psychodrama at the war's end: a comfortable imprisonment, prompting psychological stress, interpersonal tensions and growing antagonisms, reminiscent of Sartre's *Huis Clos*. Ill-assorted in background and personality, the detainees nevertheless shared in common the fact that they were scientists of different talents who had worked for the Hitler regime. They also shared in common the remarkable fact that they failed to acknowledge moral responsibility for their collusion with the Nazi regime. Apart from Hahn and Laue, there was little evidence of soul-searching among the German physicists, only self-jusfication and a preoccupation with how they would survive and thrive in a post-war Germany. A rare shred of articulated recognition of the evil of Nazism was Wirtz's remark – 'we have done things which are unique in the world . . .'

Some of the Farm Hall detainees reacted with shock after news of the Hiroshima bomb, but it took Laue to acknowledge that it was fear and loathing of the Hitler regime that led to the bomb: 'The . . . émigrés' passionate hatred of Hitler was the thing that set it all in motion,' he remarked in a letter to his son.[1] As we have seen, even Bagge, one of the two card-carrying Nazis, managed to duck responsibility for his Party membership by blaming it on his mother.

Looking to the future, and their scientific fates in the post-war era, the scientists were inclined to draw an implicit moral equivalence between Anglo-American democracy and the Soviet Union, making it clear that the crucial consideration was not which of the two power blocks represented a free and good society, where their scientific activities might flourish with integrity, but which of the two power blocks would lend them status and fund their research programmes. 'If we find we are only able to eke out a meagre existence under the Anglo-Saxons, whereas the Russians offer us a job for say 50,000 roubles, what then?' asked Weizsäcker, rhetorically. On one occasion Heisenberg said: 'Can they expect us to say: "No, we will refuse the 50,000 roubles as we are so pleased and grateful to be allowed to remain on the English side." '[2]

Later, Heisenberg said to Hahn: 'I don't want to do petty physics . . . If the final decision is that I can't do any proper physics and I go back to Germany again, naturally they, too, will realize that I am then going to consider doing physics with the Russians after all.'

Heisenberg, the Final Verdict

So what are we to make, finally, of Werner Heisenberg? Was he the brilliant hero who deprived Hitler of the atom bomb? Or was he an incompetent physicist who would have made an atom bomb had he known how? Or does he remain for ever the enigma of Michael Frayn's *Copenhagen*? It is worth recapitulating what we know about him for certain.

He was not a Nazi; he disdained to join the Party even during the period in which it looked as if Germany had won the war. He had many reasons to despise the Nazis, not least his persecution by Stark and Lenard and the investigation of his life by the SS; he expressed his loathing of the Nazis and their regime to his wife on various occasions. He was nevertheless a patriot, a nationalist, with a great love of German culture, landscape, music. His attachment to Germany was neither crudely bucolic nor based on the pseudo-history of Teutonic mythologies. He loved his discipline, physics, and his desire to protect his subject and bring on a new generation of physicists for a Germany beyond Nazism seems genuine.

There is no evidence that he was anti-Semitic. He had studied under Max Born at Göttingen, and everything indicates that he loved him. When he won the Nobel prize for physics for 1932, a prize he was convinced Born should have shared, he travelled into Switzerland in order to send a letter to Born expressing sorrow that they had not been cited together. Among his close colleagues and friends he numbered Einstein, Wolfgang Pauli, Rudolph Peierls and above all Niels Bohr – all of whom had Jewish ancestry.

He was a great physicist, one of the greatest theoretical physicists of the twentieth century. If his science had a flaw it was, as we have seen, his lack of aptitude for experiment, and he could be careless mathematically. He was clearly the wrong person to direct the Nazi atomic bomb programme and his half-hearted, fragmented and distracted leadership was almost certainly a major factor in its lack of progress. But there is no evidence that he aspired to lead in order to deprive Hitler of the atomic bomb. Like many people, he was convinced that the war could not last long and was even bold enough to suggest as much to Albert Speer, of all people.

What we know for certain too, however, is that he supported Hitler's war aims. While there is a minor mitigating factor in that he defended only the war with Russia, as opposed to the war with the West, he did not reflect on the illegality and the brutality of the entire war. Nor did he have the slightest scruple, as we have seen, about flying the Nazi flag in places like Denmark, the Netherlands and Poland, despite his certain knowledge of the atrocities committed in these places.

Finally, at Farm Hall, he evidently sided with Weizsäcker in the *Lesart*, the version aimed at presenting their stewardship of physics under Hitler as not simply apolitical, but noble. He did not attempt to portray himself as a hero, but he never expressed openly a regret for having worked for Hitler and contributing to the morale of Nazism.

The worst that can be said of Heisenberg, I am convinced, is that he was morally and politically obtuse. Born into a tradition of academia in which the great professor exists in an ambit of 'irresponsible purity', he had that 'inexcusable habit of combination', as Thomas Mann put it in *Dr Faustus*. The most damaging aspect of this fellow travelling under Hitler was its power to give the regime credibility and status in the eyes of the fence-sitters and the undecided, and especially the young. There is no evidence that Heisenberg ever regretted his wartime role, and every evidence that he falsified his memory of it.

The falsification of memory in the post-war era has driven historians, mostly in North America, to continue to examine and explore the significance of science under Hitler. As Mark Walker puts it, the German scientific community and most of its members entered into 'a Faustian pact with National Socialism, trading financial and material support, official recognition, and the illusion of professional independence for conscious or unconscious support of National Socialist policies culminating in war, the rape of Europe, and genocide'.[3] It needs to be added, moreover, that few areas of science under the Third Reich were free of taint from the employment of slave labour.

Walker's perspective is based on the distinction that should be made between scientists who embraced National Socialism with enthusiasm – the likes of Lenard and Stark – and those who simply stayed in Germany, working within science, remaining, on the face of it, morally and politically aloof from Nazism, while arguing that science is an apolitical, neutral pursuit. This distinction was routinely maintained by historians and journalists, with a measure of plausibility, during an era in which the main task was to identify the villains, and the obvious heroes, while leaving the rest – the fellow travellers – in a kind of untouchable taboo status. Against this background, moreover, it was all too easy to characterize the likes of Heisenberg and Hahn, not merely as neutral, but as heroes. While Heisenberg and Hahn were not villains in the mould of Lenard and Stark, persistent and pernicious myths have been constructed maintaining that they resisted Hitler by 'sabotaging' the bomb programme. The fullest extent of the myth-making at its origins is exposed in the Farm Hall tapes, for close attention to these conversations reveals the extent of their readiness to engage not only in false exculpations but in claims of moral superiority over the nuclear scientists (many of them expelled from Germany) in the United States.

The status of fellow travellers in the Nazi regime acquires a virtually unique distinctiveness in relation to Germany's scientists in view of the familiar appeal to value-free science. According to this view it is possible to do good science under any employer, patron, regime or auspices. And yet, it is precisely within this claim for apolitical, value-neutral, untouchable, 'taboo' status that the fellow traveller stands revealed, in Professor Walker's words, as 'often far more dangerous than either Hitler's true believers or his bitter opponents'.

It was the exceptions, few and far between, who revealed the nature of true courage and integrity. Take Hermann Weyl, the mathematician, who left Göttingen for the United States. 'I could not bear to live,' he wrote, 'under the rule of that demon who had dishonoured the name of Germany, and although the wrench was hard and the mental agony so cruel that I suffered a severe breakdown, I shook the dust of the Fatherland from my feet.'[4]

But was it possible to remain in Germany working as a scientist and retain one's integrity in opposition to the regime without being arrested? The outstanding example is Max von Laue, whose form of quiet passive resistance started with his speech in Würzburg in 1933, when he referred to Galileo as a thinly disguised cover to talk about the oppression of Einstein. Describing how his life changed in 1933, he wrote:

Often, when I awoke and remembered the terrible events of the previous day, I asked myself if they were all a dream. Unfortunately they were real . . . As often as I could I helped those affected, for example with timely warnings . . . On one occasion I took a man who thought he was being persecuted in my car into Czech territory. All this had to happen with the utmost secrecy. But my feelings I declared publicly.[5]

Gerda Freise has written, moreover, of the moving example of Professor Heinrich Wieland, whose courage is perhaps less well known. Wieland won the Nobel prize in 1928 for work in organic chemistry, and succeeded to the chair of chemistry at Munich in succession to Liebig, Baeyer and Willstätter.[6] Wieland organized his departments so as to ensure that he employed many half-Jewish students as demonstrators and arranged for them to complete their studies unofficially when they were forbidden grants and places. His laboratory was nicknamed the 'ghetto room'.

He refused to use the Hitler greeting – 'Heil Hitler' – and reprimanded Nazi students who were threatening to denounce a fellow student for fraternizing with a half-Jewish student called Hans Konrad Liepelt. In 1943 Liepelt was arrested along with six other students for collecting money for needy Jewish students and for possession of treasonous reading matter and an illegal radio. Wieland attempted to help the arrested students, visiting their families and providing lawyers. In the autumn of 1944 Liepelt faced trial and Wieland appeared for the defence. Liepelt was nevertheless found guilty and executed on 29 January 1945.

Freise, who was a student of Wieland's, concedes that most scientists who were not Nazi supporters managed to adopt a dual role of 'autonomy' and 'adaptation', whereby they kept within the bounds of state law and focused on the purity and objectivity of their research, which they deemed a higher good. The inference is that they acquiesced in the face of the political drift.[7] Wieland, she writes, was one of those who refused to see his science as an activity outside and beyond the political circumstances in which he was obliged to live and work. His notion of 'autonomy' involved the social context of his work and his loyalty to the academic institution was not absolute. Wieland was a resistance worker, but he was fortunate to have engaged in non-cooperation publicly with impunity. His Nobel award probably helped, but he had still courted danger and he was an exception in that he got away with it.

Lise Meitner's Letter

There was another kind of important exception: the scientist who served the Hitler regime and later gained insight, culminating in remorse and self-criticism. One of the most eloquent critics of the scientist as fellow traveller in the Third Reich was Lise Meitner, whose arguments are all the more telling as she includes herself in her vehement condemnation of those who worked under Hitler.

In June 1945, a month in which the truth about the concentration camps was published and broadcast in Britain, Meitner poured out her heart in a letter to Hahn, her old collaborator. It did not reach him, but the spirit of the letter was to be repeated in subsequent correspondence. Towards the end of this long and powerful letter she wrote that what she had heard 'of the uncontained horrors in the concentration camps exceeds everything that one had feared'.[8] After she had heard on the BBC 'a very factual report from the British and Americans on Belsen and Buchenwald', she 'took to howling out loud and could not sleep the whole night'. Recalling her memory of seeing people arriving from the camps in a pitiful state, she wrote that 'Heisenberg and many millions with him should be forced to see these camps and the martyred people. His appearance in Denmark in 1941 is unforgivable.'

Portions of the rest of the letter deserve quoting in all their poignant detail as one of the most important denunciations of Hitler's fellow traveller scientists by a former member of their circle:

You all worked for Nazi Germany. And you tried to offer only a passive resistance. Certainly, to buy off your conscience you helped here and there a persecuted person, but millions of innocent human beings were allowed to be murdered without any kind of protest being uttered. . I must write this to you, because so much depends for both Germany and yourselves on your recognizing what you allowed to happen . . . I and many others believe that a way for you would be to publish an open declaration that you are conscious that through your passivity you have incurred a joint responsibility for what happened . . . But many believe that it is too late for that. They say that you first betrayed your friends, then your men and children in that you let them stake their lives on a criminal war – and finally that you betrayed Germany itself, because when the war was already quite hopeless, you did not once arm yourselves against the senseless destruction of Germany. This sounds irredeemable, yet believe me, I write all this to you out of the most honourable friendship.

She ends with an admission of her own guilt, a consequence of her having left it until 1938 to leave Germany, having been forced out by events rather than a political and moral decision. 'You yourself may perhaps recall,' she wrote,

> how when I was still in Germany (and today I know that it was not only stupid, but a great wrong that I had not immediately left) I often said to you: 'As long as only we and not you have sleepless nights, things will not be better in Germany.' But you had no sleepless nights, you did not want to see, it was too uncomfortable. I could give you so many examples, great and small. I beg you to believe me that all I write here is an attempt to help you all.[9]

Three years later, Meitner returned to the question of her own guilt in remaining in Germany: 'Today it is very clear to me that it was a grave moral fault not to leave Germany in 1933, since in effect by staying there I supported Hitlerism.'[10] The admission contrasts dramatically with the self-exculpations and moral pretensions of the ten Farm Hall detainees.

The Post-war Fate of Hitler's Scientists

Despite Lise Meitner's powerful criticisms, Hitler's scientists showed little remorse, as science in West Germany accommodated itself to the new situation of a divided nation.

Professor Mark Walker claims that the leadership of the Kaiser Wilhelm Society and its successor, the Max Planck Society, 'has systematically ignored, down-played or misrepresented its consequential collaboration with National Socialism since the end of the war'.[11] The Society and its institutes became a football between the victors. The Soviets tried to take it over for East Germany and put a communist in charge. The French tried to take valuable parts of it to France. The Americans considered the society as a whole redundant, but were willing to recognize certain of the institutes as independent entities. The British, however, nurtured the society by establishing its headquarters in their zone of occupation. They insisted, however, on a name change. By 1948 the Kaiser Wilhelm Society was renamed the Max Planck Society. Hahn, who had become the Society's first post-war president, fought hard to retain the old name. He threatened to resign and to take the organization over to the Soviets. In the end he capitulated.

In its effort to appear rehabilitated and distanced from the Nazis the society severed its ties with German industry and declared itself committed exclusively to basic research. The new federal government funded the society entirely. Funding was short and the old Emergency Foundation for German Science, established after the First War, emerged once again. This act indicated, in the view of Mark Walker, 'an attempt to turn back the clock to the Weimar period and treat the Third Reich as a brief aberration, an attitude widespread in Germany after the Second World War'.[12] Hahn had attempted to set up a centralized clearing house for research and funding, called the 'German Research Council', but it was suspected of a fascist taint and a decentralized system of science policy was developed.

In 1946, the year he took over the direction of the Kaiser Wilhelm Society, the most distinguished role for a scientist in Germany, Hahn accepted the Nobel prize in Stockholm for his discovery of fission. Meitner, who had done so much to prepare for, and elucidate, Hahn's and Strassmann's result, did not share the prize.

By the 1950s Hahn had emerged as a leading German spokesperson on the ethics and politics of science. In early 1954, for example, he wrote a paper on the exploitation of nuclear energy which was widely promoted in Germany and around the world. Three years later he became an international focus of controversy when he and seventeen of his colleagues, members of a so-called 'Nuclear Physics Group', attacked Chancellor Adenauer's government for cooperating with NATO in a scheme to stockpile nuclear weapons on West German soil. The group argued the dangers of tactical nuclear weapons, pointing out that, if they were used against the Warsaw Pact countries, Germany would bear the brunt of the radioactive fallout. Among the signatories to the denunciatory manifesto, published in the three largest German newspapers, were Max Born, Walther Gerlach, Werner Heisenberg, Max von Laue,

Josef Mattach, Fritz Strassmann, Carl Friedrich von Weizsäcker and Karl Wirtz: several of them were former detainees of Farm Hall. The initiative of this band of ageing scientists was not, as it happened, successful, but the importance of their collaboration is how former Third Reich scientists lined themselves up alongside a noble figure like Max Born; and how he, in turn, was prepared to be part of it.[13]

In an ironic twist of destiny, Adenauer chose Pascual Jordan, the physicist who had equated quantum physics with National Socialism, to argue the case of the Christian Democrats in support of an expanding nuclear armament of NATO in Europe. Jordan resurrected a catalogue of his old will–to–power arguments. Hahn's group of physicists, he charged, were less capable of making judgments about fundamental political questions than the 'average democratic citizen'. The incident drew a harsh broadside from Max Born, who accused Jordan of a past that stank of compromise and self-delusion. Born went on to condemn publicly Jordan and his aptitude for manipulating people and their ideas to justify the primacy of power. Born's anger was not surprising. Jordan had been brazen enough in 1948 to seek a reference from Max Born, attaching an exculpatory apologia for his conduct under Hitler. He boasted in the document that he had done all in his power to defend quantum physics from the attacks of Lenard and Stark and congratulated himself on not having worked on the atom bomb or on rockets (which was a lie, as Peter Wegener could testify). Born simply sent by return of post a list of his friends and family members who had been killed under the Nazi regime. Heisenberg, on the other hand, was inclined to accept Jordan's defence.

In the remaining years of his life Hahn continued to be the dean of science in Germany. Eventually Meitner confirmed her reconciliation with him when she symbolically joined her name with his at the opening in 1959 of the Hahn–Meitner Institute for Atomic Research. Hahn died full of honours in July of 1968 in Göttingen.

Werner Heisenberg, who was happy to help Pascual Jordan rehabilitate himself, was no less speedily embraced by West

Germany in the post-war era. A year after he was released from Farm Hall he was appointed director of the Kaiser Wilhelm (soon to be renamed Max Planck) Institute for Physics and Astrophysics at Göttingen. In the remaining quarter century of his life (he died in 1971, aged seventy), he appears to have been reinstated as an honoured member of the scientific establishment both at home and abroad. He was made a Fellow of the Royal Society in London and was the recipient of many awards, prizes and honorary doctorates. He travelled the world attending scientific conferences, where he was on the whole well received.

Niels Bohr and Heisenberg, who had once formed one of the most famous friendships in physics, were never truly reconciled. Bohr steadfastly refused to discuss what had passed between them during the war. They met and were civil to each other, but the old friendship was never rekindled. Heisenberg cut an isolated figure for the rest of his life. The physicist John Wheeler told the story that he and his wife went out one evening in Copenhagen to a restaurant near the harbour. In the middle of their dinner a man entered and sat down to eat alone. Wheeler turned to his wife and said: 'That's Heisenberg.'[14]

The tension between Bohr and Heisenberg eventually flared when Robert Jungk's book, *Brighter than a Thousand Suns*, appeared in 1956 (1957 in English) claiming a high moral stance for Heisenberg. Bohr seems to have believed that Jungk had been persuaded to take this view by Heisenberg himself. Bohr, as we have seen, wrote a number of draft letters to Heisenberg, but never sent them. Bohr died in 1963 without ever having rehearsed with Heisenberg that famous meeting in wartime Denmark. As the years passed, there were still those who cut him dead and refused to shake his hand. He represented Germany as a delegate to the UNESCO conferences for a European atomic research centre, and in 1955 he was invited to give the prestigious Gifford Lectures on philosophy and science at the University of St Andrews in Scotland.

Heisenberg's colleague, Carl Friedrich von Weizsäcker, the author of the *Lesart* that promoted the moral superiority of the Nazi bomb physicists, became head of the theoretical section of

Heisenberg's Kaiser Wilhelm Institute in Göttingen in 1946. In 1957 he was appointed professor of philosophy at the University of Hamburg; aged fifty-eight in 1970, he was appointed head of the Max Planck Institute of Research into the Conditions of Life in the Scientific-technical World in Starnberg. He has continued to give interviews and to write, providing a sanitized account of the Nazi atom bomb effort. He continues to deny that scientists under Hitler ever intended building a bomb. In an interview in Germany in March 1996 he said that the Americans spent 1,000 times more on their atom bomb than the Germans. Germany, he declared, would not have spent such vast resources on something of such uncertain value.[15]

In contrast the exponents of *Deutsche Physik* ended their days in disgrace. Johannes Stark was put on trial as a Nazi at the end of the war (Heisenberg testifying against him). He was found guilty in July 1947 as a 'major offender' and sentenced to four years' hard labour, although the sentence was later suspended. He died in 1957 on his estate in Bavaria. Philipp Lenard, on the other hand, had reached the age of eighty-three by the war's end, and had long withdrawn from active participation in Nazi activities. The authorities were keen to try him for responsibility for Nazi crimes, but an appeal by the acting rector of Heidelberg University let him off the hook. He died in 1947.

Erich Bagge, the youngest of the Farm Hall detainees, joined Heisenberg in Göttingen before moving on to a chair of physics at Hamburg. In the mid-1950s he served as the technical-scientific project leader of a study group for the promotion of nuclear energy in maritime construction in Hamburg-Geestacht and assisted in the development of the first German nuclear-powered ship, the *Otto Hahn*. He died in 1996.

Walther Gerlach, the individual who had been thrown into a depression following the news of the Hiroshima bomb at Farm Hall, returned to Germany and to a chair in physics in Bonn. In 1948 he moved on to the Ludwig Maximilian University in Munich, where he had once held a chair in physics, to serve as rector. He died in 1979. Paul Harteck, the physical chemist who had alerted German Army Ordnance to the possibility of a nuclear weapon after reading Joliot-Curie's article in *Nature*, moved into biophysics at Hamburg after the war. In 1951 he emigrated to the United States, where he became a professor in the Polytechnic Institute in Troy, New York, until his retirement in 1974. He died in 1985.

Max von Laue helped to re-establish the German Physical Society and assumed, in his long retirement (he died in a road accident in 1960), an important role in the normalization of German culture, providing a link with an earlier era when Germany was a respected Mecca of world science.

The end of the war saw the great Max Planck aged and exhausted. No stranger to tragedy (his elder son was killed in the trenches in 1916, and his twin daughters died in childbirth), his younger son Erwin was murdered by the Nazis in 1945 for his alleged part in a plot against Hitler. Towards the end of the war he was caught in a bombing raid on Kassel and buried for hours in a shelter. Finally his home and library were destroyed in an Allied bombing raid on Berlin. He fled from the Russians, hiding out in woods until he was rescued by the Americans. He spent the two remaining years of his life living with his grand-niece in Göttingen, where he died in 1947. His grave there is marked by a simple rectangular tombstone with the equation denoting the constant in nature which carries his name.

Among other scientists in related disciplines, Konrad Lorenz suffered no repercussions for his work during the war. He was captured by the Russians and practised medicine for four years while in detention. He returned to Austria in 1949 and founded the Institute for Comparative Behavioural Studies (Ethology) in Altenberg. In 1961 he was appointed director of the Max Planck Institute for Physiology in Seewiesen near Starnberg, and in 1973 won the Nobel prize for Physiology or Medicine for his work on human and animal behaviour. In 1982, when he was seventy-nine years of age, he was appointed director of the Ethology Section of what was to become the Konrad Lorenz Institute of the Austrian Academy of Sciences. Among his published books were *On*

Aggression, which argued the deterministic nature of violence, and *The Eight Deadly Sins of Civilized Humanity*, a campaigning book against unrestricted population growth and environmental pollution. He died in 1989 aged eighty-five.

Otmar von Verschuer, who benefited from death camp experiments, was similarly rehabilitated after being classed a 'fellow traveller' and submitting to a denazification process. By 1951 he had become professor and director of the Institute for Human Genetics at the University of Münster. Ten years later he was guest of honour at the Second International Conference on Human Genetics in Rome. He died as a result of car accident in 1969.

Josef Mengele was imprisoned in Munich at the end of the war, but the authorities did not know his identity or what he had been guilty of. After his release he fled in 1949 via Rome to Buenos Aires, then travelled from country to country in South America. He became a citizen of Paraguay in 1959, where he continued to live in anonymity until his death from a stroke while swimming in the sea in 1979.

31. Scientific Plunder

A British aeronautical engineer, later a distinguished Cambridge academic, Professor Austyn Mair, relates how as a young RAF officer he flew to Germany in the early summer of 1945 to inspect the Reich's wind tunnel technology. As he sat with a German rocket engineer by the side of a river in southern Bavaria, the conversation turned to the A4 missile. The German engineer asked what it was like when a V2 exploded on a city like London. Mair remembers how the man appeared to gloat as he asked the question. 'I just stared at him and said nothing,' says Mair. 'I was astonished by his callousness.' The incident illustrated for one young British officer the difference between the Germans and the Allies as scientists and engineers.[1]

In the laboratories, workshops, foundries and testing grounds of Hitler's vulcans, the Allied armies came upon the dark horrors of slavery and sadism; they also found discipline, high technology, organization and industry. The products, inventions and discoveries of Nazi science and technology covered a huge circuit of plans, research and development programmes and production facilities. Many of the designs were buried or destroyed to evade seizure; plants and materials were blown up, abandoned or hidden; personnel attempted to flee or to go into hiding. Groups of American and British scientific intelligence officers were often profoundly impressed at what they found, especially in the case of aircraft production facilities; sometimes, as in the case of the Nazi nuclear programme, they were contemptuous of the lack of Nazi achievement. Whatever the case, if the Allied scientific officers had imbibed the notions expounded in their countries on the relationship between democracy and science – only good societies produce good science – they would have been in no doubt as to Allied superiority in science and technology.

In the early 1940s, after the fall of Poland, the Low Countries and France, two outstanding writers, the biochemist and historian of science Joseph Needham in Britain and the sociologist Robert Merton in the United States, argued that there was a marriage between good science and technology and the virtues of liberal democracy. They meant that the exercise of scientific imagination flourishes in an environment that is rational, free, objective, systematically sceptical and universalist. Their comments were understandable in the light of events. They were reacting to Hitler's onslaught on civilization throughout continental Europe. They were also declaring that the products of science and technology under democracy are put, by definition, to just and ethical use. The notion was in clear reaction to the implied Faustian bargain of Nazi science: a pact with the devil Hitler that produced flashy, initially impressive technology, profoundly flawed and dedicated to irrational goals: bad science for bad ends.

The oversimplification inherent in the Needham–Merton myth was soon made plain, however, by a scheme known as 'Project Overcast', whereby British and American scientific officers (the French and the Russians had their projects too) aimed to plunder every technological plan and product, every useful engineer and scientist to be found in the ruins of the Third Reich, whatever their depraved associations and auspices.

The original objective of the race for scientific spoils of war had been articulated in Britain not long before the D-Day landings, on 5 June 1944, when Lieutenant-General Sir Ronald Weeks, Deputy Chief of the Imperial General Staff, declared that 'German equipment is as good or better than ours'. He was convinced that the seizure of German research, design and development programmes was 'one of the most vitally important of our immediate post-war aims . . . It may be that this is the only form of reparation which [it] will be possible to exact from Germany. Everything possible to ensure that it is exacted must be carefully planned now.'[2] In 1945 the American Secretary of Commerce, Henry A. Wallace, enunciated a parallel rationale for the technological plunder of the Nazi war machine: 'The transfer of the outstanding German scientists to this country for the advancement of our science and industry,' he told President Truman, 'seems wise and logical. It is well known that there are presently under U.S. control eminent scientists whose contributions, if added to our own, would advance the frontier of scientific knowledge for national benefit.'[3] The Russians, and to a lesser extent the French, similarly competed with the Americans and the British in justifying their determination to exact reparations in the form of technology by sending intelligence teams into their zones of influence. Sometimes these Allied teams trespassed on each other's zones of interest, behaving like gangsters: even resorting to threats, blackmail and kidnap, to obtain anything of value.

The Russians dismantled whole factories and transported them along with the workers to Russia. Helped by the British, the Americans, led by Colonel Holger Toftoy (Chief of US Ordnance Rocket Branch), shipped to the United States V1 and V2 rockets and all the spares they could lay their hands on. The Americans also grabbed an entire supersonic wind tunnel from Bavaria, a submarine with an advanced propulsion system, and many different types of aircraft including jet prototypes and rocket planes. The loot transported to the Air Document Research Center at Wright Field in the United States included tons of designs.

If the Americans revealed little interest in the Nazi associations of all this material, they showed even fewer qualms of conscience at colluding with German scientists guilty of employing slave labour and other crimes. Within weeks of being rounded up mainly in Bavaria, about 120 researchers handpicked by von Braun himself were taken to Fort Bliss, near El Paso, Texas, and set to work on guided missile research.

At first the Germans were housed in barrack accommodation. They enjoyed plentiful food, salaries, some privileges and limited freedoms, such as a weekly trip to the cinema in a chartered bus. At the level of ordinary USAF personnel, there was evident indignation at the arrival of the Germans. One aeronautical engineer, Hans Amtmann, who was originally billeted with his family at Wright Field, relates how he and a group of compatriots repaired a tennis court with much hard labour and ingenuity. When they started playing on it, they were thrown off by American officers, who appeared to resent the fact that the Germans were enjoying themselves. But by 1948, still working for the Americans in the aircraft industry, Amtmann was the proud owner of a Studebaker Champion automobile and had bought a house in the Dayton area, where he settled his wife and children..

Eventually the key team of von Braun's Germans was moved to Redstone Arsenal in Huntsville, Alabama, where they were set to work in earnest on the Redstone missile, which was an advanced A4, as well the Jupiter and Pershing missiles. In 1960 the remnants of the original German group still active were transferred once again to NASA at the Marshall Space Flight Center, where the Saturn programme was in development.

Michael J. Neufeld has noted that the reconstruction of an entire technological system from one defeated country to the victors' country was probably unique. Despite American doubts about state ownership, the arsenal system of weapons development and production was part of a long tradition in the United States. At Redstone von Braun oversaw his 'under-one-roof' principle of research and development of missiles, with contracting-out restricted to sub-systems and mass production. What was missing was concentration camp labour; the input of the concentration camps, of course, had already been provided.

By March of 1946 'Project Paperclip' had replaced Project Overcast, providing a stronger legal basis for the long-term exploitation of former enemy scientists and engineers and new scope to bend President Harry Truman's rules regarding Nazi war criminals: namely, that they emphatically should not be allowed into the United States as bona fide immigrants.[4] When the US war crimes investigators launched a search for evidence and witnesses for the Mittelbau-Dora atrocities (the underground factory facility where the A4 had been in production), Toftoy protected his team of Germans from scrutiny. The Peenemünde scientists in the United States had dissociated themselves from any guilt for the atrocious treatment of slave labour at Dora, pinning all blame on the SS. Their American masters were happy to go along with this exculpation. As the Cold War went into deep freeze, the reasons for amnesia about the past required little rationalization. Peenemünde was reborn in Huntsville, an essential part of the US army's research on anti-aircraft missiles and intercontinental ballistic missiles (ICBMs); some of the Germans (the numbers of former Nazi engineers eventually rose to 500) jokingly referred to Huntsville as Peene-münde South. While it made little sense not to exploit the techno-logical know-how in the interests of the defence of the West, it was important, and still is, that we who set much store by recording the crimes of Nazi science should recognize that there was, and still is, the blood of slave labourers on the missiles that defend the West. A token of a refusal to forget occurred on the eve of the Apollo moon-shot, when von Braun strode out of a press conference on being asked by a journalist whether he could guarantee that the rocket would not land on London.

Soviet Plunder

The Soviet Union behaved no better, and in some ways worse. 'Technology transfer', forced and voluntary, to the Soviet Union in the aftermath of war is still shrouded in mystery. Estimates for numbers of German scientists, technicians and engineers who ended up in the Soviet Union range from 6,000 to 100,000. Those pressur-ized to go between May and August 1945 included rocket and nuclear scientists. Large numbers of specialists in all the armament industries were mostly forced from their beds on the night of 21–2 October 1946.[5]

Helmut Gröttrup, a key figure at Peenemünde, went over to the Russians, possibly because other Peenemünders were accusing him of betraying the location of Nazi rocket documents. The Russians set him up as head of a rocket research institute near the Mittel-werke, eventually sending him to Moscow. The Russians then forced a number of German specialists at gunpoint to depart for Russia. Gröttrup continued to exist in fairly comfortable accommo-dation, but many of his compatriots were imprisoned.

The testimony of Werner Albring, one of the leading German

experts in aerodynamics in the Third Reich, illustrates how Nazi scientists were 'recruited' by the Russians at the end of the war and how they were treated.[6] An important factor in his decision to go with the Russians was the promise to allow families to stay together (scientists who went to America were also eventually joined by wives and children). Those who went over to the Russians were told that they would be allowed to remain in Germany; but this was a lie.

After a brief period in East Germany Albring was transferred to an island called Gorodomlia on the lake of Seligersee, situated half-way between Moscow and Leningrad. He was one of 150 German scientists who along with their families formed a community of about 500. In a familiar model of Soviet camps designed for scientists working under enforced conditions, they lived in wooden huts surrounded by a forest. There was a school, a small restaurant and a general practice medical facility. The entire island was surrounded by barbed wire. Albring's incarceration lasted for six years. The scientists were required to work forty-eight hours a week. If they were late on arrival, they were fined. For relaxation they played handball and tennis; some took up gardening and painting. The girls formed a choir and there was a chamber ensemble. They attempted to form a debating club; but this was immediately banned.

Albring was convinced by the time that he wrote his book in 1957 (he continued to work on it until 1987) that the German scientists in Russia played a much less significant role in rocket science than those in the United States. He was convinced that the pressure to engage Germans came from 'on high' rather than from real need, as a means of preventing them from going over to the West. Work and research conditions were extremely primitive. They were equipped with nothing more than pencils and paper. The island group formed part of the 88th Moscow Scientific Institute, which was directed by a Soviet general. Grottrup, who was based in Moscow with other German scientists, pleaded for more freedom for his compatriots but none was forthcoming. The Germans remained separated from their Russian colleagues and never saw the fruit of their labours in applied results and tests. They were working far from appropriate aerodynamic and thermodynamic institutes, wind tunnels and test sites, and without contemporary literature. The scientists were continually evaluated by their masters in confrontational sessions; yet Albring remained convinced that the Soviets could have done everything the Germans were doing themselves; that they were talented scientists in their own right.[7] He claimed the Germans contributed little or nothing to the success of the Sputnik, but he believes that he and his colleagues contributed to anti-aircraft missiles, which needed to fly in tight curves at high speeds. He was finally allowed to return to Germany in 1952, but thereafter refused to work on armaments technology for the GDR.

PART EIGHT

Science from the Cold War to the War on Terrorism

32. Nuclear Postures

On the final day of an academic colloquium held in November 2002 in Cambridge to discuss Michael Frayn's play *Copenhagen*, a group of conference speakers drove out on a Sunday morning to Farm Hall, the former detention house of the German physicists. We were surprised to discover that the façade of the house overlooked a busy road that runs from the village of Godmanchester to the towns of St Ives and Huntingdon. Published photographs of the house have been taken most often from the back of the premises, giving the impression of remoteness. The actual circumstance of the house had the effect, for me, of breaking the spell of isolation that surrounds its story, and of bringing Farm Hall, which is a short distance from two large American air bases, into the present.

There were Mark Walker, historian of science; Paul Rose, author of a biography of Heisenberg; Jeremy Bernstein, nuclear physicist and writer; Walter Gratzer, biochemist, and author of *The Undergrowth of Science*; the playwright Michael Frayn; and his wife, the author Claire Tomalin. The current owners of the house also joined us – Professor Marcial Echenique, professor of art and architecture at Cambridge University, and his wife.

Sitting in the panelled room, with its deep Georgian windows, where the physicists had listened to the news of the Hiroshima bomb on 6 August 1945, we made a party of nine, just one short of the ten scientists. The party of scientists back in 1945 might well have felt cramped, claustrophobic. Unlike the reception rooms on the second floor, the sitting room on the first floor had low ceilings facing north. The view towards the river Ouse and England's most expansive commons, open land free for the people, was inviting and bathed in sunshine; the room itself, however, was sunless. Remembering the presence of the scientists, in 1945, prompted divided feelings.

The colloquium on the previous day had been contentious, punctuated by outbursts of emotion. Professor Rose had been shouted down for showing a photograph of Werner Heisenberg on an overhead projector, suggesting that he looked 'brutish'. Michael Frayn, on being accused of sympathizing with the Nazi scientists, had responded with icily English sang-froid; he had been 'as polite as a razor', remarked one of the participants.

As we exchanged feelings about the house and our knowledge of the circumstances of the detentions, tensions arose once again between the academics. It was difficult, evoking the moment when the news of the dropping of the first atom bomb on Japan broke, to avoid those questions of conscience that had nagged down the decades. The task of telling the story of Nazi scientists and their guilt was inseparable from considering the ethics and politics of British and American scientists, during the war and afterwards. The subterranean theme of the conference, and its intermittent eruptions of anger, had shed cold light, once again, on the Anglo–American construction of the first nuclear weapon of mass destruction and its use against the Japanese.

The most familiar argument of those who justified, and continue to justify, first use of nuclear weapons on Hiroshima and Nagasaki is that the decision shortened the war and saved countless American lives. Another compelling argument, employed, for example, by war historian John Keegan, is that the atomic bomb warned dictators that they could not, as in the past, get their way through violence. There are defenders of the act, moreover, who appeal to contrasting kill rates: 300,000 Japanese killed by atom bombs, compared with some 50 million slaughtered during World War I by conventional means, privation, labour camps and displacements.

But there are equally familiar opposing arguments to the legitimacy of first use: that Japan was on the brink of capitulation, that the Americans saw the Japanese as sub-human, and that Truman was determined to use the bomb to impress Stalin. There were prominent military leaders on the American side who opposed the use of the bombs. General Eisenhower and General MacArthur, air force generals Arnold and Le May, Admirals Leahy and King were all against first use.[1] With the vantage point of fifty years, moreover, the late John Rawls, author of *Theory of Justice*, argued in 1995 that Truman failed to end the war without nuclear weapons for sheer lack of statesmanship.[2]

The most powerful argument against the bombs, however, focuses on the proposition that the use of nuclear weapons crossed a boundary that could lead to future use with weapons a thousand times more powerful. Leo Szilard, in a petition to President Truman just three weeks before the dropping of the Hiroshima bomb, wrote precisely about that weighty consideration:

The development of atomic power will provide the nations with new means of destruction. The atomic bombs at our disposal represent only the first step in this direction, and there is almost no limit to the destructive power which will become available in the course of their future development. Thus a nation which sets the precedent of using these newly liberated forces of nature for purposes of destruction may have to bear the responsibility of opening the door to an era of devastation on an unimaginable scale.[3]

Szilard's fears for the unlimited destructive power of nuclear weapons and for proliferation were not unfounded.

The H-bomb

The physicist Freeman Dyson recounts how when he came to work at Cornell University under Hans Bethe in 1947, he embarrassed everybody at lunch by remarking in all innocence, 'It's lucky that Eddington proved it's impossible to make a bomb out of hydrogen.'[4] Dyson remembers that there was an awkward silence and the subject was changed. A colleague took him aside afterwards to inform him that a lot of work on hydrogen bombs was in progress and that he was never to mention the subject again.

Dyson reports that in those days, just two years on from Hiroshima and Nagasaki, the Los Alamos veterans were politically involved in pleading for the surrender of all nuclear science under strong international authority. Bethe, who 'was troubled by these questions', often discussed the morality of nuclear weapons with Dyson. In October 1949 Bethe was being courted by Edward Teller to join the team that was to build the hydrogen bomb, and seemed unable at that stage to make up his mind. Teller remembers Bethe turning to him after a meeting to say: 'You can be satisfied. I am still coming to Los Alamos.' Less than a week later, when Teller talked to him again, Bethe said that he had changed his mind. 'He did not give me an explanation,' remembered Teller.[5] It was not the last time that Bethe changed his mind.

In the previous year, J. Robert Oppenheimer, who was still suffering pangs of conscience over Hiroshima and Nagasaki, published his famous confession in a February issue of *Time* magazine: 'In some sort of crude sense, which no vulgarity, no humor, no overstatement can quite extinguish, the physicists have known sin; and this is a knowledge which they cannot lose.'

The stated motivation of many of the scientists at Los Alamos was to prevent Hitler from being first with a nuclear weapon. When it was discovered in December of 1944 that Hitler did not have a bomb, only one of their number, Joseph Rotblat, resigned from the project, convinced that the only moral basis on which such a weapon could be built *was* as a deterrent, and that in consequence the project should now be abandoned. He was obliged by the authorities not to tell his colleagues his reasons for withdrawing.

Rotblat stands out as an extraordinarily courageous and principled individual, prepared to resist the pressures of the prevailing tide of opinion. He later became a co-founder of Pugwash, the body dedicated to world peace and to the elimination of all nuclear weapons, and won the Nobel peace prize. After America's use of the atom bombs on Japan, scientists like Rotblat, Leo Szilard and Niels Bohr argued that nuclear knowledge should be shared with the Soviet Union; but that was a view neither Truman nor Churchill (nor Clement Attlee, Britain's post-war Labour Prime Minister) was inclined to take. Fear of Hitler had been replaced with fear of Stalin, and in the confrontation of the Cold War a new, more lethal arms race began. According to Rotblat, who learned of the remark retrospectively, General Leslie Groves, head of the Manhattan Project, said privately: 'You realise, of course, that the main purpose of the project is to subdue the Russians.'[6]

By 1949, some six years earlier than the Americans had expected, the Russians tested their first atom bomb. The United States had already begun work, as Dyson discovered in 1947, on a hydrogen bomb under the scientific leadership of Edward Teller, and eventually carried out a first test on 1 November 1952, vaporizing the Pacific island of Elugelab with a 12-megatonne bomb 1,000 times more powerful than the atom bomb that destroyed Hiroshima. The Russians followed suit just nine months later, in August 1953. It was believed at the time that the Russians had produced a lighter device than the American bomb that feasibly could be carried by an intercontinental ballistic missile; this only lent further impetus to the arms race. According to other accounts, the American analyses of the radioactivity released by the test indicated that the explosion was caused by nuclear rather than thermonuclear fusion.[7] Both sides pressed on, with the Russians, under Kruschev, aiming for a 100-megatonne H-bomb, despite the environmental dangers.

Given the ruthless politics of the Cold War, and powerful convictions on both sides about the aggressive intent of the other, there was little that individual scientists could do to restrain the determination of their leaders; the technology for a hydrogen bomb was known in theory at an early stage and there were scientists prepared to step into the shoes of those who demurred. This did not mean that, as some ethicists have argued, moral and political consciousness had no place in the lives and work of the H-bomb scientists.

The scientists' scope of influence in the development of the H-bomb can be illustrated by the contrasting careers of Edward Teller in the United States and Andrei Sakharov in the Soviet Union, both cited as 'fathers of the H-bomb'.[8]

Sakharov, who was aged twenty-four at the end of World War II, was a brilliant theoretical physicist who had begun to focus on pure research after working in a munitions factory for much of the war. He refused to join the Communist Party and twice declined invitations to get involved in nuclear weapons research. In 1949, however, he was commanded by Beria, the chief of secret police, to lead the Russian equivalent of Los Alamos: it was called the 'Installation'. Built by slave labour, it was situated in a secret place in the Urals. To have refused would have involved severe punishment. In his memoirs, however, Sakharov claims that despite the terrifying nature of this weapon of mass destruction, and the brutality of the regime under which he lived and worked, he thought his participation was essential to the balance of power between the United States and the Soviet Union, and consequent maintenance of the peace. He was elected to the Soviet Academy of Sciences and awarded the title of Hero of Socialist Labour; but a rift soon opened between him and the authorities.

Sakharov was concerned about the effect of radioactive fall-out across the planet from atmospheric tests and quarrelled with Khruschev, who wanted more tests to be carried out in order to match the number completed by the United States. At a public banquet Khruschev coarsely rebuked the scientist, telling him to stay out of politics: 'Sakharov, don't try to tell us what to do or how to behave.'[9] But that is precisely what Sakharov now decided to do:

The very moment of the explosion, the shock wave which moves along the field and which crushes the grass and flings itself at the earth . . . All of this triggers an irrational and yet very strong emotional impact. How not to start thinking of one's responsibility at this point?[10]

Sakharov continued to warn about the dangers of radioactive fall-out. He pleaded on behalf of dismissed colleagues who had been detained in the Gulags; he interceded for political prisoners, and initiated human rights campaigns. His initiatives were all the more remarkable as he was constantly courting personal danger. Cautious of throwing him in jail, the authorities vindictively docked his salary, withdrew his privileges, slashed his car tyres and subjected him to constant surveillance. When he won the Nobel peace prize in 1975 he was refused permission to travel to Norway; his second wife, Elena Bonner, received the prize and delivered his acceptance lecture on his behalf. His Nobel prize nevertheless gave him greater prominence and outreach in his campaigns for human rights, which angered and incited the Kremlin. When he vociferously opposed the invasion of Afghanistan in 1979, he was exiled for seven years to Gorky, a city off limits for foreigners. Deprived of a telephone, he continued to battle with the authorities by going on hunger strike; on several occasions his minders resorted to force-feeding him. Following Mikhail Gorbachev's rise in 1985, Sakharov was released from Gorky and returned to Moscow, where he became a member of parliament and worked tirelessly for a pluralist society and human rights until his death aged sixty-nine.

Edward Teller, a Hungarian of Jewish background, had begun work with Stanislav Ulam on a hydrogen bomb, known as the Super, early in the days of the Manhattan Project. Ulam, a brilliant mathematician, did most of the calculations based on Teller's ideas. At first, Ulam found Teller's design for the bomb mathematically unfeasible, but Teller persisted with his idea even beyond the end of 1950, when President Truman ordered that the H-bomb programme should proceed without delay.

According to Richard Rhodes, it was Stanislav Ulam who eventually came up with a design that was at last feasible.[11] Teller, adding refinements of his own, accepted. In Teller's recent memoirs, completed at the age of ninety-two, he repudiates Rhodes's thesis, dismissing Ulam's contribution and claiming for himself all the credit for inventing the H-bomb. Teller's criticism of Ulam confirms the pressures, personal rivalries and conflicts within scientific communities, even over the dubious hubris of who created the biggest weapon of mass destruction. In a footnote in his memoirs Teller writes acidulously: 'Stan [Ulam] was not at all addicted to exertion, a characteristic so marked that it became a joke . . . I should have liked to know how Stan looks when he is working. No one ever caught him red-handed.'[12]

Teller's enthusiasm for making a hydrogen bomb contrasts with J. Robert Oppenheimer's reluctance in the late 1940s and thereafter. Teller took this reluctance as an indication of disloyalty to the United States. He recalls Oppenheimer saying during the war: 'We have a real job ahead. No matter what Groves demands now, we have to cooperate. But the time is coming when we will have to do things differently and resist the military.'[13] Oppenheimer's comment was by any criterion natural for a scientist looking forward to the day when, as a researcher, he would be independent of military pressures and direction. Teller, however, professed to be 'shocked'. He comments in his memoir: 'The idea of resisting our military authorities sounded wrong to me.' Teller later accused Oppenheimer of plotting to undermine the H-bomb.

While it is true that Oppenheimer was morally against the project, this did not amount to anything approaching treason or subversion. Testifying at a hearing in Washington, DC, in the spring of 1954, Teller was instrumental in having Oppenheimer's security clearance withdrawn. He criticized his former chief for complaining about the 'exaggerated secrecy in connection with the A-bomb' and reported Oppenheimer as saying that if he undertook the thermonuclear development, it 'should be done more openly'.[14]

Teller was asked at a government hearing: 'Do you or do you not believe that Dr Oppenheimer is a security risk?' Teller responded with the remarkable weasel words that in the event destroyed Oppenheimer's reputation as a patriot and a trustworthy individual. He said:

In a great number of cases I have seen Dr Oppenheimer act – I understood that Dr Oppenheimer acted – in a way which for me was exceedingly hard to understand. I thoroughly disagreed with him in numerous issues and his actions frankly appeared to me confused and complicated. To this extent I feel that I would like to see the vital interests of this country in hands which I understand better, and therefore trust more.[15]

Many colleagues thereafter decided to have nothing further to do with Oppenheimer. Teller's reputation and responsibilities now rose inexorably; he became an adviser of presidents. He encouraged an expansion of the deployment of nuclear weapons and he was the originator of SDI, the Strategic Defence Initiative, popularly known as the Star Wars programme.

In 1988 Teller and Sakharov had an encounter at a meeting of the Ethics and Public Policy Center in Washington, DC. Sakharov used the opportunity to lambast Teller on two occasions about the Star Wars programme as a scheme that would undermine the nuclear balance and therefore offer a greater threat to peace and security. Teller has continued to defend his eagerness to develop the hydrogen bomb, and the Strategic Defence Initiative, arguing that not to have done so would have constituted a suppression of knowledge.

Joseph Rotblat comments on this rationale: 'The underlying notion that the acquisition of knowledge overrides all other considerations is unsustainable. Josef Mengele justified his "experiments" in Auschwitz on the grounds that they would provide new knowledge.'[16] Rotblat continues, 'There are other principles that override it, humanitarian principles. Scientists must always remember that they are human beings first, scientists second. And adherence to ethical principles may sometimes call for limits on the pursuit of knowledge.'

Sakharov vehemently opposed atmospheric nuclear testing in the sure knowledge of the effects on many thousands of people. Teller, in contrast, advocated such tests precisely because it would advance knowledge and aid the security of the United States. Teller rationalized his position by insisting that small doses of radiation were not harmful: 'a claim,' Joseph Rotblat comments, 'that is contradicted by the generally accepted norms on radiation exposures'.[17]

Teller has been judged harshly by his peer scientists on both sides of the East–West divide during the Cold War, to the extent of being dubbed the scientist who will go down in history as 'the real Dr Strangelove'.

Freeman Dyson records in his autobiography how one day he returned to his house to hear someone playing Bach's Prelude No. 8 in E-flat minor on the piano.

Whoever was playing it, he was putting into it his whole heart and soul. The sound floated up to us like a chorus of mourning from the depths, as if the spirits in the underworld were dancing to a slow pavane. There sitting at the piano, was Edward Teller.

Dyson's verdict: 'I decided that no matter what the judgment of history upon this man might be, I had no cause to consider him my enemy.'[18] Edward Teller, for all his civilization, represents in his urgent insistence on ever-bigger thermonuclear bombs and ballistic defence systems the apotheosis of the dark drives that impelled the Germans to force their rocket programme and the urgent suspicions that drove the Americans to produce the atomic bomb. The legacy of Peenemünde, built with the labour of slaves in the tunnels of the Harz mountains, is the array of ballistic missiles which, despite the fall of the Berlin Wall, continue to threaten the existence of the planet. The legacy of the Manhattan Project is the atomic and thermonuclear warheads which form the payloads of those missiles.

Tactical Nuclear Weapons

Even as the doctrine of massive nuclear deterrence in the form of thermonuclear bombs of ever-larger yield dominated the thinking on both sides, East and West, American scientists in the late 1940s were urging the military to take note of new developments. They were conquering the size problem to such an extent that they could offer nuclear warheads so small as to form the payload of artillery shells and smaller missiles. When in 1955 it became obvious that NATO was not going to reach its goal of opposing the Soviet armies with 100 divisions, the notion of creating a battlefield deterrent with tactical nuclear weapons took shape.

The notion was enthusiastically received by NATO generals, especially as it was believed that the American scientists had stolen a march on the Soviets. The edge did not last long. By the end of the decade the Soviet Union was also deploying tactical missiles and the advantage was lost. NATO's response was to increase the quantity of tactical nuclear weapons, which only invited the Soviets to follow suit, making the concept of deterrence ever more complex and prone to accidents. The Americans continued to develop smaller and smaller calibre guns with dual use capability, conventional and nuclear. But the Soviets again followed suit.

The same went for missiles. The immediate successors to the A4 built at Peenemünde were developed by both the Soviet Union and the United States as tactical intermediate-range nuclear missiles. The US army built identical copies of the A4 called Hermes. Subsequent missiles were the Redstone, deployed in 1958, which could deliver a 1- or 2-megaton warhead. That was succeeded in 1962 by the Pershing, which could deliver a 60-, a 200- and a 400-kiloton warhead over a range of 740 kilometres. The Soviet version of the A4 was the SS1A, which developed into the SS12, which carried a 500-kiloton nuclear warhead.[19]

The American strategic doctrine for a war against the Warsaw Pact in Europe was based on a willingness to substitute nuclear weapons for conventional forces. This, of course, implied that NATO had adopted a first-use strategy. When President Kennedy came to power, the Pentagon developed a 'flexible response' doctrine, which more or less meant that NATO would keep the Soviets guessing; but Kennedy had also decided that the West would in future attempt to oppose the Soviet Union with credible conventional forces. Had 'battlefield' tactical weapons been used on both sides in a European conflict, the continent would have been entirely devastated.

In 1974 the Soviet Union increased its armoury of intermediate-range nuclear weapons, the SS20s, in an apparent attempt to gain an advantage. The SS20 system was highly mobile and the payload included three MIRV warheads. In a move that resulted in widespread demonstrations in Europe, NATO responded by replacing its Pershings with a Ground Launched Cruise Missile system, which was deployed across Europe. A new missile arms race was in progress, a dangerous build-up of new generations of weapons that only ceased after Mikhail Gorbachev came to power, initiating retrenchments that would end in the fall of the Soviet system.

MAD

Scientists, along with historians, politicians and strategists, have joined the chorus down the years proclaiming that nothing less than the principle of mutually assured destruction has enabled the world to avoid a repetition of the two world wars. Even Niels Bohr, a man of the most delicate conscience, asked on arriving in the United States whether the atomic bomb was going to be big enough to end the war. Here is Richard Rhodes on that theme: 'If the bomb seems brutal and scientists criminal for assisting at its birth, consider: would anything less absolute have convinced institutions capable of perpetrating the First and Second World Wars, of destroying with hardware and callous privation 100 million human beings, to cease and desist?'[20] The proof of Rhodes's argument lies in the fact that the world has indeed avoided a third world war. And yet, so far so lucky. The Cuban crisis, mulled over by the media and historians on its fortieth anniversary in October 2002, left many believing that it was luck, rather than the prospect of mutually assured destruction, that avoided a nuclear holocaust on that occasion. As John Lewis Gaddis succinctly puts it in his *Now We Know*, 'Fortunately no black hole lurked at the other end, although new evidence confirms how easily one might have.'[21] The new evidence involves the fact that Soviet commanders on Cuba had been given discretion to use tactical nuclear missiles, without permission from Moscow, against an invading American force; that Soviet submarine commanders in the Atlantic were similarly given latitude to use their initiative with the use of nuclear weapons while out of contact with naval superiors.

An even more dangerous predicament arose at the end of 1983, the year in which a Korean Airlines jumbo jet had been shot down, killing 269 passengers, sixty of them Americans, after straying into Soviet air-space. In November Soviet missions in NATO countries reported that US bases had been put on alert, a mistake due to a misunderstanding of the import of a NATO exercise code-named Able Archer. In the following weeks, the cumulative impact of the Star Wars project, the collapse of arms-control talks and general suspicion of Reagan's 'evil empire' rhetoric found the Soviet Union anticipating a surprise attack, with attendant readiness to pre-empt. The hazards of nuclear war by accident persist during an era in which the current Russian deterrent – still comprising 6,000 nuclear war-heads – decays and becomes increasingly subject to under-funding and mismanagement, during an era, moreover, in which such weapons and new generations of such weapons are proliferating.

On 1 June 1997, the *Washington Post* carried a report claiming that the US government was deploying a versatile new kind of nuclear bomb intended to penetrate the earth and destroy under-ground facilities. These new nuclear capabilities were the first, according to the paper, to be added to the US arsenal since 1989 and the end of the Cold War. It was described as 'a slim, 12-foot-long weapon known as the B61 model-11 gravity bomb'. The bomb was developed and deployed without public or congressional debate and in contradiction to official assurances that no new nuclear weapons were being developed. The government, according to the *Washington Post*, contended that the B61-11 was 'merely a modification to the B61-7 gravity bomb'.

According to Greg Mello, head of the Los Alamos Study Group, an arms-control lobby group in New Mexico, the danger of the new bomb was that claims for its low-yield nuclear fall-out were liable to establish rationalizations for its use. The US government apparently threatened Libya with such a bomb over its construction of an alleged underground chemical weapons factory. The greatest danger, however, was that the new weapon violated the spirit of the 'delicately forged international ban on nuclear testing', undermining the US commitment to nuclear disarmament embodied in the Nuclear Non-Proliferation Treaty.

Six years on from this early warning, the nuclear weapons whistle-blower Ian Hofman, also associated with the Los Alamos Study Group, filed two syndicated news stories in February 2003.[22] The decision has been made, according to the study group, to push for nuclear testing of a new class of mini-nukes. According to minutes obtained of a meeting of federal defence executives and weapons scientists at the Pentagon on 10 January 2003, the preliminary groundwork is being laid to update nuclear weapons technology.

Hofman writes that advocates of a new generation of low-yield nuclear weapons argued that the current US arsenal was full of overpowered weapons designed for the Cold War rather than smaller state-terror threats. He quotes weapons physicists at Lawrence Livermore and Los Alamos laboratories as proud of developing weapons which they informally refer to as 'dial-a-yield', meaning the ability to control variable radiation yields.

Have scientists allowed themselves to be mere voiceless onlookers in the face of these developments? A bold warning comes from at least one world-class physicist, Steven Weinberg. Writing in the *New York Review of Books* on 18 July 2002, Weinberg argues that:

As the world's leader in conventional weaponry, we have a very strong interest in preserving the taboo against the use of nuclear weapons that has survived since 1945 . . . We cannot tell what countries may be tipped towards a decision to develop nuclear weapons by new US weapons programs or resumed nuclear testing.

Weinberg then reflects sombrely on a piece of history: Britain's decision in 1905 to build a new generation of battleship that made its entire fleet obsolete, enabling Britain's rivals to catch up merely by following suit. 'Other countries,' writes Weinberg, 'could now compete with Britain by building Dreadnoughts, and a naval arms race began between Britain and Germany.' Weinberg concludes that, like Dreadnoughts,

nuclear weapons can act as an equalizer between strong nations like the US, with great economic and conventional military power, and weaker countries or even terrorist organizations. It should be clear by now that national security is not always best served by building the best weapons.

The proliferation of nuclear weapons has kept pace with the spread of nuclear power for peace, reinforcing Weinberg's argument.

Atoms for Peace

The policy makers and scientists who harnessed nuclear power for peace believed that there would be a peace dividend in cheap, clean, unlimited energy. The Soviet Union built the first nuclear power station at Ominsk near Moscow in 1954. Two years later Britain followed suit with its Windscale power plant in Cumberland. Finally, in 1957, the Shippingport reactor in Pennsylvania was started up. The economics of nuclear power were inextricably linked with the production of plutonium for atomic weapons; at the Shippingport plant, moreover, tests were being conducted on the feasibility of a nuclear-powered submarine. The development of nuclear power was, and continued to be, a government-funded operation, for the costs were prodigious and the returns uncertain.

By the mid-1970s there were 514 nuclear power plants up and running, or under construction, throughout the world, but the expansion was about to lose impetus. The environmental movements of the 1970s cast doubt on the safety of nuclear plants as well as the problems of waste. There had been two accidents in the early days of nuclear power. In September 1957 an explosion occurred at a nuclear waste-disposal facility at Kyshtym in the Soviet Union, when 70 tons of radioactive material blew up, contaminating 400 square miles of agricultural land for years. A fire at Windscale the following month resulted in a large emission of radioactive gas and the plant was closed.

Any lingering doubts about the potential hazards of nuclear power plants became a shocking reality as a result of a leak at the nuclear plant on Three Mile Island near Harrisburg, Pennsylvania, on 28 March 1979. Much worse was the explosion at the power plant at Chernobyl in the Ukraine on 26 April 1986. Contamination from the incident spread throughout the Ukraine, Belorussia, Poland and Scandinavia. Very large numbers of the populations of Eastern Europe have been affected through the absorption of radioactive material into the food chain. It may be another decade before the full extent of the damage is known. It is estimated that by the end of 1988 the accident had cost the Soviet Union some

£12.5 billion. In addition to human error, it was considered that indifference to safety was a major cause of the accident.

The Chernobyl accident reminded the world that nuclear power meant nuclear fall-out and the by-product plutonium, which could be used in weapons production. In 1974 India conducted its first nuclear explosion test with plutonium produced by a reactor which had been constructed with help from Canada and the United States. Meanwhile a queue of other countries was waiting to join the nuclear power club. Through the spread of nuclear power, some forty countries had the potential to build nuclear weapons by 1980. Among them were the 'threshold states' (so-called because they had not signed the non-proliferation treaty): Brazil, India, Pakistan, Israel and Iraq. Of particular interest was the case of Israel and Iraq. It appears that Israel had the know-how for nuclear weapons by the early 1970s but announced that the components would be kept unassembled unless another power in the Middle East prompted them to change their minds. It was the French (tempted by commercial rather than political considerations) who enabled Saddam Hussein to start a nuclear power programme which led to the Israelis making a pre-emptive strike on the Iraqi reactor at Tammuz in June 1981. North Korea's first nuclear reactor, at Yongbyou, was provided by the Soviet Union. The country built a plutonium separation plant at Yongbyou in 1987.

There have been many reasons for nuclear physicists throughout the world to ponder carefully the long-term consequences of providing their governments with nuclear weapons potential. It would be an obtuse researcher who could argue that scientists have no responsibility in such circumstances.

33. Uniquely Nazi?

Have science and scientists behaved better since the fall of the Third Reich? Historians continue to ponder what was distinctively unique about science, medicine and technology under Hitler; whether there are clear margins where Nazi science begins and ends.

A unique aspect of science under Hitler was the attempt by doctors and anthropologists to supply a scientific basis for murderous 'racial hygiene'. Authentically scientific proof for the claims of an Aryan racial superiority were never found, but the regime constructed and promoted a pseudo-scientific racist biology that led to the death camps, placing Nazi science in a sphere all of its own.

Human experiments without consent reached horrifying depths of depravity and cruelty under the Nazis; but experimentation on groups and individuals without their consent has occurred before, during and since Nazi rule. During the war the Japanese ran a network of camps through the occupied countries of the Far East in which teams of researchers (notably the notorious Unit 731) conducted experiments involving lethal pathogens and chemical poisons on thousands of human guinea pigs.

Slave labour was a feature of Nazi technology and industry, and employed even by German university departments during the war.[1] But murderous slave labour was also used by the Soviet Union before, during and after World War II. The building of the White Sea–Baltic Canal with enforced Kulak labour cost as many as 150,000 lives. Links with Nazi slave labour, moreover, continued beyond the war's end. German manufacturers like Daimler-Benz, which had used forced labour under National Socialism, started building automobiles for the civilian market within weeks of the war's end, their machine tools still usable in dispersed factory workshops. The transition to a profitable peacetime production,

based originally on collusion with the racial barbarism of the Third
Reich and active cruelty towards the regime's victims, explodes
the myth that the German economic miracle arose from a post-Nazi
'zero-hour'.[2]

Even the Nazi suicide mission tactics towards the end of the
regime have not proved exclusive to Nazi rule. And the 'prone'
position required of pilots in the design of later prototypes of
German fighter planes, taken by some historians to be a peculiarly
Nazi feature, were copied by the Americans in a modified Lockheed
F-80E after the war.[3]

Michael J. Neufeld, pondering what was 'specifically Nazi' in
character other than slave labour in the Nazi missile programme,
suggests that, at first sight, there is 'not very much'.[4] The emergence
of the military-industrial complex shared much in common with
the Manhattan Project, including secrecy, prodigious state funding,
the scaling up from modest beginnings to massive plant and workforce. Yet what is widely neglected, he claims, is the unique
historical circumstance of the Third Reich. Without the Nazi
predicament of competing power centres combined with the
dilettantism and amateurism of Hitler and Goering, it would have
been impossible for the modest endeavours of von Braun and his
associates to have developed into the massive Peenemünde project
without an industrial base. 'As a result,' writes Neufeld, 'the rocket
programme built an institution and a weapon which made little
sense, given the Reich's limited research resources and industrial
capacity – a perfect symbol of the Nazi regime's pursuit of irrational
goals with rational, technocratic means.'[5]

This focus on the unique historical context that attends the
corruption of science under Hitler should make us vigilant for the
rapidly changing social, economic and political circumstances in
which science has been operating since 1945.

Science in the Cold War

Pondering what was specifically Nazi in character about science under the Third Reich is to probe the ethics and politics of scientists in a wider ambit than Germany: how the scientists of the Western democracies and the Soviet Union behaved towards their disciplines, towards their societies and towards nature itself during the Cold War.

World War II brought about widespread changes in the conduct of science and technology which became permanent features in the post-war era. Many of the standards of scientific freedom and exchange of knowledge were suspended by all the belligerents. The reduction of scientists to the status of faceless team members was also a result of the consequent growth of Big Science through state, industrial and military auspices from 1940 onwards.

The most dramatic alteration was in the West. The Office of Scientific Research and Development under the government science chief Vannevar Bush commissioned more than 2,000 research programmes in the course of World War II. The projects involved industrial research and development units employing tens of thousands of scientists and technicians in companies such as Du Pont and General Electric, as well as major university laboratories like MIT and Caltech. Bush, however, intended that in the post-war era the state would fund science on a grand scale, while leaving scientists free to decide the direction of basic research.

Late in 1944 Vannevar Bush was commissioned by President Franklin Roosevelt to write a proposal about the future of science and technology. Roosevelt died before Bush had completed the document, so he presented President Truman with his visionary tract, entitled *Science, the Endless Frontier*, in July of 1945. He proposed a remarkable new relationship between government and science funding, to be run by a body he called the National Research Foundation (NRF). The NRF would work in the interests of American military might, to be sure, but also for economic prosperity and towards the intellectual and cultural flourishing of the entire nation. His proposal for a barrier between government and military funding and civilian control of the choice and direction of basic research would prove, however, a vain hope.

Any complacency about the freedom of science in the West in the post-war era must be qualified, therefore, by the overwhelming and combined power of industry and the military in the United States, a combined hegemony over science that was barely questioned through the 1950s. The growth of Big Science was also evident in the smaller developed countries like Britain and France. Government funding for research and development expanded in Britain, for example, from 44 per cent of total R and D expenditure in the 1930s to 57 per cent in the 1960s.[6]

During this period the predicament of science in the United States came to resemble the state grip on science in authoritarian regimes. Science came under surveillance, never more evident than in the McCarthyite communist witch-hunt trials, when many distinguished figures were deprived of funding and driven into exile.

In 1961 the retiring President Dwight D. Eisenhower made his famous 'industrial-military complex' speech and went on to warn that:

this conjunction of an immense military establishment and a large arms industry is new in the American experience. The total influence – economic, political, even spiritual – is felt in every city, every State house, every office of the Federal government . . . we must not fail to comprehend its grave implications.

Other critics would extend the combination to the 'university-industrial-military complex'.

The results of the wartime synergy of the military and science were by no means deleterious in every case. Particle physics, astrophysics, astronomy and space exploration benefited from the research done at Los Alamos, as well as the rocket developments that were heir to Peenemünde. Many might feel, however, that the vast sums of money spent on space programmes and Big Science physics might have been better spent elsewhere.

Work on molecular biology in the United States and Britain benefited from the field of biophysics and flourished as a result of knowledge and expertise gained in wartime. Work on radioactivity, radio, radar, the electron microscope, supersonic and vacuum techniques, electronics and chemotherapy contributed to the expanding field of biology for medicine.

Yet not every endeavour in the field of biology for medicine turned out to be irreproachable, despite the reaction in the democracies to the medical crimes of the Nazis. Several prominent cases in the United States reveal the extent to which the standards invoked at the Nuremberg doctors' trial were violated from the very outset by denizens of the country of the code's authors. The Tuskegee Syphilis Experiment involved a decision not to treat 400 Afro-Americans suffering from syphilis between 1932 and 1972 in order to determine the course of the disease.[7] Between 1950 and 1969 the United States government carried out a range of tests involving chemical and biological agents, radioactive materials and drugs on unsuspecting civilians, or subjects with very little political clout – people in the military, prison inmates, mental patients – involving chemical and biological agents.[8] Other scandals involved the exposure of people to radiation.[9] In the 1960s scientists at the Hanford Nuclear Reservation in Washington state contaminated a district of some 8,000 square miles by releasing quantities of radioactive substances into the atmosphere to assist the military in assessing contamination from Soviet plutonium factories.[10] In a series of nuclear tests conducted by Britain during the 1950s on Australian islands and the Australian mainland, military personnel were deliberately exposed to the blasts to establish the effects of radiation. In the Soviet Union, German, Italian and Spanish prisoners of war, along with officers of the Vlassov army, incarcerated on Wrangel Island, were being subjected to experiments as late as 1962 to determine the effects of radiation on the human body. Other experiments involved 'the effects of prolonged submersion at great depths'.[11] In 1975 Amnesty International reported 'torture' as a component of the treatment of political prisoners detained in Soviet psychiatric hospitals for indeterminate periods.[12]

Parallel with the issue of non-consent to dangerous human experiments was the failure of industries in the democratic West to inform the general public, workers and consumers of the dangers of toxic contaminants in a wide range of waste materials and products. Rachel Carson revealed the dangers of pesticides in her book *Silent Spring* (1962), and Ralph Nader exposed the priorities of profit over safety in the motor industry. Authors Gerald Markowitz and David Rosner have shown in their book *Deceit and Denial: The Deadly Politics of Industrial Pollution* (2002) how the lead and vinyl industries did not merely withhold or suppress information but sought to mislead the public by issuing reassuring propaganda, containing manipulated scientific research, claiming that dangerous products were beneficial.

Less than a year after the reunification of East and West Germany, it was revealed that East German scientists and physicians had been using men, women and children as human guinea pigs in a state-sponsored research programme intended to perfect hormone drugs in an effort to develop compounds that would boost the performance of East German athletes.[13] It appears that the research was conducted on children without their consent (or that of their parents) and that athletes were permanently injured.

Information Revolution

Meanwhile the wheel of science and technology has continued to turn, giving rise to novel forms of ethical and political pressures on scientists. The innovations in telecommunications and computer technology from the 1960s had inevitable and far-reaching consequences for the providers as well as the recipients.

The deregulation of broadcast media and telephone systems, against the background of satellite technology and coaxial and fibre-optic cables, saw a growing challenge to national broadcast systems from commercial media companies. The military had led the way with satellite technology, but civilian application was not far behind. The new technology of communications was leading to a shrinking world – the global village. At the same time, cheap and extensively improved communications were meeting the needs of information transmission, expanded by the power, capacity and speed of modern computers.

The proliferation of chip technology in the development of microprocessing saw a shift from technology dominated by the military to the enthusiasm of a market led by civilian consumers.

The transition to mass consumerism involved the determination of IBM to enter the market. IBM's project involved the contracting-out of software operating systems to Microsoft, a small company founded by Bill Gates and Paul Allen in Seattle. In 1978 computer sales revenues were $10 billion; by 1984 they had risen to $22 billion, of which more than half were in the small business and home market sector.[14]

The information revolution had arrived; its next quantum leap was interaction, which emerged from a project to link the participating computers of the Pentagon's Advanced Research Projects Agency (ARPA) throughout North America. Lawrence G. Roberts devised a 'packet switching' system which broke up messages for transmission by the best available routing, to be reassembled at the other end. This system of computer-to-computer networks spread through government institutions, universities and businesses, and by 1983 ARPA designed a system through the same 'packet switching' for interaction between the different existing networks. As the market for PCs grew, so did the Net, as interactive communications came to be known. The Net meant e-mail (electronic mail), enabling vast numbers of individuals, homes and businesses to communicate interactively. The Net also meant, in time, a vast and ever-growing source of information of every kind, uncontrolled, unrefereed and mostly free. A means of ordering this mass of information was provided by the World Wide Web, which developed out of a system created by Tim Berners-Lee for CERN, the international physics laboratory in Geneva.

The advent of the computer had from its earliest days enabled scientists and engineers in every discipline to create sophisticated models, to number-crunch and analyse at ever greater speeds. From modelling receptor sites in cells, to building artificial intelligence architectures, to calculating the behaviour of Black Holes and entire universes, the advent of small, powerful and relatively cheap computers was speeding up the rate of generation of knowledge and discovery. With the advent of e-mail, scientists began creating research groups without frontiers, ideological, national or geographical.

IT, Publishing and Biotechnology

During an era in which computer technology was transforming the ease of collaboration and exchange of information between scientists, economic and political developments in the West were imposing market criteria on the world of academia. The natural sciences, along with other disciplines in universities, were being forced to adapt to a culture of accountability and assessment according to 'productivity' in terms of publication. The proliferation of papers and articles has not necessarily increased the quality of published work: it has certainly led to information overload. At the same time, academic publishers, which once operated on the basis of subsidies, have come under increasing pressure to be profitable or cease publishing. Many of the great academic imprints in the United States, Britain, France and Germany have ceased to publish learned monographs in the natural sciences, preferring trade titles that will pay their way. Meanwhile, the publishers of many learned journals have become increasingly commercial, while abandoning hard copy distribution and opting to offer online packages to academic libraries at exorbitant rates: this in turn has created crises within journal publishing, which impacts on the ever-expanding need for scientists to be published in order to justify their funding, tenure and promotion prospects.

The 'globalization' that resulted from the worldwide webs and nets of information technology though the 1990s had its optimists and its prophets of gloom, with convincing arguments on both sides. The borderless economy in which money, technology, industry and goods move without hindrance throughout the world promises a future in which every home, in the words of Alvin Toffler, will become an 'electronic cottage'. But globalization also has its losers, its economic have-nots, destined to be tranquillized by digitized trivial entertainment or to nourish hatreds that threaten to break out in violence.

An ominous consequence of the IT revolution for academia, and science especially, is the circumstance foretold, and up to a point fulfilled, by the philosopher Jean-François Lyotard in his essay *The Postmodern Condition* (1979), subtitled 'a Report on Knowledge'. Lyotard saw that the extent of the domination and penetration of IT would be a globalized information domain in which the only kinds of knowledge to survive would be those amenable to translation (meaning cultural translatability as well translatability in terms of information technology). Lyotard's vision is best explained by imagining a world of scientific and educational publishing similar to the *Reader's Digest* magazine. Articles chosen for publication are those suitable for translation and editing for syndication to all the different national and language editions. The result is a kind of highly bland product that serves maximum editorial economy of scale.

Lyotard's insight into the repercussions of IT forms part of his theory of the collapse of the great modern ideals: the French pursuit of liberty and equality; the German Enlightenment project of free knowledge for all. Information technology in Lyotard's view is now inseparable from commodity capitalism, scientific research being dependent on investment linked with improved performance which itself is dependent on technological advances. As a result, science ceases to celebrate the pursuit of truth and the dissemination of free knowledge, and handles knowledge as a commodity traded for profit. Lyotard's dark vision of the future of science already shows signs of ominous realization in the realms of biotechnology.

The period of respite from military priorities which began in 1989 after the collapse of the Berlin Wall, with the promise of substantial peace dividends for medicine, agriculture and the environment, in the form of massive funding, had not meant a relaxation of political and ethical pressures on scientists. If anything, those pressures have increased.

R and D in the medical sciences in the United States, including molecular biology, genetics, neuroscience and biochemistry, is dominated by the National Institutes of Health at Bethesda near Washington, DC, whose annual budget by 2000 reached a prodigious $18 billion and whose goals were, clearly, the applied science of cures and palliatives in close collaboration with the American pharmaceutical industry and its markets. That synergy has raised many questions about the relationship between medical science and business.

The Human Genome Project, costing $3 billion, was a scheme to sequence the estimated 100,000 (now scaled down to 35,000) human genes by the turn of the new century. Reaching for parallels with Big Science, big budgets and space-age outreach, some called it 'biology's equivalent of the Apollo space programme'. Mapping the human genome not only promised a transformation of our understanding of what makes us tick biologically, but tantalized with the prospect of altering, for better or for worse, what humans might become. The rhetoric proclaimed cures for some 4,000 diseases, including Alzheimer's, multiple sclerosis, many forms of cancer and cystic fibrosis. But there were also remarkable financial rewards awaiting those who would win the race to exploit the new knowledge commercially.

The architects of the project had not been insensitive to the need for control. Nobel laureate James Watson, the director, had insisted on a budget of $90 million to be spent on research into bioethics. Yet in 1992, just a year into the project, signs of future ethical and political problems began to emerge. Watson opposed the decision of the NIH to file for patent on a batch of brain genes produced under the auspices of the project. Watson felt so strongly that he resigned.

In April of 1992 Bernadine Healey, director of the US National Institutes of Health, filed for patent on nearly 3,000 gene fragments, known as cDNA (complementary DNA), genetic information obtained by a breakneck 'mechanical' sifting method. The process enables researchers to isolate large numbers of potentially useful genetic fragments swiftly, before their precise function is specified. The process was developed by Dr Craig Venter, then employed by a Washington-based federal-funded genetics laboratory. Venter's research was applauded by many scientists, since it promised to hasten the discovery of therapies for major diseases. Watson complained that Bernadine Healy had acted precipitately in championing mechanized short-cuts; moreover, he described her attempt to copyright unspecified genetic fragments as 'lunacy'.[15]

The issue of patenting goes to the heart of tensions between those who see science as free, objective, universalist, and those who argue that science can only flourish when discoveries are patented or copyrighted and profits from its success ploughed back. Bernadine Healey defended the patenting of cDNA as a decisive choice between science as 'pro-competitive', as she put it, as opposed to science as 'anti-competitive'.[16] She was, of course, wedded to the 'pro-competitive'.[17]

For his part, Watson's unhappiness over the brain genes and the NIH was not only a reluctance to have forms of nature patented, but also stemmed from his conviction that such issues should involve the views of scientists and not just politicians, bureaucrats and business people. Scientists who claim that science is apolitical, value-free and morally neutral might not agree with him.

In a world of academic and commercial research in which knowledge is increasingly seen in terms of 'intellectual property rights', organized and disseminated in the interests of efficiency and profit, the undermining of scientific freedom increases apace. On 4 October 2001 the science weekly *Nature* reported a rebellion throughout Europe against a patent on a gene for breast cancer held by the American company Myriad Genetics. In a situation claimed by many professionals to be unprecedented, Myriad was also attempting to enforce the provision of a diagnostic service in relation to the gene as well as claim property rights over the genetic information. Myriad had requested that all test samples be sent to its labs in Salt Lake City; but scores of geneticists and dozens of diagnostic labs have continued to perform their own BRCA1 testing.[18]

The Ethics of Cloning

The Human Genome Project is subject to continuing controversy relating to patenting and intellectual property rights; at the same time, it has provoked polemic over the relationship between genetics, diagnostics and the issue of cloning.

Do we have the right actively to encourage the birth of cloned animals or humans with a high rate of certainty that they will suffer abnormalities as a result of reproductive technology? James Watson ran into trouble in the late 1990s in Germany by publicly promoting early identification of genetic diseases such as Fragile X and Tay Sachs so as to give the mother the option to abort. He was accused of lining himself up with Nazi scientists who had advocated 'euthanasia' of 'lives not worth living'. The Germans, sensitive to the fact that Tay Sachs has a high rate in Jewish populations, had a point: who, after all, is to say when a life is not worth living? Cloning, however, puts a new slant on the same issue.

The cloning experiments of I. Wilmut et al., published early in 1997,[19] showed that sheep embryonic eggs (ocytes) can re-programme the nuclei of differentiated cells, enabling the cells to develop into any type. The implications for medicine appear remarkable, not least the possibility of the eradication of certain heritable diseases and developments in stem cell research using human embryos. As for cloning entire human beings, as opposed to specific tissues for therapeutic reasons, questions have been posed by ethicists about the possible loss of the uniqueness of such a human being.

The problem, according to biologists, is not so much the possibility of reproducing 'identical personalities'. The brain and central nervous system develop to a significant extent epigenetically (that is, not solely on the basis of genes), and hence exhibit differentiation in both cloned animals and monozygotic twins. The real problem, as described by many specialists who counsel caution, is the possibility of introducing abnormal chromosomes. Richard Lewontin explains:

[In the cloning process] the nucleus containing the egg's chromosomes is removed and the egg cell is fused with a cell containing a nucleus from the donor that already contains a full duplicate set of chromosomes. These chromosomes are not necessarily in the resting state and so they may divide out of synchrony with the embryonic cells. The result will be extra and missing chromosomes so that the embryo will be abnormal and will usually, but not necessarily, die.[20]

This is why Wilmut and his colleagues produced only one successful Dolly out of 277 tries. In other words the cloning process is not safe. What it means for the cloned embryo brought to term is that telomeres, which control the ageing process in the donor DNA, could be deteriorated, leading to early abnormalities. More ominous still is the prospect of thus affecting germ line cells in the offspring of such a cloned animal or human being. In other words, it has taken nature aeons of evolution to establish the process of reproduction, and, as yet, we are not entirely sure of the consequences of interfering with germline cells. If there were to be a better rate of success in producing cloned humans than Wilmut's success with sheep, what rate of consequent abnormalities would we consider acceptable?

The new and continuing developments in biotechnology have bestowed extraordinary insights into the relationship between evolution, the building blocks of life and prospects for advances in medicine. The advances in knowledge have presented new dilemmas for scientists that threaten to outstrip our ethical standards and traditions. But historical circumstances have also occurred, as they did with Hitler's rise to power, that will put scientists to the test in new and unforeseen ways.

34. Science at War Again

On 11 September 2001, the United States was traumatized by an act of mass murder, committed by a terrorist group that had turned the power of science and technology against innocent civilians going about their everyday lives. Within the space of several hours the Western world became a far more vulnerable, fragile and perilous place. Virtually every aspect of science, in the view of the American government, had become a potential weapon in the consequent war on terror.

The attacks, followed by anthrax terrorism, fears of biological weapons and 'dirty bombs' delivered by suicide bombers, have had a profound impact on the government management of science and technology in the United States and in other Western countries. Scientists may be experiencing the impact, counting the cost, for decades to come.

September 11 saw an end to the delivery of the peace dividends for science that were supposed to follow the fall of the Berlin Wall. President George W. Bush supervised the shift of priorities back towards intelligence, defence systems and a newly shaped doctrine of pre-emptive strikes, including the new generation of low-yield 'mininukes'. The relationship between the American administration and scientific research, already strained by the President's scheme for a trillion-dollar ballistic missile shield and his refusal to sign the international Kyoto Protocol for protecting the environment, became more tense. The National Institutes of Health at Bethesda, Washington, DC, formerly an open, collegial place, became ring-fenced with high security. The National Science Foundation, charged with supporting pure science, came under pressure to scrutinize the relevance of its funding targets. Awards were being made a month after the attacks for projects that would study 'human, social responses' to the terrorism; even its much-vaunted earlier promotion of more funding for mathematics has been rewritten to attract grant applications that can be applied to secret intelligence.[1]

Now a whole range of pathogens are deemed to have dual uses, for medical benefits and for terrorist outrages. There are government calls for biologists to be aware that their work can be exploited to develop weapons of mass destruction. Biological weapons, we have been warned, present a serious and growing threat that poses similar dangers to nuclear proliferation. A calculation made by the US Congress Office of Technology Assessment suggests that 100 kilograms of anthrax spores spread effectively on the wind and allowed to drift over Washington, DC, could, in the right conditions, cause 3 million deaths. George Poste, who chairs a Ministry of Defence task force on bio-terrorism in Britain, told a pharmaceuticals industry conference in London on 6 November 2002 that biology must 'lose its innocence', face up to implicit dangers of certain kinds of biological data even in legitimate projects and accept the need for more regulation of information. The sort of project Poste had in mind was the creation of so-called 'stealth' viruses that evade the immune system. A month later, the United States moved to block new means for verifying compliance with the Biological and Toxin Weapons Convention (originally established in 1972). In December 2001 an international conference seeking to press for compliance broke up in disarray and without agreement.

Since then we have seen confirmation of President Bush's commitment to a new generation of nuclear weapons, with clear indications that such weapons will no longer be regarded as deterrence but for pre-emptive use against nuclear and non-nuclear powers alike. We have entered an era in which the United States, the lone superpower, has assumed the role of world policeman – a circumstance that Leo Szilard prophesied for either America or Russia, whichever should win the Third World War.

Washington is convinced that future conflicts will be conducted with superior American science and technology: surveillance, information, smart bombs, smart electronics. War will be waged via computer-generated imaging from a safe and distant environment: virtual warfare with all too real consequences on its targets. The American interventions in Afghanistan and Iraq seemed to confirm the view that the Pentagon can win wars without significant casualties among its own military. This can only encourage America to believe that it can, and must, tackle perceived threats as they arise. Where will such threats start and finish? Will all of science, technology and medicine now be seen as either for or against America in the war on terrorism? How do scientists retain their integrity against this background?

Doing good science at any time and in any country in the century described in the span of this narrative has involved familiar, time-honoured ethical principles. Researchers who falsify data, massage statistics, appropriate other people's discoveries or ideas cannot be said to be 'good scientists', in either a moral or a professional sense: certainly their results are unlikely to be any good.

The events of World War II, the widespread militarization and industrialization of science and technology, the expansion of Big Science, the obstacles to freedom of information, the shroud of secrecy and the development of weapons of mass destruction undermined for ever the notion of basic science driven by individual talent divorced from consideration of consequences and social and political auspices. The infiltration of science in the Cold War by industry, business, the military, government planning, intellectual property rights and the scramble for funding has blurred the separation between basic and applied research, reducing the individuality of scientists and exposing them to an array of moral and political scruples. The tendency has been for scientists to ignore these scruples, as did Hitler's scientists, by withdrawing into a cocoon of 'irresponsible purity'. In the post–September 11 era the scruples have increased and the temptation to withdraw politically and ethically is even greater.

Doing good science today involves a principled vigilance for consequences, an awareness of the impact of scientific discovery on society, on the environment, on nature. The good scientist evidently does not place dangerous knowledge or techniques into the hands of the untrustworthy. The good scientist attempts to publicize by any means possible the social and environmental consequences of potentially dangerous knowledge.

The good scientist, moreover, rejects the use of people as instruments, as means to an end. That principle is increasingly complex, as we have seen in the expanding fields of biotechnology and genetics, where proliferation of knowledge and techniques involves difficult ethical choices. The imperative calls for ever more subtle distinctions to be made between persons and things in an era of rapidly advancing reproductive technology and potential for genetic enhancement.

Resolving moral dilemmas in science, however, involves not merely isolated choices but a committed pattern of behaviour, long-term resistance to compromise leading to feelings of self-respect: integrity.[2]

The greatest pressures on the integrity of scientists are exerted at the interface between the professional practice of science and the demands of the fund-awarding patrons. At every stage of this narrative, from Fritz Haber's decision to promote poison gas to Max Planck's decision to raise his arm in a Nazi salute, to Paul Harteck's acceptance of a chair made vacant by a dismissed Jew, to Heisenberg's decision to accept Hans Frank's hospitality in Cracow, to Wernher von Braun's use of slave labour, we have seen the pressures of hubris, loyalty, competition and dependence leading to compromise. In the final analysis the temptation was a preparedness to do a deal with the Devil in order to continue doing science.

The Faustian bargains lurk within routine grant applications, the pressure to publish for the sake of tenure and the department's budget, the treatment of knowledge and discovery as a commodity that can be owned, bought and sold. Handling these pressures and realities is inseparable from the difficult task of being a good scientist today.

Are these pressures largely resolved by the auspices of a more or less democratic government and a benign constitution? As we have seen, the view was widely promoted during the early 1940s by the British biochemist and historian Joseph Needham, and by Robert Merton, that science reaffirms the values of Western liberal democracy – rationality, objectivity, freedom, internationalism, systematic scepticism – values that were then under threat from European totalitarian regimes. There have been interesting shades of difference on this theme. While broadly agreeing with the thesis, J. D. Bernal proved soft on Stalin, as if to judge some authoritarians more acceptable than others. And whereas the physicist Samuel Goudsmit believed that democracy encourages good science, Merton also believed that good science actually encourages democracy.

At first sight the symbiosis between science and democracy presents a powerful alternative to the irresponsible purity not only found among German scientists of the 1920s and the Third Reich, but widely familiar since the end of World War II. The danger is, however, that scientists working within more or less democratic societies are inclined to abdicate responsibility by assuming that their democratically elected governments always know best. That assumption is naive in the extreme. As with truth in the media, the first casualty of war, and the threat of war – whether we are talking Cold War, or, as now, a war on terrorism – is freedom of information and free access to knowledge within science and technology. Should a scientist in a democracy keep silent on demand, or surrender knowledge on demand, hoping that the government knows best?

Scientists, as this narrative has sought to demonstrate, may not have political power, but neither do they exist in a moral, social and political vacuum. Scientists have the power of knowledge and expertise, which they can choose to employ, disseminate or withdraw. They are not obliged to work under auspices of, or impart knowledge to, those they distrust or believe to be reprehensible.

Norbert Wiener provides us with a telling example. In 1947 he was approached by an American aircraft corporation requesting a

technical account of his guidance research. Refusing to pass over
the information. Wiener invoked some basic principles relating to
scientific responsibility in weapons creation. 'The policy of the
government itself during and after the war,' he began, 'say in the
bombing of Hiroshima and Nagasaki, has made it clear that to
provide scientific information is not a necessarily innocent act,
and may entail the gravest consequences.'[3] In these circumstances,
Wiener continues, the tradition of free exchange of scientific information requires careful scrutiny 'when the scientist becomes an
arbiter of life and death'. Wiener goes on:

The experience of the scientists who have worked on the atomic bomb
has indicated that in any investigation of this kind the scientist ends by
putting unlimited powers in the hands of the people whom he is least
inclined to trust with their use. It is perfectly clear also that to disseminate
information about a weapon in the present state of our civilization is to
make it practically certain that the weapon will be used. In that respect
the controlled missile represents the still imperfect supplement to the
atom bomb and to bacterial warfare.

In a further reflection, as chillingly relevant today as when he
wrote it half a century ago, Wiener argues:

The practical use of guided missiles can only be to kill foreign civilians
indiscriminately, and to furnish no protection whatsoever to civilians in
this country. I cannot conceive a situation in which such weapons can
produce any effect other than extending the kamikaze way of fighting to
whole nations.

Scientists possess specialist knowledge of the 'laws of nature'
and specialist technology, as well as specialist knowledge of the
consequences of that technology. Scientists can affect social and
political decisions by choosing not to share or advance their knowledge, or, on the other hand, by refusing to limit its dissemination.
Witness, for example, Dr Jeffrey Wigand, the tobacco executive
who decided to expose at great cost to himself the tobacco industry's

efforts to downplay knowledge of the health hazards of smoking. Wigand said: 'I felt that the industry as a whole had defrauded the American people. And there were things that I felt needed to be said.'[4]

Wigand's story gives the lie to the notion that scientists have no voice, as they are invariably mere cogs in corporate machines. The circumstances of the communications revolution have made all the difference. The gesture of a lone hero, like Rotblat, made no difference back in 1945, when he resigned from the Manhattan Project. He was forbidden under pain of imprisonment from communicating the reason for his departure from Los Alamos to his colleagues, let alone the media. Today scientists have the ability to collaborate with like-minded, responsible practitioners, making the widest possible use of new information technology and the media.

Can scientists exert greater political and ethical responsibility in the face of the overwhelming power of those who own and control most of science and technology? Science today is already influenced by non-specialist groups of citizens operating outside of official government and commercial organizations.[5] The movement can be broadened and deepened, influencing public opinion on an array of social issues – health, poverty, education, criminal justice, human rights, the environment, nutrition.

The blurring of the margins between markets and industry, the academic world and government institutions, as well as greater access to the media through the internet, create opportunities for multi-disciplinary, non-hierarchical groups to make their views felt. It is possible for scientists and non-scientists, together, to question the impact of new developments in science, medicine and technology, from genetically modified crops to SSRI drugs, from embryonic stem cell research to attempts to clone human beings, from the use of organophosphates in farming to patenting that hampers rather than aids the application of medical science. The most important issue, however, remains the panoply of new generations of weapons of mass destruction.

Nearly sixty years after Vannevar Bush's dream for a civilian-administered funding agency, basic science is still largely shaped and managed by government and military patronage in the United States, the lone superpower. The National Science Foundation manages some $4 billion (of which only an estimated $2.5 billion is spent on basic science) of a federal R and D budget that spends annually some $75 billion, half of which is administered by the Pentagon.[6] Even so, research supported by the National Science Foundation is available to the Pentagon. The concentration of such prodigious and unscrutinized power over science and technology is extremely dangerous, calling for scientists themselves to insist on an increasing critique of the direction, the choices and the bias of new weapons.

Will scientists today, in an increasingly crisis-ridden world, in which they are ever more dependent on paymasters to pursue their vocations, behave like the fellow travellers under Hitler – the Heisenbergs, Weizsäckers and the von Brauns – taking benefits from the government and the military, while claiming that as individuals they are aloof from society and politics? Will they argue that they are not in any way responsible for the uses to which their knowledge and discoveries are put? Or will they take part in bringing down the barriers that insulate defence research from public scrutiny, criticism and influence?

There is an urgent need today for scientists who are not only skilled practitioners in their disciplines but who possess a highly developed grasp of politics and ethics, who are prepared to question, probe, expose and criticize the trends of military-dominated science. Under Hitler, the dissident scientist risked imprisonment and death; but at least, in the early days of the regime, it was possible to emigrate in the hope of doing science under more benign political auspices. Today the dissident does not risk imprisonment or death, but in the globalized domains of science and technology there are no oases of irresponsible purity into which a scientist can retreat. The best defence against the prostitution and abuse of science is for scientists to unite in small and large unofficial constituencies, to create communicating communities of scientists who, in Joseph Rotblat's words, are 'human beings first and scientists second'. These constituencies could provide the pluralist checks and balances that alert the public to irresponsible exploitation of science that poses threats not just to the American 'homeland', but to societies and peoples everywhere; to the environment, to peace, to human rights and to nature itself.

Notes

Introduction: Understanding the Germans

1 See Ulrich Schwarz, *Der Spiegel* (20 January 2003), pp. 82–8; and Jörg Friedrich, *Der Brand: Deutschland im Bombenkrieg 1940–1945* (Munich, 2002). See also the posthumous essays of the late W. G. Sebald, *On the Natural History of Destruction*, trans. by Anthea Bell (London, 2003).

2 Stockholm International Peace Research Institute [SIPRI], *Warfare in a Fragile World: Military Impact on the Human Environment* (Stockholm, 1980), p. 87.

3 Joseph Haberer, *Politics and the Community of Science* (New York, 1969).

4 Geoffrey Cocks, *Psychotherapy in the Third Reich: The Göring Institute* (New York, 1985), p. 19.

5 See Michael Wildt, *Generation des Unbedingten: Das Führungskorps des Reichssicherheitshauptamtes* (Hamburg, 2002).

6 See FIAT Review of German Science 1939–1946, published by the Office of Military Government for Germany Field Agencies Technical (Wiesbaden, 1947).

7 Robert Jungk, *Brighter than a Thousand Suns: A Personal History of the Atomic Scientists*, trans. by James Cleugh (London, 1960), p. 102.

8 Thomas Powers, *Heisenberg's War: The Secret History of the German Bomb* (Boston, 1993), p. 482.

9 Lewis Wolpert, *The Unnatural Nature of Science* (London, 1992), p. 170.

10 Quoted in French in Jean-Jacques Salomon, *Le Scientifique et le guerrier* (Paris, 2001), p. 127.

Chapter 1: Hitler the Scientist

1 Walter Dornberger, *V2*, trans. by James Cleugh and Geoffrey Halliday (London, 1954), pp. 70–73.

2 Quoted in Michael J. Neufeld, *The Rocket and the Reich: Peenemünde and the Coming of the Ballistic Missile Era* (Cambridge, MA, 1999), p. 118.

3 Albert Speer, *The Slave State: Heinrich Himmler's Masterplan for SS Supremacy*, trans. by Joachim Neugroschel (London, 1981), p. 84.

4 Paul Weindling, *Health, Race and German Politics Between National Unification and Nazism 1870–1945* (Cambridge, 1993), p. 492.

5 Quoted in Michael H. Kater, *Doctors under Hitler* (Chapel Hill, 1989), p. 178.

6 Quoted ibid., p. 178.

7 Quoted ibid., p. 178.

8 Speer, *Slave State*, p. 83.

9 See for example Geoffrey Brooks, *Hitler's Nuclear Weapons: The Development and Attempted Deployment of Radiological Armaments by Nazi Germany* (London, 1992).

10 Adolf Hitler, *Mein Kampf*, trans. by Ralph Manheim (London, 1992), p. 10, note 1.

11 Brian McGuinness, *Wittgenstein: A Life*, vol. I: *Young Ludwig, 1889–1921* (London, 1988), p. 51.

12 Albert Speer, *Inside the Third Reich*, trans. by Richard and Clara Winston (London, 1995), p. 317.

13 Adolf Hitler, *Table Talk, 1941–1944: His Private Conversations*, trans. by Norman Cameron and R. H. Stevens, ed. by H. R. Trevor-Roper (London, 2000), p. 509.

14 Quoted in Wolfgang Wagner, *The History of German Aviation: The First Jet Aircraft* (Atglen, 1998), p. 121.

15 See Weindling, *Health, Race and Politics*, p. 492.

16 Quoted in Robert N. Proctor, *The Nazi War on Cancer* (Princeton, 1999), pp. 134–35.

17 Speer, *Third Reich*, p. 463.

18 Hitler, *Mein Kampf*, p. 379.

19 Ibid., p. 380.

20 Ibid., p. 384.

21 Ibid., p. 387.

22 Hitler, *Table Talk*, p. 720.

23 Ibid., p. 323.

24 See Brigitte Nagel, 'Die Welteislehre: Ihre Geschichte und ihre Bedeutung im "Dritten Reich"', in *Medizin, Naturwissenschaft, Technik und Nationalsozialismus: Kontinuitäten und Diskontinuitäten*, ed. by Christoph Meinel and Peter Voswinckel (Stuttgart, 1994), pp. 166–72.

25 See Klaus Hentschel, *The Einstein Tower: An Intertexture of Dynamic Construction, Relativity Theory, and Astronomy*, trans. by Ann M. Hentschel (Stanford, 1997), p. 137.

26 Hitler, *Table Talk*, p. 182.

27 Ibid., p. 3.

28 Ibid., pp. 59–60.

29 Ibid., p. 84.

30 Ibid., p. 85.

31 Ibid., p. 308.

Chapter 2: Germany the Science Mecca

1 See Diana Kormos Barkan, *Walther Nernst and the Transition to Modern Physical Science* (Cambridge, 1999), pp. 104ff.

2 Ibid., p. 105.

3 Quoted ibid., p. 106.

4 Ibid., p. 107.

5 Constance Reid, *Hilbert* (New York, 1996), p. 73.

6 Ibid., pp. 82–3.

7 For the story of Perkin, see Simon Garfield, *Mauve: How One Man Invented a Colour that Changed the World* (London, 2000).

8 Wolfgang Wetzel, *Naturwissenschaften und chemische Industrie in Deutschland: Voraussetzungen und Mechanismen ihres Aufstiegs im 19. Jahrhundert* (Stuttgart, 1991), pp. 50–51.

9 See Jeffrey Allen Johnson, 'The Academic-Industrial Symbiosis in German Chemical Research, 1905–1939', in *The German Chemical*

Industry in the Twentieth Century, ed. by John E. Lesch (Dordrecht, 2000), pp. 15ff.

10 Roy Porter, *The Greatest Benefit to Mankind: A Medical History of Humanity from Antiquity to the Present* (London, 1997), p. 325.

11 Thomas Mann, *Doctor Faustus* (London, 1999), p. 18.

Chapter 3: Fritz Haber

1 See Fritz Stern's essay 'Together and Apart: Fritz Haber and Albert Einstein', in Fritz Stern, *Einstein's German World* (London, 2000), pp. 59–164.

2 The best biographies of Haber are those by Dietrich Stoltzenberg, *Fritz Haber: Chemiker, Nobelpreisträger, Deutscher, Jude: Eine Biographie* (Weinheim, 1998); and Margit Szöllösi-Janze, *Fritz Haber 1868–1934: Eine Biographie* (Munich, 1998).

3 Stoltzenberg, *Fritz Haber*, p. 28; Szöllösi-Janze, *Fritz Haber*, p. 44.

4 Quoted in Stoltzenberg, *Fritz Haber*, pp. 352–3.

5 For the foundation of the Kaiser Wilhelm Society and its institutes see Kristie Macrakis, *Surviving the Swastika: Scientific Research in Nazi Germany* (New York, 1993), pp. 12ff.

6 See von Harnack's memorandum to the Kaiser of 21 November 1909. For a critique of this see ibid., pp. 15ff.

7 Fritz Haber, *Briefe an Richard Willstätter, 1910–1934*, ed. by Petra Werner and Angelika Irmscher (Berlin, 1995), p. 10.

8 See Vaclav Smil, 'Millennium Essay', *Nature* (29 July 1999), p. 415.

9 Stern, *Einstein's German World*, p. 119.

10 Jonathan Steinberg, *Yesterday's Deterrent: Tirpitz and the Birth of the German Battle Fleet* (London, 1965), p. 18.

11 I am grateful for the Richards quote to Steven Weinberg: see his 'The Growing Nuclear Danger', *New York Review of Books* (18 July 2002), p. 20. Weinberg draws an interesting parallel between the arms race of the Dreadnought and the nuclear arms race: 'As a scientist, I can recognize a kind of technological restlessness at work from the building of the Dreadnought to this year's Nuclear Posture Review.'

12 For details and background to the British battleship programme, see

Roger D. Thomas, *Dreadnoughts in Camera: Building the Dreadnoughts, 1905–1920* (Stroud, 1998).

13 See J. T. Sumida, 'British Capital Ship Design and Fire Control into the Dreadnought Era', *Journal of Modern History*, 51 (1979), pp. 205–30. See also Peter Padfield, *The Battleship Era* (London, 1972), especially the introduction.

14 Quoted in Walter Gratzer, *The Undergrowth of Science: Delusion, Self-Deception and Human Frailty* (Oxford, 2000), p. 171. Gratzer usefully cites the following on chauvinism and bigotry in science: Pierre Duhem's lectures, translated under the title *German Science* (La Salle, IL, 1991); and Harry W. Paul, *The Sorcerer's Apprentice: The French Scientist's Image of German Science 1840–1919* (Gainsville, FA, 1972).

15 Gratzer, *Undergrowth*, p. 162.

16 Quoted ibid., p. 162.

17 Quoted ibid., p. 163.

18 Quoted ibid., p. 167.

19 Quoted ibid., p. 170.

20 Quoted in Helge Kragh, *Quantum Generations: A History of Physics in the Twentieth Century* (Princeton, 1999), p. 145.

21 Quoted in Gratzer, *Undergrowth*, p. 174.

Chapter 4: The Poison Gas Scientists

1 Quoted in Jungk, *Thousand Suns*, p. 16.

2 See Christoph Gradmann, ' "Vornehmlich beängstigend" – Medizin, Gesundheit und chemische Kriegführung im deutschen Heer 1914–1918', in *Die Medizin und der Erste Weltkrieg*, ed. by Wolfgang U. Eckart and Christoph Gradmann (Pfaffenweiler, 1996), pp. 131–2.

3 Quoted in Malcolm Brown, *The Imperial War Museum Book of the Western Front* (London, 2001), p. 77.

4 Ibid., p. 77.

5 See SIPRI, *The Problems of Chemical and Biological Warfare*, vol. I: *The Rise of CB Weapons* (Stockholm, 1971), pp. 30ff.

6 Otto Hahn, *My Life*, trans. by Ernst Kaiser and Eithne Wilkins (London, 1970), p. 120.

7 Brown, *Western Front*, p. 248.

8 Quoted in G. B. Carter, *Chemical and Biological Defence at Porton Down 1916–2000* (London, 2000), p. 2.

9 Edward M. Spiers, *Chemical Warfare* (Chicago, 1986), p. 22.

10 Quoted in Adolf Wild von Hohenborn, *Briefe und Tagebuchaufzeichnungen des preußischen Generals als Kriegsminister und Truppenführer im Ersten Weltkrieg*, ed. by Helmut Reichold (Boppard am Rhein, 1986), p. 167, note 2.

11 Quoted in Dieter Martinetz, *Der Gaskrieg 1914/18: Entwicklung, Herstellung und Einsatz chemischer Kampfstoffe* (Bonn, 1996), p. 21.

12 See Stoltzenberg, *Fritz Haber*, p. 356.

13 Hahn, *My Life*, p. 122.

14 Ibid., pp. 122–3.

15 Ibid., p. 130.

16 Ibid., p. 130.

17 Ibid., pp. 130–31.

18 Ruth Lewin Sime, *Lise Meitner: A Life in Physics* (Berkeley, 1996), p. 58.

19 Quoted in Szöllösi-Janze, *Fritz Haber*, p. 479.

20 Ibid., p. 449.

21 Ibid., p. 450.

22 Quoted ibid., p. 450.

23 Quoted in Stoltzenberg, *Fritz Haber*, p. 317.

24 Stern, *Einstein's German World*, p. 124.

25 Quoted in Lewis S. Feuer, *Einstein and the Generations of Science* (New York, 1974), p. 49.

26 Quoted in Vaclav Smil, *Enriching the Earth: Fritz Haber, Carl Bosch and the Transformation of World Food Production* (Cambridge, MA, 2001) p. 231.

27 Quoted in Stoltzenberg, *Fritz Haber*, p. 575.

Chapter 5: The 'Science' of Racial Hygiene

1 For a discussion of the Krupp prize see Weindling, *Health, Race and Politics*, pp. 112–18.

2 Ibid., p. 86.

3 See Anthony Pagden, *The Fall of Natural Man: The American Indian and the Origins of Comparative Ethnology* (Cambridge, 1982), p. 104.

4 Robert N. Proctor, *Racial Hygiene: Medicine under the Nazis* (Cambridge, MA, 1988), p. 13.

5 Alfred Ploetz, *Die Tüchtigkeit unserer Rasse und der Schutz der Schwachen* (Berlin, 1895).

6 Quoted in Proctor, *Racial Hygiene*, p. 17.

7 Quoted ibid., p. 25.

8 Quoted ibid., pp. 23–4.

9 Quoted ibid., p. 28.

10 Quoted ibid., p. 29.

11 Niels C. Lösch, *Rasse als Konstrukt: Leben und Werk Eugen Fischers* (Frankfurt am Main, 1997), pp. 37–8.

12 Quoted in Ute Deichmann, *Biologists under Hitler*, trans. by Thomas Dunlap (Cambridge, MA, 1996), p. 323.

13 Proctor, *Racial Hygiene*, p. 36.

Chapter 6: Eugenics and Psychiatry

1 Quoted in Stefan Kühl, 'The Relationship between Eugenics and the so-called "Euthanasia Action" in Nazi Germany: A Eugenically Motivated Peace Policy and the Killing of the Mentally Handicapped during the Second World War', in *Science in the Third Reich*, ed. by Margit Szöllösi-Janze (Oxford, 2001), p. 188.

2 Ibid., p. 189.

3 Quoted in ibid., p. 201.

4 Quoted in Gratzer, *Undergrowth*, p. 292.

5 Michael Burleigh, *Ethics and Extermination: Reflections on Nazi Genocide* (Cambridge, 1997), p. 114.

6 A critical discussion of, and quotes from, Binding and Hoche's tract are provided in Rolf Winau, 'Die Freigabe der Vernichtung "lebensunwerten Lebens"', in *Medizin im 'Dritten Reich'*, ed. by Johanna Bleker and Norbert Jachertz (Cologne, 1993), pp. 162–74.

Chapter 7: Physics after the First War

1　For an assessment of classical physics in the First War see Kragh, *Quantum Generations*, pp. 130ff.

2　Quoted in Jungk, *Thousand Suns*, p. 15.

3　Russell McCormmach, *Night Thoughts of a Classical Physicist* (Cambridge, MA, 1982), p. 70.

4　E. J. Hobsbawm, *The Age of Empire 1875–1914* (London, 1987), p. 243.

5　Kurt Mendelssohn, *The World of Walther Nernst: The Rise and Fall of German Science* (London, 1973), p. 118.

6　Quoted in Barkan, *Walther Nernst*, p. 192.

7　Quoted ibid., p. 200.

8　Michael White and John Gribbin, *Einstein: A Life in Science* (London, 1993), p. 33

9　Ibid., p. 33.

10　Ibid., p. 105.

11　*The Times*, 7 November 1919.

12　Quoted in Kragh, *Quantum Generations*, p. 102.

13　Quoted in Kragh, *Quantum Generations*, p. 99.

14　Quoted ibid., p. 101.

15　Alan D. Beyerchen, *Scientists under Hitler: Politics and the Physics Community in the Third Reich* (New Haven, 1977), p. 83.

16　27 August 1920; anthologized in Klaus Hentschel and Ann M. Hentschel, eds, *Physics and National Socialism: An Anthology of Primary Sources* (Basel, 1996), pp. 1–2.

17　Mark Walker, *Nazi Science: Myth, Truth and the German Atomic Bomb* (New York, 1995), p. 9.

18　Beyerchen, *Scientists under Hitler*, p. 124.

19　Ibid., p. 102.

20　Quoted in ibid., p. 104.

21　Walker, *Nazi Science*, p. 6.

22　Ibid., p. 12.

23　Hentschel and Hentschel, *Physics and National Socialism*, p. 7.

24　Ibid., p. 8.

Chapter 8: German Science Survives

1 Quoted in Kragh, *Quantum Generations*, p. 146.

2 Quoted ibid., p. 140.

3 Quoted ibid., pp. 140–41.

4 See Helmut Werner, *From the Aratus Globe to the Zeiss Planetarium*, trans. by H. Degenhardt (Stuttgart, 1957): I am grateful to Dr Paul Murdin for alerting me to the Zeiss initiative.

5 Quoted in Richard Willstätter, *Aus meinem Leben: Von Arbeit, Muße und Freunden* (Weinheim, 1958), p. 235.

6 Ibid., p. 340.

7 Ibid., pp. 338–40.

8 Quoted in Abraham Pais, *Neils Bohr's Times: In Physics, Philosophy, and Polity* (Oxford, 1991), p. 4.

9 Quoted ibid., p. 11.

10 Otto Robert Frisch, *What Little I Remember* (Cambridge, 1979), p. 19.

11 Helmut Rechenberg and Gerald Wiemers, eds, *Werner Heisenberg 1901–1976: Schritte in die neue Physik* (Beucha, 2001), p. 25.

12 Quoted in David C. Cassidy, *Uncertainty: The Life and Science of Werner Heisenberg* (New York, 1992), p. 146.

13 Werner Heisenberg, *Encounters with Einstein and Other Essays on People, Places, and Particles* (Princeton, 1989), p. 38.

14 Quoted in Jungk, *Thousand Suns*, p. 18.

15 John Polkinghorne, *The Quantum World* (London, 1986), p. 14.

16 Victor F. Weisskopf, *The Joy of Insight: Passions of a Physicist* (New York, 1991), p. 31.

17 Ibid., p. 32.

18 Frisch, *What Little*, pp. 29ff.

19 Quoted in Richard Rhodes, *The Making of the Atomic Bomb* (New York, 1986), p. 318.

20 Frisch, *What Little*, p. 34.

Chapter 9: The Dismissals

1 Quoted in Ute Deichmann, 'The Expulsion of German-Jewish Chemists and Biochemists and their Correspondence with Colleagues in Germany after 1945: The Impossibility of Normalization?', in Szöllösi-Janze, *Science*, p. 247.

2 Quoted in Wilfried van der Will, 'Culture and the Organization of National Socialist Ideology 1933 to 1945', in *German Cultural Studies: An Introduction*, ed. by Rob Burns (Oxford, 1995), p. 111.

3 William L. Shirer, *The Rise and Fall of the Third Reich: A History of Nazi Germany* (London, 1961), p. 248.

4 Donald Prater, *Thomas Mann: A Life* (Oxford, 1995), p. 198.

5 Quoted in Sime, *Lise Meitner*, p. 137.

6 Quoted ibid., p. 138.

7 Quoted ibid., p. 139.

8 Max Born, *The Born–Einstein Letters: Correspondence between Albert Einstein and Max and Hedwig Born from 1916 to 1955, with Commentaries by Max Born*, trans. by Irene Born (London, 1971), p. 114.

9 Hans Krebs (with Anne Martin), *Reminiscences and Reflections* (Oxford, 1981), p. 61.

10 Ibid., p. 61.

11 Quoted in www.spartacus.schoolnet.co.uk/GERrust.htm.

12 Quoted in van der Will, 'Culture', p. 113.

13 Quoted in Nicholas Boyle, *Who Are We Now? Christian Humanism and the Global Market from Hegel to Heaney* (Edinburgh, 1998), p. 200.

14 Alfred Rosenberg, *Selected Writings*, ed. by Robert Pois (London, 1970), p. 89.

15 Ibid., p. 87.

16 Quoted in Sime, *Lise Meitner*, p. 143.

17 Quoted in Deichmann, 'Expulsion', in Szöllösi-Janze, *Science*, p. 250.

18 Quoted ibid., p. 251.

19 Gratzer, *Undergrowth*, p. 277.

20 Ibid., p. 277.

21 Hentschel and Hentschel, *Physics and National Socialism*, p. xliii.

22 Hahn, *My Life*, p. 145.

23 Quoted in Beyerchen, *Scientists under Hitler*, p. 1.

24 Helmuth Albrecht, 'Max Planck "Mein Besuch bei Adolf Hitler":
Anmerkungen zum Wert einer historischen Quelle', in Helmut
Albrecht, ed., *Naturwissenschaft und Technik in der Geschichte* (Stuttgart,
1993), p. 49.

25 Jane Caplan, *Government without Administration: State and Civil Service
in Weimar and Nazi Germany* (Oxford, 1988), p. 324.

26 Planck's account of his interview with Hitler, published in *Physikal-
ische Blätter* in 1947, is reprinted in full in Albrecht, 'Max Planck', in
Albrecht, *Naturwissenschaften*, pp. 41–2.

27 Ibid., p. 42.

28 Quoted in Stern, *Einstein's German World*, p. 157.

29 Quoted ibid., p. 159.

30 Quoted ibid., p. 159.

31 Anthologized in Léon Poliakov and Joself Wulf, eds, *Das Dritte Reich
und seine Denker: Dokumente* (Berlin, 1959), pp. 306–7.

32 Stern, *Einstein's German World*, p. 163.

33 For the exodus see Kragh, *Quantum Generations*, pp. 230 ff.; and
Beyerchen, *Scientists under Hitler*, pp. 40ff.

34 Hentschel and Hentschel, *Physics and National Socialism*, p. 145.

Chapter 10: Engineers and Rocketeers

1 Holger H. Herwig, 'Innovation Ignored: The Submarine Problem –
Germany, Britain, and the United States, 1919–1939', in *Military
Innovation in the Interwar Period*, ed. by Williamson Murray and Allan
R. Millett (Cambridge, 1998), p. 232.

2 Allison W. Sakville, 'The Development of the German U-boat Arm
1919–1935', unpublished Ph.D. thesis, University of Washington,
1963, pp. 173–4; cited in Herwig, 'Innovation Ignored', p. 233.

3 Williamson Murray, 'Strategic Bombing: The British, American, and
German Experiences', in Murray and Millett, *Military Innovation*,
p. 110.

4 Ulrich Albrecht, 'Military Technology and National Socialist Ideology', in *Science, Technology and National Socialism*, ed. by Monika
Renneberg and Mark Walker (Cambridge, 1994), p. 92.

5 Quoted in Andreas Heinemann-Grüder, '"Keinerlei Untergang":

German Armaments Engineers During the Second World War and in the Science of the Victorious Powers', in Renneberg and Walker, *Science, Technology*, p. 32.

6 For the background to this see Heinemann-Grüder, ' "Keinerlei Untergang" ', Renneberg and Walker, *Science, Technology*, pp. 30–50.

7 Ulrich Albrecht, Andres Heinemann-Grüder and Arend Wellmann, *Die Spezialisten: Deutsche Naturwissenschaftler und Techniker in der Sowjetunion nach 1945* (Berlin, 1992), pp. 19–20.

8 Quoted in Heinrich Adolf, 'Technikdiskurs und Technikideologie im Nationalsozialismus', *Geschichte in Wissenschaft und Unterricht*, 48 (7 August 1997), p. 442.

9 Peter P. Wegener, *The Peenemünde Wind Tunnels: A Memoir* (New Haven, 1996), p. 22.

10 Quoted in Freeman Dyson, *Disturbing the Universe* (New York, 1979), p. 108.

11 Ibid., p. 108.

12 Quoted in Neufeld, *The Rocket*, p. 22.

13 Quoted in Michael J. Neufeld, 'The Guided Missile and the Third Reich: Peenemünde and the Forging of a Technological Revolution', in Renneberg and Walker, *Science, Technology*, p. 57.

Chapter 11: Medicine under Hitler

1 Kater, *Doctors under Hitler*, p. 56.

2 Ibid., p. 57.

3 Klaus-Dieter Thomann, 'Dienst am Deutschtum: Der medizinische Verlag J. F. Lehmanns und der Nationalsozialismus', in Bleker and Jachertz, *Medizin*, p. 65.

4 Kater, *Doctors under Hitler*, p. 25.

5 Michael Hubentorf, 'Von der "freien Auswahl" zur Reichsärzteordnung – Ärztliche Standespolitik zwischen Liberalismus und Nationalsozialismus', in Bleker and Jachertz, *Medizin*, p. 45.

6 Quoted in Hendrik van den Bussche, 'Ärztliche Ausbildung und medizinische Studienreform im Nationalsozialismus', in ibid., p. 119.

7 Quoted in Kater, *Doctors under Hitler*, p. 183.

8 Werner Friedrich Kummel, 'Antisemitismus und Medizin im 19/20

Jahrhundert', in *Menschenverachtung und Opportunismus: Zur Medizin im Dritten Reich*, ed. by Jürgen Peiffer (Tübingen, 1992), p. 44.

9 Kater, *Doctors under Hitler*, p. 185.

10 Ibid., p. 188.

11 Hertha Nathorff, *Das Tagebuch: Berlin–New York, Aufzeichnungen 1933 bis 1945* (Munich, 1987). The following material on Nathorff's experiences is all taken from this diary.

12 Albert Einstein and Sigmund Freud, 'Why War?', in *An International Series of Open Letters*, vol. 2 (Paris, 1933), p. 47.

13 Quoted in Ronald W. Clark, *Freud: The Man and the Cause* (London, 1980), p. 490.

14 Ibid., p. 490.

15 *Zentralblatt für Psychotherapie*, January 1934; quoted and trans. in Clark, *Freud*, p. 493.

16 See Käthe Dräger, 'Bemerkungen zu den Zeitumständen und zum Schicksal der Psychoanalyse und der Psychotherapie in Deutschland zwischen 1933 and 1949', in *Psychoanalyse und Nationalsozialismus: Beiträge zur Bearbeitung eines unbewältigten Traumas*, ed. by Hans-Martin Lohmann (Frankfurt am Main, 1994), pp. 41–53.

17 Nathan G. Hale, *The Rise and Crisis of Psychoanalysis in the United States: Freud and the Americans, 1917–1985* (Oxford, 1995), p. 125.

18 Ibid., p. 115.

19 Quoted in Clark, *Freud*, p. 511.

20 Mitchell G. Ash, *Gestalt Psychology in German Culture 1890–1967: Holism and the Quest for Objectivity* (Cambridge, 1998), p. 332.

21 Quoted ibid., p. 334.

22 Quoted ibid., p. 335.

23 Ibid., p. 340.

Chapter 12: The Cancer Campaign

1 For a detailed assessment of basic biology in the Third Reich see Deichmann, *Biologists*.

2 Proctor, *Nazi War on Cancer*, p. 4.

3 Ibid., p. 36.

4 Ibid., p. 111.

5 Ibid., p. 46.
6 Quoted ibid., p. 124.
7 Ibid., p. 153.
8 Ibid., p. 173.
9 Ibid., p. 195.
10 Quoted ibid., p. 196.
11 Ibid., p. 249.

Chapter 13: Geopolitik *and* Lebensraum

1 G. Kearns, 'Halford Mackinder', *Geographers: Biobibliographical Studies*, 9 (1985), pp. 71–86.
2 See Henning Heske, 'German Geographical Research in the Nazi Period', *Political Geography Quarterly*, 5 (July 1986), pp. 267–81.
3 Hitler, *Mein Kampf*, p. 590.
4 Quoted in Robert Strausz-Hupé, *Geopolitics: The Struggle for Space and Power* (New York, 1972), p. 7.
5 Quoted in Geoffrey Parker, *Geopolitics: Past, Present and Future* (London, 1998), p. 30.
6 Quoted ibid., p. 30.
7 See Mechtild Rössler, 'Geography and Area Planning under National Socialism', in Szöllösi-Janze, *Science*, p. 62.
8 Quoted in David Thomas Murphy, *The Heroic Earth: Geopolitical Thought in Weimar Germany 1918–1933* (Kent, OH, 1997), p. 242.

Chapter 14: Nazi Physics

1 Hentschel and Hentschel, *Physics and National Socialism*, p. 100.
2 13 May 1933.
3 Hentschel and Hentschel, *Physics and National Socialism*, p. 49.
4 Walker, *Nazi Science*, p. 20.
5 Cassidy, *Uncertainty*, p. 336.
6 Ibid., p. 330.
7 Quoted in ibid., p. 349.
8 Quoted in Powers, *Heisenberg's War*, p. 41.
9 Quoted in Cassidy, *Uncertainty*, p. 379.

10 Ibid., p. 381.

11 Quoted ibid., p. 384.

12 Quoted ibid., p. 386.

13 Ibid., p. 389.

14 Quoted ibid., p. 390.

15 Quoted ibid., p. 393.

16 Ibid., pp. 391–92.

17 Quoted in M. Norton Wise, 'Pascual Jordan: Quantum Mechanics, Psychology, National Socialism', in Renneberg and Walker, *Science, Technology*, p. 233.

18 Peter P. Wegener, *The Peenemünde Wind Tunnels*, p. 28.

19 Quoted in Wise, 'Pascual Jordan', in Renneberg and Walker, *Science, Technology*, p. 224.

20 Quoted in Jungk, *Thousand Suns*, p. 20.

21 Quoted in Wise, 'Pascual Jordan', Renneberg and Walker, *Science, Technology*, p. 226.

22 Quoted ibid., p. 238.

23 Quoted ibid., p. 226.

24 Quoted ibid., p. 250.

Chapter 15: Himmler's Pseudo-science

1 Anthologized in Helmut Heiber, ed., *Reichsführer!* . . . *Briefe an und von Himmler* (Stuttgart, 1968), p. 41.

2 Quoted in Margit Szöllösi-Janze, 'National Socialism and the Sciences: Reflections, Conclusions and Historical Perspectives', in Szöllösi-Janze, *Science*, p. 1.

3 I am indebted to Simon Schama for the link between Tacitus and the Ahnenerbe: see his *Landscape and Memory* (London, 1995), pp. 82–3.

4 Szöllösi-Janze, 'National Socialism', *Science*, p. 4.

5 Quoted in Deichmann, *Biologists*, p. 255.

6 Gratzer, *Undergrowth*, p. 233.

7 Quoted in Szöllösi-Janze, 'National Socialism', *Science*, p. 4.

8 Heiber, *Reichsführer!*, p. 57.

9 Ibid., p. 57.

10 Quoted in Szöllösi-Janze, 'National Socialism', *Science*, p. 3.

11 Quoted in Proctor, *Nazi War on Cancer*, p. 138.
12 Heiber, *Reichsführer!*, p. 80.

Chapter 16: Deutsche Mathematik

1 Quoted in Helmut Lindner, '"Deutsche" und "gegentypische" Mathematik: Zur Begründung einer "arteigenen Mathematik" im "Dritten Reich" durch Ludwig Bieberbach', in *Naturwissenschaft, Technik und NS-Ideologie: Beiträge zur Wissenschaftsgeschichte des Dritten Reichs*, ed. by Herbert Mehrtens and Steffen Richter (Frankfurt am Main, 1980), p. 88.
2 Quoted by Karl Sabbagh, *Dr Riemann's Zeros* (London, 2002), p. 100.
3 Quoted in Ash, *Gestalt Psychology*, p. 335.
4 Léon Poliakov and Josef Wulf (eds.), *Das Dritte Reich und Seine Denker* (Berlin, 1959), p. 316.
5 Ibid., pp. 312–13.
6 Ibid., pp. 314–15.
7 Quoted in Reid, *Hilbert*, p. 208.
8 The story of these extraordinary escapes is told by John W. Dawson Jr in *Notices of the AMS*, 49:9 (October 2002), pp. 1,068ff.

Chapter 17: Fission Mania

1 Quoted in Sime, *Lise Meitner*, p. 136.
2 Quoted ibid., p. 136.
3 Leo Szilard, *His Version of the Facts: Selected Recollections and Correspondence*, ed. by Spencer R. Weart and Gertrud Weiss Szilard (Cambridge, MA, 1978), p. 14.
4 Ibid., p. 14.
5 Ibid., p. 17.
6 Quoted in Jungk, *Thousand Suns*, p. 53.
7 Quoted in Sime, *Lise Meitner*, p. 310.
8 Jungk, *Thousand Suns*, p. 65.
9 See Patricia Rife, *Lise Meitner and the Dawn of the Nuclear Age* (Boston, 1999), p. 163.
10 Quoted ibid., p. 165.

11 Quoted ibid., p. 166.

12 Quoted in Sime, *Lise Meitner*, p. 233.

13 Quoted ibid., p. 234.

14 Quoted ibid., p. 235.

15 *Die Naturwissenschaften*, 27:1 (1939), p. 15. Translation and parentheses supplied by the late Max Perutz, *I Wish I'd Made You Angry Earlier: Essays on Science, Scientists, and Humanity* (Oxford, 1998), p. 23.

16 Frisch, *What Little*, pp. 115–16.

17 Ibid., p. 116.

18 *New York Review of Books*, 20 February 1997.

19 *Nature*, 11 February 1939.

20 *Nature*, 18 February 1939.

21 Quoted in Rife, *Lise Meitner*, p. 202.

22 Pais, *Niels Bohr's Times*, p. 454.

23 Quoted in William Lanouette and Bela Szilard, *Genius in the Shadows: A Biography of Leo Szilard* (Chicago, 1992), p. 179.

24 J. A. Wheeler, 'The Discovery of Fission', *Physics Today*, 20 (November 1967), pp. 49–52; quoted with a corrective footnote [9] in Paul Lawrence Rose, *Heisenberg and the Nazi Atomic Bomb Project: A Study in German Culture* (Berkeley, 1998), pp. 83–4.

25 Szilard, *His Version*, pp. 69–70 and note 24.

26 The connections of the tale are skilfully drawn together by Jeremy Bernstein in *Hitler's Uranium Club: The Secret Recordings at Farm Hall* (New York, 2001), pp. 1–2, 14.

27 Quoted in Rhodes, *Atomic Bomb*, p. 290.

28 Harteck divulged his thinking on this matter in an interview with Joseph J. Ermenc, which is published in Ermenc's book *Atomic Bomb Scientists: Memoirs 1939–1945* (Westport, 1967), p. 97; see also Bernstein, *Uranium Club*, pp. 1–2, who adds the comment: 'His motivation appears to have been not too different from that of Willie Sutton, who was famous for saying that he robbed banks because that was where the money was.'

29 Quoted in Rhodes, *Atomic Bomb*, p. 305.

30 Quoted ibid., p. 314.

31 Quoted ibid., p. 314.

Chapter 18: World War II

1 See Bernstein, *Uranium Club*, p. 2.
2 Quoted in Cassidy, *Uncertainty*, p. 414.
3 Quoted ibid., p. 413.
4 Mark Walker, *German National Socialism and the Quest for Nuclear Power 1939–1949* (Cambridge, 1989), p. 17.
5 Bernstein, *Uranium Club*, p. 26.
6 Walker, *German National Socialism*, p. 18.
7 Ibid., pp. 19–20.
8 See Michael Schaaf, 'Der Physiochemiker Paul Harteck (1902–1985)', in CENSIS-REPORT–33–99 (Hamburg, 1999), p. 96.
9 See Bernstein, *Uranium Club*, p. 127; and comment in Horst Kant, 'Werner Heisenberg and the German Uranium Project', preprint for the Max-Planck-Institut für Wissenschaftsgeschichte (2001), p. 6.
10 For the original of the specific passage see Kant, 'Werner Heisenberg', p. 6; see also Cassidy, *Uncertainty*, p. 421.
11 Quoted in Cassidy, *Uncertainty*, p. 422.
12 See Kant, 'Werner Heisenberg', p. 7.
13 *Die Naturwissenschaften*, 27 (1939), pp. 402–10.
14 Rose, *Heisenberg*, pp. 97–98.
15 Frisch, *What Little*, p. 125.
16 Ibid., p. 126.
17 Ibid., p. 126.
18 Quoted in Pais, *Niels Bohr's Times*, p. 494.
19 Quoted ibid., p. 494.

Chapter 19: Machines of War

1 Monika Renneberg and Mark Walker, 'Scientists, Engineers and National Socialism', in Renneberg and Walker, *Science, Technology*, pp. 1–2.
2 Ibid., p. 2.
3 Neufeld, *The Rocket*, p. 120.
4 See for example Richard Overy, *Why the Allies Won* (London, 1995),

pp. 208ff.; and Gerhard L. Weinberg, *A World at Arms: A Global History of World War II* (Cambridge, 1994), pp. 536ff.

5 Overy, *Why the Allies Won*, pp. 217–18.

6 Ibid., p. 212.

7 For extensive details of the *Bismarck* see Ulrich Elfrath and Bodo Herzog, *The Battleship Bismarck: A Documentary in Words and Pictures*, trans. by Edward Force (West Chester, 1989).

8 *Völkischer Beobachter*, 15 February 1939.

9 See discussion in Alan Beyerchen, 'From Radio to Radar: Interwar Military Adaptation to Technological Change in Germany, the United Kingdom, and the United States', in Murray and Millett, *Military Innovation*, p. 295.

10 Quoted in Wagner, *German Aviation*, p. 22.

11 Clay Blair, *Hitler's U-Boat War: The Hunted 1942–1945* (London, 1999), p. 512.

12 Ibid., p. 512.

13 Quoted ibid., p. 314.

14 Neufeld, *The Rocket*, p. 70.

15 Quoted ibid., p. 126.

16 Quoted ibid., p. 137.

Chapter 20: Radar

1 David Zimmerman, *Britain's Shield: Radar and the Defeat of the Luftwaffe* (Stroud, 2001), p. xiii.

2 An authoritative account of early German radar can be found in Harry von Kroge, *GEMA: Birthplace of German Radar and Sonar*, trans. and ed. by Louis Brown (Bristol, 2000).

3 GEMA is an acronym for *Gesellschaft für elektroakustische und mechanische Apparate*.

4 Zimmerman, *Britain's Shield*, p. 57.

5 Robert Watson-Watt's entry on radar in the *Chambers Encyclopaedia* (1973), vol. XI, p. 423.

6 Kroge, *GEMA*, p. 159.

7 Norbert Wiener, *Atlantic Monthly*, January 1947.

8 Robert Buderi, *The Invention that Changed the World: The Story of Radar from War to Peace* (London, 1998), p. 202.

9 Zimmerman, *Britain's Shield*, p. xiii.

10 For Rosbaud see Arnold Kramish, *The Griffin* (London, 1987).

11 Winston S. Churchill, *The Second World War* (London, 1949–54), vol. I, p. 645.

12 Ibid., vol. II, p. 16.

13 Ibid., vol. II, p. 149.

14 Quoted in Robert Harris and Jeremy Paxman, *A Higher Form of Killing: The Secret History of Gas and Germ Warfare* (London, 1982), p. 112.

15 Ibid., p. 112.

16 Churchill, *The Second World War*, vol. II, pp. 407–8.

17 Buderi, *Invention*, p. 91.

18 Kroge, *GEMA*, p. 43.

19 Buderi, *Invention*, p. 85.

20 Ibid., p. 86.

21 Quoted ibid., p. 211.

Chapter 21: Codes

1 Quoted in Simon Singh, *The Code Book: The Secret History of Codes and Codebreaking* (London, 2000), p. 142.

2 Quoted in David Kahn, *Seizing the Enigma: The Race to Break the German U-Boat Codes, 1939–1943* (London, 1996), p. 31.

3 Ibid., p. 32.

4 The intriguing story of the Polish decoders has been told in a variety of books. See Singh's *Code Book* and also Rudolf Kippenhahn, *Code Breaking: A History and Exploration*, trans. by Ewald Osers (London, 1999), pp. 172–80.

5 Singh, *Code Book*, p. 155.

6 Erich Hüttenhain, 'Erfolge und Mißerfolge der deutschen Chiffrier-dienste im Zweiten Weltkrieg', in *Die Funkaufklärung und ihre Rolle im Zweiten Weltkrieg*, ed. by Jürgen Rohwer and Eberhard Jäckel (Stuttgart, 1979), pp. 101–2.

7 R. A. Haldane, *The Hidden World* (London, 1976), p. 119.

8 For a remarkable history of German radio intelligence see Heinz Bonatz, *Die deutsche Marine-Funkaufklärung 1914–1945* (Darmstadt, 1970).

9 Stephen Budiansky, *Battle of Wits: The Complete Story of Codebreaking in World War II* (London, 2000), pp. 286ff.

10 Quoted in Andrew Hodges, *Alan Turing: The Enigma* (London, 1992), p. 165.

11 Singh, *Code Book*, p. 192; also Michael Smith and Ralph Erskine, *Action this Day* (New York, 2001), p. 371.

12 Interview with Dr Derek Taunt, former Bletchley Park cryptanalyst, 12 February 2003.

13 Konrad Zuse, 'Some Remarks on the History of Computing in the Twentieth Century', in *A History of Computing in the Twentieth Century: A Collection of Essays*, ed. by N. Metropolis, J. Howlett and Gian-Carlo Rota (New York, 1980), p. 611.

14 Ibid., p. 611.

15 Friedrich L. Bauer, 'The Early Development of Digital Computing in Central Europe', in Metropolis et al., *History of Computing*, p. 518.

Chapter 22: Copenhagen

1 The letter, in the keeping of Professor Helmut Rechenberg of the Max Planck Institute for Physics in Munich, was referred to by Michael Frayn at the Jesus College 'Copenhagen Colloquium', 22–24 November 2002.

2 This differs from the account of Powers, who, with no accompanying source, writes that Heisenberg left Berlin on the Sunday and arrived in Copenhagen at 6.15 on the Monday evening (*Heisenberg's War*, p. 121).

3 Walker, *Nazi Science*, pp. 147–8.

4 Cassidy, using an account by Stefan Rozental, recounted in Walker, *German National Socialism*, p. 224, places the incident in March 1941, whereas Rose places it in September 1941; see Rose, *Heisenberg*, p. 272, note 9.

5 Walker, *Nazi Science*, pp. 149–50.

6 Werner Heisenberg, *Physics and Beyond: Encounters and Conversations*, trans. by Arnold J. Pomerans (London, 1971), p. 201.

7 Pais, *Niels Bohr's Times*, p. 484.

8 Quoted in Powers, *Heisenberg's War*, p. 124.

9 Quoted in Jungk, *Thousand Suns*, p. 101.

10 Powers, *Heisenberg's War*, p. 124; for the source quoted see p. 510, note 17.

11 Quoted ibid, p. 126.

12 Rose, *Heisenberg*, p. 281.

13 Quoted in Jungk, *Thousand Suns*, p. 101.

14 Niels Bohr Archive (NBA), document 1.

15 NBA document 7.

16 NBA document 9.

17 NBA document 11a.

18 NBA document 11a.

19 Walker, *German National Socialism*, p. 222.

Chapter 23: Speer and Heisenberg

1 Quoted in Cassidy, *Uncertainty*, p. 443.

2 Quoted ibid., p. 443.

3 Quoted in David Irving, *The German Atomic Bomb: The History of Nuclear Research in Nazi Germany* (New York, 1967), p. 108.

4 Quoted ibid., p. 110.

5 Quoted in Cassidy, *Uncertainty*, p. 450.

6 Quoted ibid., pp. 450–51.

7 Speer, *Third Reich*, p. 272.

8 Joachin C. Fest, *Speer: The Final Verdict*, trans. by Ewald Osers and Alexandra Dring (London, 2001), p. 129.

9 Speer, *Third Reich*, p. 279.

10 Ibid., p. 276.

11 Ibid., p. 278.

12 Ibid., p. 33.

13 Ibid., p. 296.

14 Ibid., p. 297.

15 Ibid., p. 297.

16 Ibid., p. 315.

17 Quoted in Rose, *Heisenberg*, p. 180; see also Irving, *German Atomic Bomb*, p. 120.

18 Speer, *Third Reich*, p. 316.

19 Ibid., p. 316.

20 Ibid., p. 317.

21 Leslie R. Groves, *Now It Can Be Told: The Story of the Manhattan Project* (New York, 1983), pp. 296–7.

22 Speer, *Third Reich*, p. 319.

23 Quoted in Rhodes, *Atomic Bomb*, p. 405.

24 Ermenc, *Scientists: Memoirs*, p. 115.

25 Rhodes, *Atomic Bomb*, p. 567.

26 Cassidy, *Uncertainty*, p. 464.

27 Ibid., p. 472.

28 Quoted ibid., p. 473.

29 Hendrik Casimir, *Haphazard Reality: Half a Century of Science* (New York, 1983), p. 208.

30 M. Dresden, *H. A. Kramers: Between Tradition and Revolution* (New York, 1987), p. 485.

31 For the main details of this story I am obliged to Jeremy Bernstein's unpublished paper 'Heisenberg in Poland' (2003).

32 Quoted ibid.

33 Quoted ibid.

34 Quoted in Cassidy, *Uncertainty*, p. 470.

Chapter 24: Haigerloch and Los Alamos

1 Heisenberg, *Physics and Beyond*, pp. 187ff.

2 Quoted in Jungk, *Thousand Suns*, p. 155.

3 See Powers, *Heisenberg's War*, notes for chapter 34, pp. 563ff., citing OSS and CIA papers, including Berg's own notes. See also Louis Kaufman et al., *Moe Berg: Athlete, Scholar, Spy* (New York, 1974).

4 Quoted in Powers, *Heisenberg's War*, p. 400.

5 Ibid., p. 565, notes 21, 22.

6 Quoted ibid., p. 402.

7 Letter from Martin Heisenberg to Powers quoted ibid., p. 404.

8 Haakon Chevalier, *Oppenheimer: The Story of a Friendship* (London, 1966), p. 11. I am grateful to Jeremy Bernstein for drawing attention to this description; Bernstein says that it accorded with his memory of the man twenty years later.

9 Groves, *Now It Can Be Told*, p. 207.

10 Samuel A. Goudsmit, *Alsos* (New York, 1996), p. 15.

11 Ibid., pp. 47–8.

12 Ibid., pp. 48–9.

13 Antony Beevor, *Berlin: The Downfall 1945* (London, 2002), pp. 324–5.

14 Quoted in Goudsmit, *Alsos*, p. 113.

Chapter 25: Slave Labour at Dora

1 Quoted in Neufeld, *The Rocket*, p. 187.

2 Ibid., p. 187.

3 Quoted ibid., p. 192.

4 Martin Middlebrook, *The Peenemünde Raid 17–18 August 1943* (London, 2000), p. 223.

5 Quoted in Neufeld, *The Rocket*, p. 210.

6 Quoted ibid., p. 211.

7 Quoted ibid., p. 212.

8 Ibid., p. 212.

9 Ibid., pp. 207–8.

Chapter 26: The 'Science' of Extermination and Human Experiment

1 Burleigh, *Ethics*, p. 117.

2 Ibid., p. 116.

3 Christiane Rothmaler, 'Zwangssterilisationen nach dem Gesetz zur Verhütung erbkranken Nachwuchses', in Bleker and Jachertz, *Medizin*, p. 137.

4 For a recent scholarly account of the Wannsee Protocol see Mark Roseman, *The Villa, the Lake, the Meeting: Wannsee and the Final Solution* (London, 2002).

5 Jean-Claude Pressac and Robert Jan van Pelt, 'The Machinery of Mass Murder at Auschwitz', in *Anatomy of the Auschwitz Death Camp*,

ed. by Yisrael Gutman and Michael Berenbaum (Bloomington, 1994), pp. 183–245. See also Debórah Dwork and Robert Jan van Pelt, *Auschwitz* (New York, 2002), pp. 176–7; and Robert Jan van Pelt and Carroll William Westfall, *Architectural Principles in the Age of Historicism* (New Haven, 1993), pp. 150–51.

6   Pressac and van Pelt, 'Machinery', in Gutman and Berenbaum, *Anatomy*, p. 209.

7   T. C. Bridges and H. Hessell Tiltman, *Master Minds of Modern Science* (London, 1930), p. 104.

8   Alexander Mitscherlich and Fred Mielke, *The Death Doctors*, trans. by James Cleugh (London, 1962), p. 17.

9   Christian Pross, 'Nazi Doctors, German Medicine and Historical Truth', in *The Nazi Doctors and the Nuremberg Code: Human Rights in Human Experimentation*, ed. by George J. Annas and Michael A. Grodin (New York, 1992), p. 36.

10 See Benno Müller-Hill, *Murderous Science: Elimination by Scientific Selection of Jews, Gypsies, and Others in Germany, 1933–1945*, trans. by George R. Fraser (Oxford, 1988), p. 20 and passim; see also Deichmann, *Biologists*, p. 229.

11 Robert N. Proctor, 'Nazi Doctors, Racial Medicine, and Human Experimentation', in Annas and Grodin, *Nazi Doctors*, p. 18.

12 See for example Robert Jay Lifton, *The Nazi Doctors: Medical Killing and the Psychology of Genocide* (New York, 1986); Kater, *Doctors under Hitler*; Hendrik van den Bussche, ed., *Medizinische Wissenschaft im 'Dritten Reich': Kontinuität, Anpassung und Opposition an der Hamburger Medizinischen Fakultät* (Berlin, 1989).

13 *Trials of War Criminals Before the Nuremberg Military Tribunals under Control Council Law 10*, vol. I (Washington, DC: Superintendent of Documents, U.S. Government Printing Office, 1950); Military Tribunal, Case 1, United States v. Karl Brandt et al., October 1946–April 1949, pp. 27–74.

14 Details are outlined in ibid. passim.

15 Telford Taylor, 'Opening Statement of the Prosecution December 9, 1946', in Annas and Grodin, *Nazi Doctors*, pp. 72–3.

16 Deichmann, *Biologists*, p. 257.

17 Quoted in Paul Hoedeman, *Hitler or Hippocrates: Medical Experiments*

and Euthanasia in the Third Reich, trans. by Ralph de Rijke (Lewes, 1991), p. 168.

18 Mitscherlich and Mielke, *Death Doctors*, p. 71, note 1; see also Taylor, 'Opening Statement', p. 75.

19 Hoedeman, *Hitler or Hippocrates*, p. 23.

20 Taylor, 'Opening Statement', p. 78.

21 Quoted in Hoedeman, *Hitler or Hippocrates*, p. 181.

22 Quoted ibid., p. 147.

23 Taylor, 'Opening Statement', p. 81.

24 Till Bastian, *Furchtbare Ärzte: Medizinische Verbrechen im Dritten Reich* (Munich, 1995), pp. 78–9.

25 Quoted in Taylor, 'Opening Statement', p. 81.

Chapter 27: The Devil's Chemists

1 Primo Levi, *If This is a Man; The Truce*, trans. by Stuart Woolf (London, 1990), p. 111.

2 Ibid., p. 15.

3 Ibid., p. 142.

4 Ibid., p. 145.

5 The story that follows is derived from Anthony N. Stranges, 'Germany's Synthetic Fuel Industry, 1930–1945', in Lesch, *The German Chemical Industry*, pp. 147ff.

6 Quoted in Dwork and van Pelt, *Auschwitz*, p. 205.

7 Quoted ibid., p. 232.

8 Quoted ibid., p. 233.

9 Stranges, 'Synthetic', in Lesch, *The German Chemical Industry*, p. 210.

10 Carole Angier, *The Double Bond: Primo Levi, a Biography* (London, 2002), p. 303.

11 Quoted in Joseph Borkin, *The Crime and Punishment of I.G. Farben* (London, 1979), p. 138.

12 Quoted ibid., p. 143.

13 Quoted in Dwork and van Pelt, *Auschwitz*, p. 234.

14 Quoted in Borkin, *Crime and Punishment*, p. 155.

15 Levi, *If This is a Man*, p. 47.

16 Borkin, *Crime and Punishment* p. 162.

Chapter 28: Wonder Weapons

1 See for example Friedrich Hansen, *Biologische Kriegsführung im Dritten Reich* (Frankfurt am Main, 1993), pp. 126–7.

2 Dieter Herwig and Heinz Rode, *Luftwaffe: Secret Projects – Strategic Bombers 1935–1945*, trans. by Elke and John Weal (Leicester, 2000).

3 Geoffrey Brooks, *Hitler's Terror Weapons: From VI to Vimana* (London, 2002).

4 Speer, *Slave State*, p. 146.

5 Ibid., p. 147.

6 Ibid., p. 148.

7 Ulrich Albrecht, 'Military Technology and National Socialist Ideology', in Renneberg and Walker, *Science, Technology*, p. 107.

8 Quoted ibid., p. 111.

9 Ibid., p. 113.

10 Ibid., p. 120.

11 Quoted in Ute Deichmann and Benno Müller-Hill, 'Biological Research at Universities and Kaiser Wilhelm Institutes in Nazi Germany', in ibid., p. 180.

12 See Geoffrey Brooks, *Hitler's Nuclear Weapons: The Development and Attempted Deployment of Radiological Armaments by Nazi Germany* (London, 1992), pp. 121–6.

13 Quoted ibid., p. 121.

14 Quoted ibid., p. 121.

15 The story of Gerhard Schrader and the development of tabun and sarin is told in Harris and Paxman, *A Higher Form of Killing*, pp. 53ff.

16 Ibid., p. 62.

17 For precise numbers and analysis of both the V1 and V2 campaigns see Roy Irons, *Hitler's Terror Weapons: The Price of Vengeance*, with a foreword by Richard Overy (London, 2002). See also Ken Wakefield et al., *The Blitz: Then and Now*, vol. 3 (London, 1990).

18 Quoted in Irons, *Hitler's Terror Weapons*, p. 161.

19 Quoted in ibid., pp. 161–2.

Chapter 29: Farm Hall

1 The definitive edition of the secret recordings is Bernstein, *Uranium Club*. See also the earlier edition, *Operation Epsilon: The Farm Hall Transcripts*, with an introduction by Sir Charles Frank (Berkeley, 1993).
2 Unpublished interview with Professor Marcial Echenique, 24 November 2002.
3 Bernstein, *Uranium Club*, p. 78.
4 Ibid., p. 229.
5 Ibid., pp. 115ff.
6 BBC Written Archives Centre; quoted ibid., p. 357.
7 Jungk, *Thousand Suns*, p. 102.
8 Bernstein, *Uranium Club*, p. 122, note 53.
9 Ibid., p. 137.
10 Ibid., p. 138.
11 Ibid., pp. 352–3.
12 The English and German texts are reproduced in ibid., pp. 169ff., along with a technical commentary.
13 Ibid., p. 185.
14 Powers, *Heisenberg's War*, p. 451.
15 Sime, *Lise Meitner*, p. 320.
16 Bernstein, *Uranium Club*, p. 185.

Chapter 30: Heroes, Villains and Fellow Travellers

1 Quoted in Sime, *Lise Meitner*, pp. 321–2.
2 Quoted in Bernstein, *Uranium Club*, p. 214.
3 Walker, *Nazi Science*, p. 269.
4 Quoted in Rose, *Heisenberg*, p. 300.
5 Gerhard Hildebrandt, 'Max von Laue, der "Ritter ohne Furcht und Tadel"', in *Berlinische Lebensbilder*, vol. I: *Naturwissenschaftler*, ed. by Wilhelm Treue and Gerhard Hildebrandt (Berlin, 1987), p. 233.
6 Gerda Freise, 'Der Nobelpreisträger Professor Dr Heinrich Wieland: Zivilcourage in der Zeit des Nationalsozialismus', in *Hochverrat?: Die 'Weiße Rose' und ihr Umfeld*, ed. by Rudolf Lill and Michael Kißener (Constance, 1993), pp. 135–57.

7 Ibid., pp. 152–4.

8 Meitner to Hahn, 27 June 1945. Printed in F. Krafft, *Im Schatten der Sensation* (Weinheim, 1981), p. 181ff.; quoted in Rose, *Heisenberg*, pp. 301–2.

9 Quoted ibid., pp. 301–2.

10 Quoted ibid., p. 302.

11 Mark Walker, 'Twentieth Century German Science', in *Science in the Twentieth Century*, ed. by John Krige and Dominique Pestre (Amsterdam, 1997), p. 809.

12 Ibid., p. 810.

13 Lawrence Badash, 'Otto Hahn, Science, and Social Responsibility', in *Otto Hahn and the Rise of Nuclear Physics*, ed. by William R. Shea (Dordrecht, 1983), p. 176.

14 Thomas Powers's interview with John A. Wheeler, Princeton, 5 March 1990: *Heisenberg's War*, p. 459.

15 Michael Schaaf, *Heisenberg, Hitler und die Bombe: Gespräche mit Zeitzeugen* (Berlin, 2001), p. 128.

Chapter 31: Scientific Plunder

1 Interview with Professor Austyn Mair, Cambridge, 28 November 2002.

2 Quoted in Tom Bower, *The Paperclip Conspiracy: The Battle for the Spoils and Secrets of Nazi Germany* (London, 1987), p. 66.

3 Quoted in Burghard Ciesla and Helmuth Trischler, 'Legitimation through Use: Rocket and Aeronautic Research in the Third Reich and the USA', in *Science and Ideology: A Comparative History*, ed. by Mark Walker (London, 2003), p. 157.

4 Bower, *Paperclip Conspiracy*, passim; and Neufeld, *The Rocket*, pp. 270ff.

5 Albrecht et al., *Die Spezialisten*, p. 12.

6 Werner Albring, *Gorodomlia: Deutsche Raketenforscher in Rußland* (Hamburg, 1991).

7 Ibid., pp. 126–7.

Chapter 32: Nuclear Postures

1 Sean Dennis Cashman, *America, Roosevelt, and World War II* (New York, 1993), pp. 368–69.

2 John Rawls, 'Fifty Years after Hiroshima', in John Rawls, *Collected Papers*, ed. by Samuel Freeman (Cambridge, MA, 1999), pp. 565–72.

3 Szilard, *His Version*, p. 211.

4 Dyson, *Disturbing the Universe*, p. 51.

5 Edward Teller (with Judith L. Schoolery), *Memoirs: A Twentieth-Century Journey in Science and Politics* (Cambridge, MA, 2001), p. 283.

6 Sir Joseph Rotblat Lecture at the Royal United Services Institute for Defence Studies, January 2003; see www.guardian.co.uk/nuclear.

7 Perutz, *I Wish I'd Made You Angry Earlier*, p. 58.

8 See Teller's *Memoirs*; also Richard Lourie, *Sakharov: A Biography* (Brandeis, 2002); and Andrei Sakharov, *Memoirs*, trans. by Richard Lourie (London, 1990).

9 Quoted in Joseph Rotblat, 'The Tale of Dr Strangelove vs. St Sakharov', *Times Higher Education Supplement*, 26 July 2002.

10 Quoted in David Holloway, *Stalin and the Bomb: The Soviet Union and Atomic Energy, 1939–1956* (New Haven, 1994), pp. 316–17.

11 Richard Rhodes, *Dark Sun: The Making of the Hydrogen Bomb* (New York, 1996).

12 Teller, *Memoirs*, p. 257, note 40.

13 Quoted ibid., p. 163.

14 Ibid., p. 581.

15 Ibid., p. 572.

16 Rotblat, 'Tale'.

17 Ibid.

18 Dyson, *Disturbing the Universe*, p. 92.

19 See David Miller, *The Cold War: A Military History* (London, 2001), pp. 349ff.

20 Rhodes, *Atomic Bomb*, p. 784.

21 John Lewis Gaddis, *Now We Know: Rethinking Cold War History* (Oxford, 1997), p. 261.

22 'U.S. Arsenal Stocked with 1,700 Mininukes', *Oakland Tribune*, 25

February 2003; 'Development of New Nukes Receives OK', *Tri-valley Herald*, 25 February 2003.

Chapter 33: Uniquely Nazi?

1 *Nature*, 30 November 2000, p. 504.

2 See for example Neil Gregor, *Daimler-Benz in the Third Reich* (New Haven, 1998).

3 Hans H. Amtmann, *The Vanishing Paperclips: America's Aerospace Secret* (Boston, 1988), p. 94.

4 Neufeld, 'The Guided Missile', p. 71.

5 Ibid., p. 71.

6 David Reynolds, *One World Divisible: A Global History Since 1945* (London, 2000), p. 497.

7 Sources: *Encyclopaedia Britannica*; Centers for Disease Control and Prevention; The Associated Press.

8 William Blum, *Rogue State: A Guide to the World's Only Superpower* (London, 2002).

9 Robert N. Proctor, *Value-Free Science? Purity and Power in Modern Knowledge* (Cambridge, MA, 1991), p. 239.

10 John May, *The Greenpeace Book of the Nuclear Age: The Hidden History, the Human Cost* (London, 1989), pp. 82–3.

11 Willem A. Veenhoven et al., eds, *Case Studies on Human Rights and Fundamental Freedoms: A World Survey*, vol. I (The Hague, 1975), p. 31.

12 Ibid., p. 30.

13 George J. Annas and Michael A. Grodin, 'Introduction', in Annas and Grodin, *Nazi Doctors*, p. 5.

14 Reynolds, *One World Divisible*, p. 513.

15 Interview with Brenner in John Cornwell, 'Gene Spleen', *Sunday Times Magazine*, 26 July 1992, p. 19.

16 Ibid., p. 20.

17 For the monoclonal antibodies 'scandal' see Soraya de Chadarevian, *Designs for Life: Molecular Biology after World War II* (Cambridge, 2002), pp. 353–5.

18 *Sunday Times*, 13 October 2002, p. 9.

19 *Nature*, February 1997, p. 810.
20 Richard Lewontin, *It Ain't Necessarily So: The Dream of the Human Genome and Other Illusions* (London, 2000), pp. 286–7.

Chapter 34: Science at War Again

1 *Nature*, November 2001, p. 386.
2 J. J. Smart and Bernard Williams, *Utilitarianism: For and Against* (Cambridge, 1998), pp. 98ff.
3 *Atlantic Monthly*, January 1947.
4 Apart from the Disney movie *The Insider* there is a wealth of online material on Wigand. This quote is taken from a University of Houston, Texas, website: www.uh.edu/admin/media/nr/102002/whistleblower/102302.html.
5 Helga Nowotny, Peter Scott and Michael Gibbons, *Re-Thinking Science: Knowledge and the Public in an Age of Uncertainty* (Cambridge, 2001).
6 Daniel S. Greenberg, *Science, Money, and Politics: Political Triumph and Ethical Erosion* (Chicago, 2001), p. 50.

Select Bibliography

Adolf, Heinrich, 'Technikdiskurs und Technikideologie im Nationalsozialismus', *Geschichte in Wissenschaft und Unterricht*, 48 (7 August 1997)

Albrecht, Helmuth, ed., *Naturwissenschaft und Technik in der Geschichte* (Stuttgart, 1993)

Albrecht, Ulrich, Andreas Heinemann-Grüder and Arend Wellmann, *Die Spezialisten: Deutsche Naturwissenschaftler und Techniker in der Sowjetunion nach 1945* (Berlin, 1992)

Albring, Werner, *Gorodomlia: Deutsche Raketenforscher in Rußland* (Hamburg, 1991)

Amtmann, Hans H., *The Vanishing Paperclips: America's Aerospace Secret* (Boston, 1988)

Angier, Carole, *The Double Bond: Primo Levi, a Biography* (London, 2002)

Annas, George J., and Michael A. Grodin, *The Nazi Doctors and the Nuremberg Code: Human Rights in Human Experimentation* (New York, 1992)

Ash, Mitchell G., *Gestalt Psychology in German Culture 1890–1967: Holism and the Quest for Objectivity* (Cambridge, 1998)

Barkan, Diana Kormos, *Walther Nernst and the Transition to Modern Physical Science* (Cambridge, 1999)

Bastian, Till, *Furchtbare Ärzte: Medizinische Verbrechen im Dritten Reich* (Munich, 1995)

Beevor, Antony, *Berlin: The Downfall 1945* (London, 2002)

Bernstein, Jeremy, ed., *Hitler's Uranium Club: The Secret Recordings at Farm Hall* (New York, 2001)

Beyerchen, Alan D., *Scientists under Hitler: Politics and the Physics Community in the Third Reich* (New Haven, 1977)

Blair, Clay, *Hitler's U-Boat War: The Hunted 1942–1945* (London, 1999)

Bleker, Johanna, and Norbert Jachertz, eds, *Medizin im 'Dritten Reich'* (Cologne, 1993)

Blum, William, *Rogue State: A Guide to the World's Only Superpower* (London, 2002)

Bonatz, Heinz, *Die deutsche Marine-Funkaufklärung 1914–1945* (Darmstadt, 1970)

Borkin, Joseph, *The Crime and Punishment of I. G. Farben* (London, 1979)

Born, Max, *The Born–Einstein Letters: Correspondence between Albert Einstein and Max and Hedwig Born from 1916 to 1955, with Commentaries by Max Born*, trans. by Irene Born (London, 1971)

Bower, Tom, *The Paperclip Conspiracy: The Battle for the Spoils and Secrets of Nazi Germany* (London, 1987)

Boyle, Nicholas, *Who Are We Now? Christian Humanism and the Global Market from Hegel to Heaney* (Edinburgh, 1998)

Bridges, T. C., and H. Hessell Tiltman, *Master Minds of Modern Science* (London, 1930)

Brooks, Geoffrey, *Hitler's Nuclear Weapons: The Development and Attempted Deployment of Radiological Armaments by Nazi Germany* (London, 1992)

——, *Hitler's Terror Weapons: From V1 to Vimana* (London, 2002)

Brown, Malcolm, *The Imperial War Museum Book of the Western Front* (London, 2001)

Buderi, Robert, *The Invention that Changed the World: The Story of Radar from War to Peace* (London, 1998)

Budiansky, Stephen, *Battle of Wits: The Complete Story of Codebreaking in World War II* (London, 2000)

Bullock, Alan, *Hitler: A Study in Tyranny* (London, 1954)

——, *Hitler and Stalin: Parallel Lives* (London, 1991)

Burleigh, Michael, *Death and Deliverance: 'Euthanasia' in Germany c. 1900–1945* (Cambridge, 1994)

——, *Ethics and Extermination: Reflections on Nazi Genocide* (Cambridge, 1997)

——, *The Third Reich: A New History* (London, 2000)

——, and Wolfgang Wippermann, *The Racial State: Germany 1933–1945* (Cambridge, 1991)

Caplan, Jane, *Government without Administration: State and Civil Service in Weimar and Nazi Germany* (Oxford, 1988)

Carter, G. B., *Chemical and Biological Defence at Porton Down 1916–2000* (London, 2000)

Cashman, Sean Dennis, *America, Roosevelt, and World War II* (New York, 1993)

Casimir, Hendrik, *Haphazard Reality: Half a Century of Science* (New York, 1983)

Cassidy, David C., *Uncertainty: The Life and Science of Werner Heisenberg* (New York, 1992)

Chadarevian, Soraya de, *Designs for Life: Molecular Biology after World War II* (Cambridge, 2002)

Chevalier, Haakon, *Oppenheimer: The Story of a Friendship* (London, 1966)

Churchill, Winston S., *The Second World War*, 6 vols (London, 1949–54)

Clark, Ronald W., *Freud: The Man and the Cause* (London, 1980)

Cocks, Geoffrey, *Psychotherapy in the Third Reich: The Göring Institute* (New York, 1985)

Deichmann, Ute, *Biologists under Hitler*, trans. by Thomas Dunlap (Cambridge, MA, 1996)

Delaforce, Patrick, *Churchill's Secret Weapons: The Story of Hobart's Funnies* (London, 1998)

Dornberger, Walter, *V2*, trans. by James Cleugh and Geoffrey Halliday (London, 1954)

Dresden, Max, *H. A. Kramers: Between Tradition and Revolution* (New York, 1987)

Dwork, Debórah, and Robert Jan van Pelt, *Auschwitz* (New York, 2002)

Dyson, Freeman, *Disturbing the Universe* (New York, 1979)

Eckart, Wolfgang U., and Christoph Gradmann, eds, *Die Medizin und der Erste Weltkrieg* (Pfaffenweiler, 1996)

Einstein, Albert and Sigmund Freud, 'Why War?', in *An International Series of Open Letters*, vol. 2 (Paris, 1933)

Elfrath, Ulrich, and Bodo Herzog, *The Battleship Bismarck: A Documentary in Words and Pictures*, trans. by Edward Force (West Chester, 1989)

Ermenc, Joseph J., *Atomic Bomb Scientists: Memoirs 1939–1945* (Westport, 1967)

Fest, Joachim C., *Hitler*, trans. by Richard and Clara Winston (Harmondsworth, 1977)

——, *Speer: The Final Verdict*, trans. by Ewald Osers and Alexandra Dring (London, 2001)

Feuer, Lewis S., *Einstein and the Generations of Science* (New York, 1974)

FIAT Review of German Science 1939–1946, published by the Office of Military Government for Germany Field Agencies Technical (Wiesbaden, 1947)

Fischer, Ernst Peter, *Werner Heisenberg: Das selbstvergessene Genie* (Munich, 2001)

Fischer, Hans, *Völkerkunde im Nationalsozialismus: Aspekte der Anpassung, Affinität und Behauptung einer wissenschaftlichen Disziplin* (Berlin, 1990)

Flechtner, Hans-Joachim, *Carl Duisberg: Vom Chemiker zum Wirtschaftsführer* (Düsseldorf, 1960)

Freise, Gerda, 'Der Nobelpreisträger Professor Dr Heinrich Wieland: Zivilcourage in der Zeit des Nationalsozialismus', in *Hochverrat?: Die 'Weiße Rose' und ihr Umfeld*, ed. by Rudolf Lill and Michael Kißener (Constance, 1993)

Friedrich, Jörg, *Der Brand: Deutschland im Bombenkrieg 1940–1945* (Munich, 2002)

Frisch, Otto Robert, *What Little I Remember* (Cambridge, 1979)

Gaddis, John Lewis, *Now We Know: Rethinking Cold War History* (Oxford, 1997)

Garfield, Simon, *Mauve: How One Man Invented a Colour that Changed the World* (London, 2000)

Goudsmit, Samuel A., *Alsos* (New York, 1996)

Gratzer, Walter, *The Undergrowth of Science: Delusion, Self-Deception and Human Frailty* (Oxford, 2000)

Greenberg, Daniel S., *Science, Money, and Politics: Political Triumph and Ethical Erosion* (Chicago, 2001)

Greene, Brian R., *The Elegant Universe: Superstrings, Hidden Dimensions, and the Quest for the Ultimate Theory* (London, 1999)

Gregor, Neil, *Daimler-Benz in the Third Reich* (New Haven, 1998)

Gröttrup, Irmgard, *Rocket Wife*, trans. by Susi Hughes (London, 1959)

Groves, Leslie R., *Now It Can Be Told: The Story of the Manhattan Project* (New York, 1983)

Gutman, Yisrael, and Michael Berenbaum, eds, *Anatomy of the Auschwitz Death Camp* (Bloomington, 1994)

Haber, Fritz, *Briefe an Richard Willstätter, 1910–1934*, ed. by Petra Werner and Angelika Irmscher (Berlin, 1995)

Haberer, Joseph, *Politics and the Community of Science* (New York, 1969)

Hahn, Otto, *My Life*, trans. by Ernst Kaiser and Eithne Wilkins (London, 1970)

Haldane, Robert A., *The Hidden World* (London, 1976)

Hale, Nathan G., *The Rise and Crisis of Psychoanalysis in the United States: Freud and the Americans, 1917–1985* (Oxford, 1995)

Hansen, Friedrich, *Biologische Kriegsführung im Dritten Reich* (Frankfurt am Main, 1993)

Harris, Robert, and Jeremy Paxman, *A Higher Form of Killing: The Secret History of Gas and Germ Warfare* (London, 1982)

Heiber, Helmut, ed., *Reichsführer! . . . Briefe an und von Himmler* (Stuttgart, 1968)

Heisenberg, Werner, *Physics and Beyond: Encounters and Conversations*, trans. by Arnold J. Pomerans (London, 1971)

——, *Encounters with Einstein and Other Essays on People, Places and Particles* (Princeton, 1989)

Hentschel, Klaus, *The Einstein Tower: An Intertexture of Dynamic Construction, Relativity Theory, and Astronomy*, trans. by Ann M. Hentschel (Stanford, 1997)

——, and Ann M. Hentschel, eds, *Physics and National Socialism: An Anthology of Primary Sources* (Basel, 1996)

Herwig, Dieter, and Heinz Rode, *Luftwaffe: Secret Projects – Strategic Bombers 1935–1945*, trans. by Elke and John Weal (Leicester, 2000)

Heske, Henning, 'German Geographical Research in the Nazi Period', *Political Geography Quarterly*, 5 (July 1986)

——, 'Karl Haushofer: His Role in German Geopolitics and in Nazi Policies', *Political Geography Quarterly*, 6 (April 1987)

Hildebrandt, Gerhard, 'Max von Laue, der "Ritter ohne Furcht und Tadel"', in *Berlinische Lebensbilder*, vol. I: *Naturwissenschaftler*, ed. by Wilhelm Treue and Gerhard Hildebrandt (Berlin, 1987)

Hitler, Adolf, *Mein Kampf*, trans. by Ralph Manheim (London, 1992)

——, *Table Talk, 1941–1944: His Private Conversations*, trans. by Norman Cameron and R. H. Stevens, ed. by H. R. Trevor-Roper (London, 2000)

Hobsbawm, E. J., *The Age of Empire 1875–1914* (London, 1987)

Hodges, Andrew, *Alan Turing: The Enigma* (London, 1992)

Hoedeman, Paul, *Hitler or Hippocrates: Medical Experiments and Euthanasia in the Third Reich*, trans. by Ralph de Rijke (Lewes, 1991)

Holloway, David, *Stalin and the Bomb: The Soviet Union and Atomic Energy, 1939–1956* (New Haven, 1994)

Irons, Roy, *Hitler's Terror Weapons: The Price of Vengeance* (London, 2002)

Irving, David, *The German Atomic Bomb: The History of Nuclear Research in Nazi Germany* (New York, 1967)

Jungk, Robert, *Brighter than a Thousand Suns: A Personal History of the Atomic Scientists*, trans. by James Cleugh (London, 1960)

Kahn, David, *Seizing the Enigma: The Race to Break the German U-Boat Codes, 1939–1943* (London, 1996)

Kant, Horst, 'Werner Heisenberg and the German Uranium Project', preprint for the Max-Planck-Institut für Wissenschaftsgeschichte (2001)

Kater, Michael H., *Doctors under Hitler* (Chapel Hill, 1989)

——, *Das 'Ahnenerbe' der SS 1935–1945: Ein Beitrag zur Kulturpolitik des Dritten Reiches* (Munich, 2001)

Kaufman, Louis et al., *Moe Berg: Athlete, Scholar, Spy* (New York, 1974)

Kearns, G., 'Halford Mackinder', *Geographers: Biobibliographical Studies*, 9 (1985)

Kershaw, Ian, *Hitler 1889–1936: Hubris* (London, 1998)

——, *Hitler 1936–45: Nemesis* (London, 2000)

Kippenhahn, Rudolf, *Code Breaking: A History and Exploration*, trans. by Ewald Osers (London, 1999)

Kragh, Helge, *Quantum Generations: A History of Physics in the Twentieth Century* (Princeton, 1999)

Kramish, Arnold, *The Griffin* (London, 1987)

Krebs, Hans (with Anne Martin), *Reminiscences and Reflections* (Oxford, 1981)

Kroge, Harry von, *GEMA: Birthplace of German Radar and Sonar*, trans. and ed. by Louis Brown (Bristol, 2000)

Lanouette, William, and Bela Szilard, *Genius in the Shadows: A Biography of Leo Szilard* (Chicago, 1992)

Leitner, Gerit von, *Der Fall Clara Immerwahr: Leben für eine humane Wissenschaft* (Munich, 1993)

Lesch. John E., ed., *The German Chemical Industry in the Twentieth Century* (Dordrecht, 2000)

Levi, Primo, *If This is a Man; The Truce*, trans. by Stuart Woolf (London, 1990)

Lewontin, Richard, *It Ain't Necessarily So: The Dream of the Human Genome and Other Illusions* (London, 2000)

Lifton, Robert Jay, *The Nazi Doctors: Medical Killing and the Psychology of Genocide* (New York, 1986)

Lohmann, Hans-Martin, ed., *Psychoanalyse und Nationalsozialismus: Beiträge zur Bearbeitung eines unbewältigten Traumas* (Frankfurt am Main, 1994)

Lösch, Niels C., *Rasse als Konstrukt: Leben und Werk Eugen Fischers* (Frankfurt am Main, 1997)

Lourie, Richard, *Sakharov: A Biography* (Brandeis, 2002)

McCormmach, Russell, *Night Thoughts of a Classical Physicist* (Cambridge, MA, 1982)

McGuinness, Brian, *Wittgenstein: A Life*, vol I: *Young Ludwig, 1889–1921* (London, 1988)

MacIntyre, Alasdair, *After Virtue: A Study in Moral Theory* (London, 1982)

Macrakis, Kristie, *Surviving the Swastika: Scientific Research in Nazi Germany* (New York, 1993)

Mann, Thomas, *Doctor Faustus* (London, 1999)

Martinetz, Dieter, *Der Gaskrieg 1914/18: Entwicklung, Herstellung und Einsatz chemischer Kampfstoffe* (Bonn, 1996)

May, John, *The Greenpeace Book of the Nuclear Age: The Hidden History, the Human Cost* (London, 1989)

Mehrtens, Herbert, and Steffen Richter, eds, *Naturwissenschaft, Technik und NS-Ideologie: Beiträge zur Wissenschaftsgeschichte des Dritten Reiches* (Frankfurt am Main, 1980)

Meinel, Christoph, and Peter Voswinckel, eds, *Medizin, Naturwissenschaft, Technik und Nationalsozialismus: Kontinuitäten und Diskontinuitäten* (Stuttgart, 1994)

Mendelssohn, Kurt, *The World of Walther Nernst: The Rise and Fall of German Science* (London, 1973)

Metropolis, N., J. Howlett and Gian-Carlo Rota, eds, *A History of*

Computing in the Twentieth Century: A Collection of Essays (New York, 1980)

Middlebrook, Martin, *The Peenemünde Raid 17–18 August 1943* (London, 2000)

Miller, David, *The Cold War: A Military History* (London, 2001)

Mitscherlich, Alexander, and Fred Mielke, *The Death Doctors*, trans. by James Cleugh (London, 1962)

Müller-Hill, Benno, *Murderous Science: Elimination by Scientific Selection of Jews, Gypsies, and Others in Germany, 1933–1945*, trans. by George R. Fraser (Oxford, 1988)

Murphy, David Thomas, *The Heroic Earth: Geopolitical Thought in Weimar Germany 1918–1933* (Kent, OH, 1997)

Murray, Williamson, and Allan R. Millett, eds, *Military Innovation in the Interwar Period* (Cambridge, 1998)

Nathorff, Hertha, *Das Tagebuch: Berlin–New York, Aufzeichnungen 1933 bis 1945* (Munich, 1987)

Neufeld, Michael J., *The Rocket and the Reich: Peenemünde and the Coming of the Ballistic Missile Era* (Cambridge, MA, 1999)

Nowotny, Helga, Peter Scott and Michael Gibbons, *Re-Thinking Science: Knowledge and the Public in an Age of Uncertainty* (Cambridge, 2001)

Oppenheimer, Robert, *Letters and Recollections*, ed. by Alice Kimball Smith and Charles Weiner (Cambridge, MA, 1980)

Overy, Richard, *Why the Allies Won* (London, 1995)

Padfield, Peter, *The Battleship Era* (London, 1972)

Pagden, Anthony, *The Fall of Natural Man: The American Indian and the Origins of Comparative Ethnology* (Cambridge, 1982)

Pais, Abraham, *Niels Bohr's Times: In Physics, Philosophy and Polity* (Oxford, 1991)

Parker, Geoffrey, *Geopolitics: Past, Present and Future* (London, 1998)

Peiffer, Jürgen, ed., *Menschenverachtung und Opportunismus: Zur Medizin im Dritten Reich* (Tübingen, 1992)

Perutz, Max, *I Wish I'd Made You Angry Earlier: Essays on Science, Scientists, and Humanity* (Oxford, 1998)

Ploetz, Alfred, *Die Tüchtigkeit unserer Rasse und der Schutz der Schwachen* (Berlin, 1895)

Poliakov, Léon, and Josef Wulf, eds, *Das Dritte Reich und seine Denker: Dokumente* (Berlin, 1959)

Polkinghorne, John, *The Quantum World* (London, 1986)

Porter, Roy, *The Greatest Benefit to Mankind: A Medical History of Humanity from Antiquity to the Present* (London, 1997)

Powers, Thomas, *Heisenberg's War: The Secret History of the German Bomb* (Boston, 1993)

Prater, Donald, *Thomas Mann: A Life* (Oxford, 1995)

Proctor, Robert N., *Racial Hygiene: Medicine under the Nazis* (Cambridge, MA, 1988)

——, *Value-Free Science? Purity and Power in Modern Knowledge* (Cambridge, MA, 1991)

——, *The Nazi War on Cancer* (Princeton, 1999)

Rawls, John, *Collected Papers*, ed. by Samuel Freeman (Cambridge, MA, 1999)

Rechenberg, Helmut, and Gerald Wiemers, eds, *Werner Heisenberg 1901–1976: Schritte in die neue Physik* (Beucha, 2001)

Reid, Constance, *Hilbert* (New York, 1996)

Renneberg, Monika, and Mark Walker, eds, *Science, Technology and National Socialism* (Cambridge, 1994)

Reynolds, David, *One World Divisible: A Global History since 1945* (London, 2000)

Rhodes, Richard, *The Making of the Atomic Bomb* (New York, 1986)

——, *Dark Sun: The Making of the Hydrogen Bomb* (New York, 1996)

Rife, Patricia, *Lise Meitner and the Dawn of the Nuclear Age* (Boston, 1999)

Rohwer, Jürgen, and Eberhard Jäckel, eds, *Die Funkaufklärung und ihre Rolle im Zweiten Weltkrieg* (Stuttgart, 1979)

Rose, Paul Lawrence, *Heisenberg and the Nazi Atomic Bomb Project: A Study in German Culture* (Berkeley, 1998)

Roseman, Mark, *The Villa, the Lake, the Meeting: Wannsee and the Final Solution* (London, 2002)

Rosenberg, Alfred, *Selected Writings*, ed. by Robert Pois (London, 1970)

Sabbagh, Karl, *Dr Riemann's Zeros* (London, 2002)

Sakharov, Andrei, *Memoirs*, trans. by Richard Lourie (London, 1990)

Salomon, Jean-Jacques, *Le Scientifique et le guerrier* (Paris, 2001)

Schaaf, Michael, 'Der Physiochemiker Paul Harteck (1902–1985)', in CENSIS-REPORT-33–99 (Hamburg, 1999)

——, *Heisenberg, Hitler und die Bombe: Gespräche mit Zeitzeugen* (Berlin, 2001)

Schama, Simon, *Landscape and Memory* (London, 1995)

Sebald, W. G., *On the Natural History of Destruction*, trans. by Anthea Bell (London, 2003)

Shea, William R., ed., *Otto Hahn and the Rise of Nuclear Physics* (Dordrecht, 1983)

Shirer, William L., *The Rise and Fall of the Third Reich: A History of Nazi Germany* (London, 1961)

Sime, Ruth Lewin, *Lise Meitner: A Life in Physics* (Berkeley, 1996)

Singh, Simon, *The Code Book: The Secret History of Codes and Codebreaking* (London, 2000)

SIPRI, *The Problems of Chemical and Biological Warfare*, vol. I: *The Rise of CB Weapons* (Stockholm, 1971)

——, *Warfare in a Fragile World: Military Impact on the Human Environment* (Stockholm, 1980)

Smart, J. J. C., and Bernard Williams, *Utilitarianism: For and Against* (Cambridge, 1998)

Smil, Vaclav, *Enriching the Earth: Fritz Haber, Carl Bosch and the Transformation of World Food Production* (Cambridge, MA, 2001)

Smith, Michael and Ralph Erskine, *Action this Day* (New York, 2001)

Speer, Albert, *The Slave State: Heinrich Himmler's Masterplan for SS Supremacy*, trans. by Joachim Neugroschel (London, 1981)

——, *Inside the Third Reich*, trans. by Richard and Clara Winston (London, 1995)

Spiers, Edward M., *Chemical Warfare* (Chicago, 1986)

Steinberg, Jonathan, *Yesterday's Deterrent: Tirpitz and the Birth of the German Battle Fleet* (London, 1965)

Stern, Fritz, *Einstein's German World* (London, 2000)

Stoltzenberg, Dietrich, *Fritz Haber: Chemiker, Nobelpreisträger, Deutscher, Jude: Eine Biographie* (Weinheim, 1998)

Strausz-Hupé, Robert, *Geopolitics: The Struggle for Space and Power* (New York, 1972)

Sumida, J. T., 'British Capital Ship Design and Fire Control into the Dreadnought Era', *Journal of Modern History*, 51 (1979)

Szilard, Leo, *His Version of the Facts: Selected Recollections and Correspondence*, ed. by Spencer R. Weart and Gertrud Weiss Szilard (Cambridge, MA, 1978)

Szöllösi-Janze, Margit, *Fritz Haber 1868–1934: Eine Biographie* (Munich, 1998)

——, ed., *Science in the Third Reich* (Oxford, 2001)

Teller, Edward (with Judith L. Schoolery), *Memoirs: A Twentieth-Century Journey in Science and Politics* (Cambridge, MA, 2001)

Thomas, Roger D., *Dreadnoughts in Camera: Building the Dreadnoughts, 1905–1920* (Stroud, 1998)

van den Bussche, Hendrik, ed., *Medizinische Wissenschaft im 'Dritten Reich': Kontinuität, Anpassung und Opposition an der Hamburger Medizinschen Fakultät* (Berlin, 1989)

van der Will, Wilfried, 'Culture and the Organization of National Socialist Ideology 1933 to 1945', in *German Cultural Studies: An Introduction*, ed. by Rob Burns (Oxford, 1995)

van Pelt, Robert Jan, and Carroll William Westfall, *Architectural Principles in the Age of Historicism* (New Haven, 1993)

Veenhoven, Willem A., et al., eds, *Case Studies on Human Rights and Fundamental Freedoms: A World Survey*, vol. I (The Hague, 1975)

Wagner, Wolfgang, *The History of German Aviation: The First Jet Aircraft* (Atglen, 1998)

Wakefield, Ken, *The Blitz: Then and Now*, vol. 3 (London, 1990)

Walker, Mark, *German National Socialism and the Quest for Nuclear Power 1939–1949* (Cambridge, 1989)

——, *Nazi Science: Myth, Truth and the German Atomic Bomb* (New York, 1995)

——, 'Twentieth Century German Science', in *Science in the Twentieth Century*, ed. by John Krige and Dominique Pestre (Amsterdam, 1997)

——, ed., *Science and Ideology: A Comparative History* (London, 2003)

Wegener, Peter P., *The Peenemünde Wind Tunnels: A Memoir* (New Haven, 1996)

Weigert, Hans W., *Generals and Geographers: The Twilight of Geopolitics* (New York, 1942)

Weinberg, Gerhard L., *A World at Arms: A Global History of World War II* (Cambridge, 1994)

Weindling, Paul, *Health, Race and German Politics between National Unification and Nazism, 1870–1945* (Cambridge, 1993)

Weisskopf, Victor F., *The Joy of Insight: Passions of a Physicist* (New York, 1991)

Werner, Helmut, *From the Aratus Globe to the Zeiss Planetarium*, trans. by H. Degenhardt (Stuttgart, 1957)

Wetzel, Wolfgang, *Naturwissenschaften und chemische Industrie in Deutschland: Voraussetzungen und Mechanismen ihres Aufstiegs im 19. Jahrhundert* (Stuttgart, 1991)

White, Michael, and John Gribbin, *Einstein: A Life in Science* (London, 1993)

Wild von Hohenborn, Adolf, *Briefe und Tagebuchaufzeichnungen des preußischen Generals als Kriegsminister und Truppenführer im Ersten Weltkrieg*, ed. by Helmut Reichold (Boppard am Rhein, 1986)

Wildt, Michael, *Generation des Unbedingten: Das Führungskorps des Reichssicherheitshauptamtes* (Hamburg, 2002)

Willstätter, Richard, *Aus meinem Leben: Von Arbeit, Muße und Freunden* (Weinheim, 1958)

Wilson, Thomas, *Churchill and the Prof* (London, 1995)

Wolpert, Lewis, *The Unnatural Nature of Science* (London, 1992)

Zimmerman, David, *Britain's Shield: Radar and the Defeat of the Luftwaffe* (Stroud, 2001)

Index

Hitler's Pope
The Secret History of Pius XII

The shocking untold story of Pope Pius XII that "redefines the entire history of the twentieth century" (*The Washington Post*).

Pope Pius XII has long been the subject of controversy over his failure to speak out against Hitler's Final Solution. *Hitler's Pope* describes how, even well before the Holocaust, Pope Pius XII was instrumental in negotiating an accord that helped the Nazis rise to unhindered power—and sealed the fate of the Jews in Europe. Drawing upon secret Vatican archives to which he had exclusive access, Cornwell tells the full, tragic story of how Pius became the most dangerous churchman in history. A firm and final indictment of Pius XII's papacy, *Hitler's Pope* is also a searing exploration of its lingering consequences for the Catholic Church today.   *ISBN 0-14-029627-1*

Breaking Faith
Can the Catholic Church Save Itself?

"A provocative, deeply personal, and intelligent book."
—*Library Journal*

As scandals continue to erupt and John Paul II's papacy draws to a close, the Catholic Church, its policies, and its future are being scrutinized around the globe. In this penetrating overview of an institution at the crossroads, bestselling author John Cornwell brings into play his eloquent disagreement with—and deeply felt commitment to—the Catholic Church. He probes beyond the statistics, addressing issues ranging from the centralization of power in Rome to the watering-down of the liturgy, from views on gays to the role of women, *Breaking Faith* is at once a deeply personal work and a disturbing, eye-opening analysis.
ISBN 0-14-219608-8

Hitler's Scientists
Science, War, and the Devil's Pact

"Wide-ranging and accessible . . . [a] disturbing and important account of the life of science and scientists under the Nazis."
—*The Economist*

In *Hitler's Scientists*, John Cornwell explores German scientific genius in the first half of the twentieth century, showing how Germany's early lead in the new physics led to the discovery of atomic fission, and how the ideas of Darwinism were hijacked to create the lethal doctrine of racial cleansing. By the war's end, almost every aspect of Germany's scientific culture had been tainted by the exploitation of slave labor, human experimentation, and mass killings. Cornwell argues that German scientists should be held accountable for the uses to which their knowledge was put—an issue with wide-ranging implications for the continuing unregulated pursuit of scientific progress.

ISBN 0-14-200480-4

FOR THE BEST IN PAPERBACKS, LOOK FOR THE 🐧

In every corner of the world, on every subject under the sun, Penguin represents quality and variety—the very best in publishing today.

For complete information about books available from Penguin—including Penguin Classics, Penguin Compass, and Puffins—and how to order them, write to us at the appropriate address below. Please note that for copyright reasons the selection of books varies from country to country.

In the United States: Please write to *Penguin Group (USA), P.O. Box 12289 Dept. B, Newark, New Jersey 07101-5289* or call 1-800-788-6262.

In the United Kingdom: Please write to *Dept. EP, Penguin Books Ltd, Bath Road, Harmondsworth, West Drayton, Middlesex UB7 0DA*.

In Canada: Please write to *Penguin Books Canada Ltd, 10 Alcorn Avenue, Suite 300, Toronto, Ontario M4V 3B2*.

In Australia: Please write to *Penguin Books Australia Ltd, P.O. Box 257, Ringwood, Victoria 3134*.

In New Zealand: Please write to *Penguin Books (NZ) Ltd, Private Bag 102902, North Shore Mail Centre, Auckland 10*.

In India: Please write to *Penguin Books India Pvt Ltd, 11 Panchsheel Shopping Centre, Panchsheel Park, New Delhi 110 017*.

In the Netherlands: Please write to *Penguin Books Netherlands bv, Postbus 3507, NL-1001 AH Amsterdam*.

In Germany: Please write to *Penguin Books Deutschland GmbH, Metzlerstrasse 26, 60594 Frankfurt am Main*.

In Spain: Please write to *Penguin Books S. A., Bravo Murillo 19, 1° B, 28015 Madrid*.

In Italy: Please write to *Penguin Italia s.r.l., Via Benedetto Croce 2, 20094 Corsico, Milano*.

In France: Please write to *Penguin France, Le Carré Wilson, 62 rue Benjamin Baillaud, 31500 Toulouse*.

In Japan: Please write to *Penguin Books Japan Ltd, Kaneko Building, 2-3-25 Koraku, Bunkyo-Ku, Tokyo 112*.

In South Africa: Please write to *Penguin Books South Africa (Pty) Ltd, Private Bag X14, Parkview, 2122 Johannesburg*.